Yuzuru Matsuoka

Sept. 10th 2005

Springer
Tokyo
Berlin
Heidelberg
New York
Hong Kong
London
Milan
Paris

M. Kainuma, Y. Matsuoka, T. Morita (Eds.)

Climate Policy Assessment

Asia-Pacific Integrated Modeling

With 169 Figures, Including 25 in Color

 Springer

Mikiko Kainuma
Head, Integrated Assessment Modeling Section
Social and Environmental Systems Division
National Institute for Environmental Studies
16-2 Onogawa, Tsukuba 305-8506, Japan

Yuzuru Matsuoka
Professor, Kyoto University
Yoshida-Hommachi, Sakyo-ku, Kyoto 606-8501, Japan

Tsuneyuki Morita
Director, Social and Environmental Systems Division
National Institute for Environmental Studies
16-2 Onogawa, Tsukuba 305-8506, Japan

ISBN 4-431-70264-4 Springer-Verlag Tokyo Berlin Heidelberg New York
Library of Congress Cataloging-in-Publication Data applied for.

Printed on acid-free paper

Typesetting: Camera-ready by the editors and authors
Printing and binding: Hicom, Japan
SPIN:10652883

Preface

With more than half of the world population living in the region, it is not surprising that Asia has been attracting significant attention from all over the world. While it has the highest potential for economic growth, Asia also has the largest environmental loads and the most serious damage due to deterioration of the environment. This is why a number of environmental researchers have focused on Asia, and it is now recognized that sustainable development of this region is the key to the solution of global environmental issues such as global climate change.

However, there are several reasons why the search for a solution is extremely difficult. The first is that the problems arise from huge, complex systems. There are numerous factors that are interrelated and hard to predict. Thus it is difficult to understand the problem comprehensively, making it hard to define what the problem really is, where it comes from, what we should do to address it, and what the consequences of our actions will be. Second, it will require more than a century to solve the problem. Taking the example of climate change, it is estimated that it may take more than 30 years before severe damage will appear, and it may take more than 100 years of action to stabilize the world climate. The third reason is that to implement actions it is necessary to address the differences between the North and the South, and the different value systems in the world. How should we grapple with such a problem?

We began the process of developing large-scale computer simulation models 12 years ago. This process was also intended to offer a platform for policy makers and researchers. Using the models, we attempted to enhance communication among these communities, and systematically to bring out the latest scientific findings to the policy-making processes at both international and regional levels. We tried to forge a community where researchers and policymakers from various backgrounds and nationalities discuss these long-term problems while respecting differences in opinions and values. One essential feature of our model development process is "to develop an interface between science and the policymaking process by researchers in Asia," that is, to develop a common set of models that can be used in Asia by collaborating with researchers in the region, and to offer an opportunity where policymakers and researchers think together while using the models.

The model development process was first promoted for evaluating policy options to prevent climate change, and the first basic model was completed in 1992. Since then, researchers from Asian countries such as China, India, Korea, Indonesia, and Thailand have joined this process and several country models have been applied. More than 20 models have been developed to date. Although the models focus mainly on Asia, the Pacific region is also included to analyze the Asia–Pacific region intensively. Additionally, world models have been developed to analyze international economic relationships and climate impacts for evaluating policy options from a global viewpoint. These models have been used as a single model or in combination, depending on policy needs, and have been used in various fields such as climate change mitigation, air pollution abatement, ecosystem preservation, water resource management, land use policy, energy policy, promotion of environmental industries, and integrated environmental management policy. Climate models have contributed to the establishment of a communication platform for various policy-making processes in the above fields. Because it is necessary to integrate different models to analyze complicated problems, especially related to climate change and environmental management issues, the family of models is called the Asia–Pacific Integrated Model (AIM).

Ten years have passed since the first model was released. Because significant amounts of work and experience have accumulated during this period, we decided to publish a series of books concerning the development of the model and its application. This book is the first in the Asia–Pacific Integrated Modeling series.

The book focuses on climate change issues and the evaluation of policy options to stabilize the global climate. It contains an overview of the models developed so far, their structure, and the results and analyses presented to policymakers and researchers at the levels of each Asian country, the Asia–Pacific region, and the world. AIM has contributed not only to the governments in Asian countries, but also to international organizations that have actually used the model such as the IPCC, UNEP, and OECD. Some of these contributions have been summarized in this book. Also covered are detailed country models for China, India, Korea, Vietnam, and Japan; an Asia–Pacific regional model comprising 42 countries; and regional and world models. The contents of the book vary in scope from local to global and discuss the effects of climate policies in each country and the Asia-Pacific region as a whole, cost analyses of climate policies with trade effects, and global scenario analyses. Also included are impact analyses and the effects of promoting environmental technologies.

The authors of this book are specialists who have devoted themselves to the development of AIM. Some are senior researchers who have been developing and using the model for more than 10 years, and some are young researchers who have recently become involved with the model building and application process. We are proud and pleasantly surprised by the fact that so many experts from various Asian countries have been working together for so long on the AIM model development and application process. Through this joint effort, it has been proven that Asian researchers can produce excellent results, some of which could not have been achieved by American or European researchers. We are determined to extend our collaboration and develop further models for analyzing sustainable development policies in Asia. We must leave the judgment of this and subsequent books in this series to our readers.

We sincerely appreciate the support we have received over the past 12 years from the Ministry of the Environment, Japan and we are grateful for the facilities made available by the National Institute for Environmental Studies, Japan.

Mikiko Kainuma
Yuzuru Matsuoka
Tsuneyuki Morita

Contents

Authors

Fujino, Junichi, National Institute for Environmental Studies, 16-2 Onogawa, Tsukuba 305-8506, Japan

Garg, Amit, Winrock International India, 7, Poorvi Marg, Vasant Vihar, New Delhi 110 057, India

Harasawa, Hideo, National Institute for Environmental Studies, 16-2 Onogawa, Tsukuba 305-8506, Japan

Hibino, Go, Fuji Research Institute Corporation, 2-3 Kandanishiki-cho, Chiyoda-ku, Tokyo 101-8443, Japan

Hijioka, Yasuaki, National Institute for Environmental Studies, 16-2 Onogawa, Tsukuba 305-8506, Japan

Hu, Xiulian, Energy Research Institute, State Development Planning Commission, Guohong Building, A-11 Muxidi Beili, Xicheng District, Beijing 100038, China

Ishii, Hisaya, Fuji Research Institute Corporation, 2-3 Kandanishiki-cho, Chiyoda-ku, Tokyo 101-8443, Japan

Jeon, Song Woo, Korea Environment Institute, 613-2, Bulkwang-Dong Eunpyung-Gu, Seoul 122-706, Korea

Jiang, Kejun, Energy Research Institute, State Development Planning Commission, Guohong Building, A-11 Muxidi Beili, Xicheng District, Beijing 100038, China

Jung, Tae Yong, Institute for Global Environmental Strategies, 1560-39 Kamiyamaguchi, Hayama 240-0115, Japan

Kainuma, Mikiko, National Institute for Environmental Studies, 16-2 Onogawa, Tsukuba 305-8506, Japan

Kapshe, Manmohan, Indian Institute of Management, Vastrapur, Ahmedabad 380015, India

Lal, Murari, Indian Institute of Technology, New Delhi 110016, India

Lee, Dong Kun, Sangmyung University, San 98-20, An So-Dong, Chon An, Chung Nam, 330-720, Korea

Li, Zehui, Institute of Geographical Sciences and Natural Resources Research, Chinese Academy of Science, No.3 Datun Road, Chaoyang District, Beijing 100101, China

Masui, Toshihiko, National Institute for Environmental Studies, 16-2 Onogawa, Tsukuba 305-8506, Japan

Matsui, Shigekazu, Fuji Research Institute Corporation, 2-3 Kandanishiki-cho, Chiyoda-ku, Tokyo 101-8443, Japan

Matsuoka, Yuzuru, Kyoto University, Yoshida-honmachi, Sakyo-ku, Kyoto 606-8501, Japan

Morita, Tsuneyuki, National Institute for Environmental Studies, 16-2 Onogawa, Tsukuba 305-8506, Japan

Munesue, Yosuke, Tokyo Institute of Technology, 2-12-1 Ookayama, Meguro-ku, Tokyo 152-8552, Japan

Pandey, Rahul, National Institute for Environmental Studies, 16-2 Onogawa, Tsukuba 305-8506, Japan; Indian Institute of Management, Prabandh Nagar, Off Sitapur Road, Lucknow 226013, India

Rana, Ashish, National Institute for Environmental Studies, 16-2 Onogawa, Tsukuba 305-8506, Japan

Shimada, Koji, Kyoto University, Yoshida-honmachi, Sakyo-ku, Kyoto 606-8501, Japan; Ministry of the Environment, 1-2-2 Kasumigaseki, Chiyoda-ku, Tokyo 100-8975, Japan

Shimada, Yoko, Kobe Institute of Health, 4-6 Minatojima-nakamachi, Cyuo-ku, Kobe 650-0046, Japan

Shrestha, Ram M., Asian Institute of Technology, P.O.Box 4, Klong Luang, Pathumthani 12120, Thailand

Shukla, Priyadarshi R., Indian Institute of Management, Vastrapur, Ahmedabad 380015, India

Sun, Jiulin, Institute of Geographical Sciences and Natural Resoureces Research, Chinese Academy of Science, No.3 Datun Road, Chaoyang District, Beijing 100101, China

Takahashi, Kiyoshi, National Institute for Environmental Studies, 16-2 Onogawa, Tsukuba 305-8506, Japan

Tung, Le Thanh, Asian Institute of Technology, P.O.Box 4, Klong Luang, Pathumthani 12120, Thailand

Yang, Hongwei, Energy Research Institute, State Development Planning Commission, Guohong Building, A-11 Muxidi Beili, Xicheng District, Beijing 100038, China

You, Songcai, Institute of Geographical Sciences and Natural Resources Research, Chinese Academy of Science, No.3 Datun Road, Chaoyang District, Beijing 100101, China

(Affiliation as of August 2002)

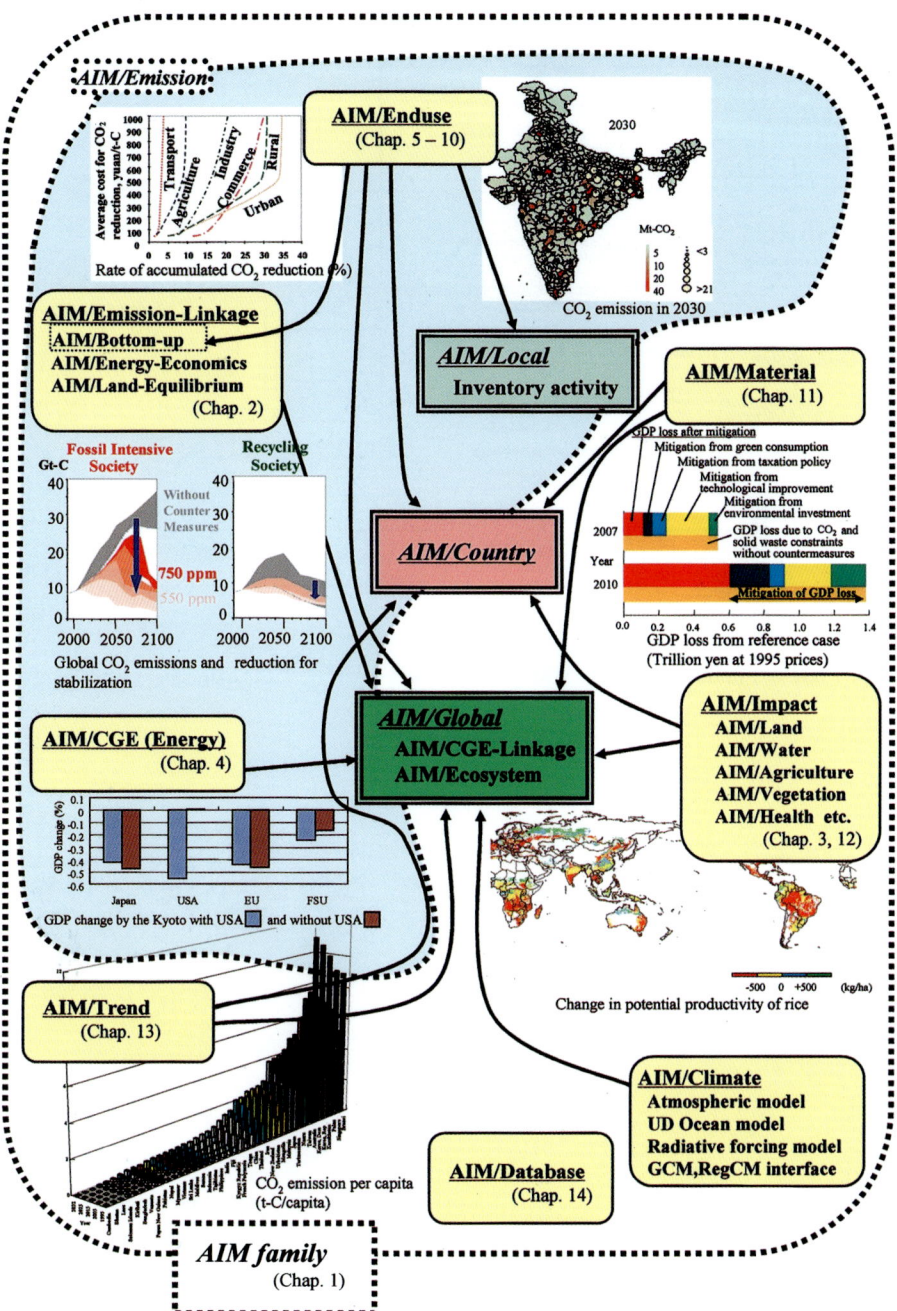

Color Plate 1. Roadmap of AIM family (see Fig. 1 of Chap. 1)

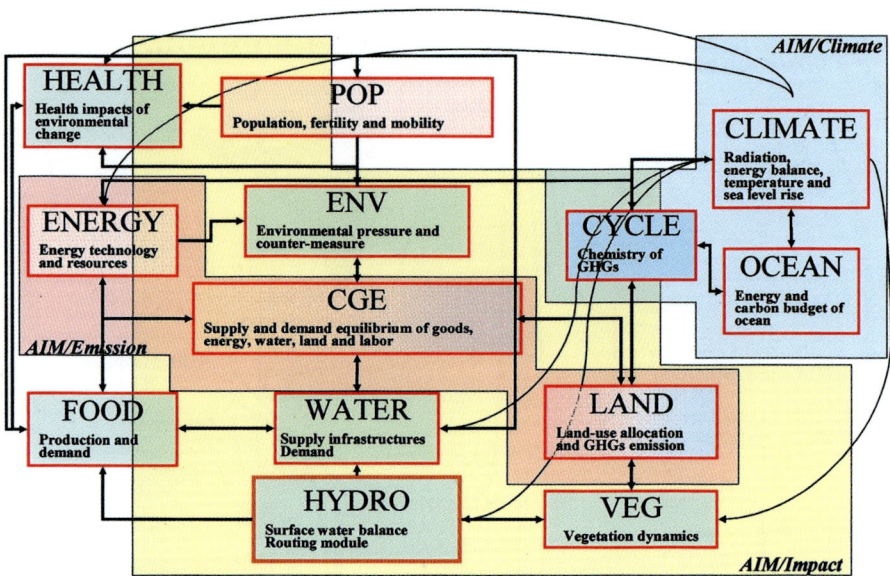

Color Plate 2. Framework of the AIM/Impact model (see Fig. 1 of Chap. 3)

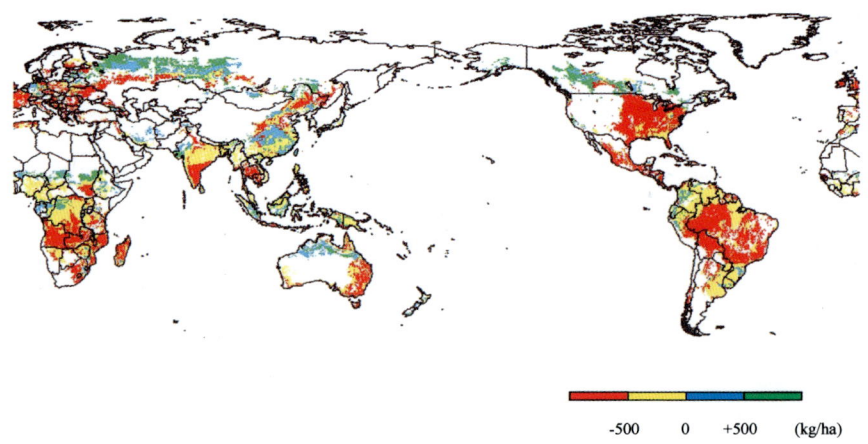

Color Plate 3. Change in the potential productivity of rice from 1990 to 2050 under the climatic conditions projected using the CCSR/NIES GCM (see Fig. 4 of Chap. 3)

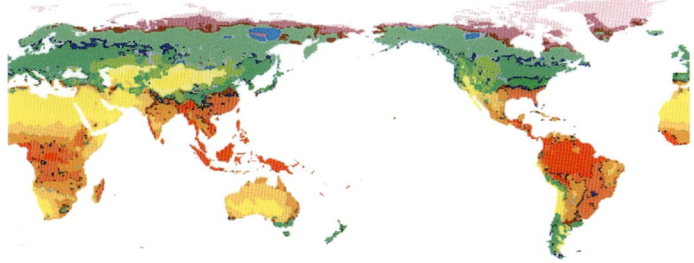

IS92c scenario with low climate sensitivity

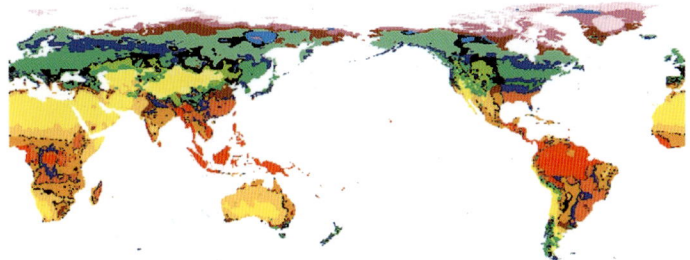

IS92a scenario with medium climate sensitivity

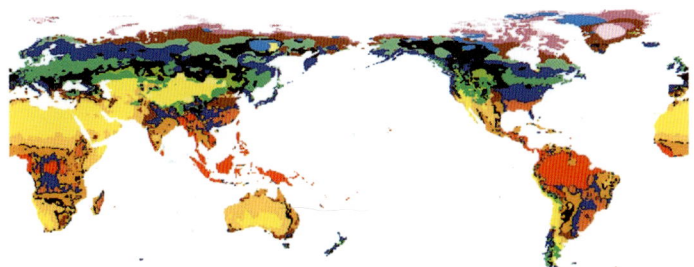

IS92e scenario with high climate sensitivity

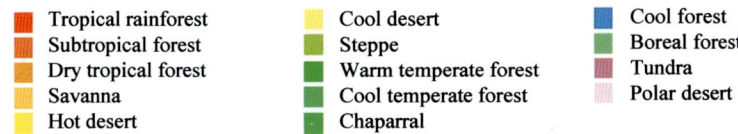

Tropical rainforest	Cool desert	Cool forest
Subtropical forest	Steppe	Boreal forest
Dry tropical forest	Warm temperate forest	Tundra
Savanna	Cool temperate forest	Polar desert
Hot desert	Chaparral	

Cells are classified to a forest type under the present climate and to a non-forest type under the future climate (diminishment of forest).

Cells are classified to a forest type under the present climate and to a different forest type under the future climate, but the shift cannot occur smoothly (replacement of the forest type with the risk of diminishment).

Cells are classified to a non-forest type under the present climate and to a forest type under the future climate, but the shift cannot occur smoothly.

Color Plate 4. Forest collapse under low, mid, high-climate change scenarios (IS92c emissions with low climate sensitivity, IS92a emissions with medium climate sensitivity, IS92e emissions with high climate sensitivity) (see Fig. 6 of Chap. 3)

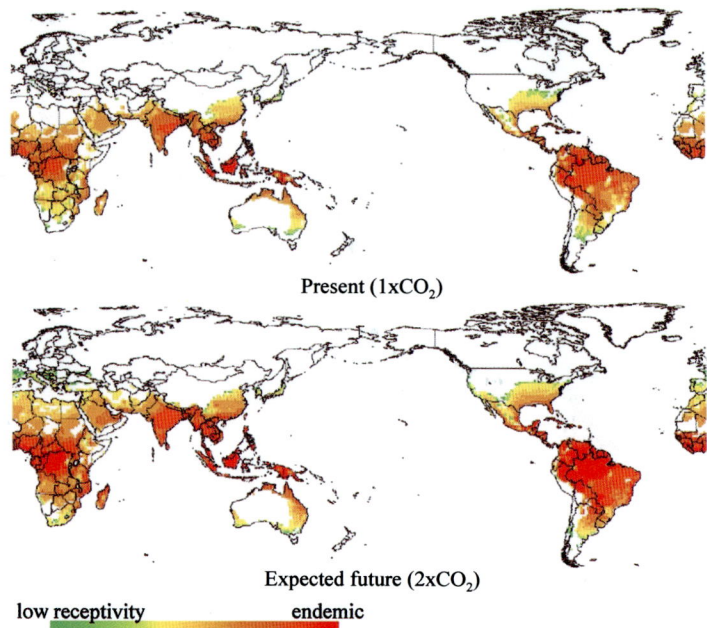

Present (1xCO$_2$)

Expected future (2xCO$_2$)

low receptivity   endemic

Color Plate 5. Expansion of the area affected by malaria (Upper: current climate, Lower: 2xCO$_2$ climate change, GCM: GFDL q-flux equilibrium experiment) (see Fig. 8 of Chap. 3)

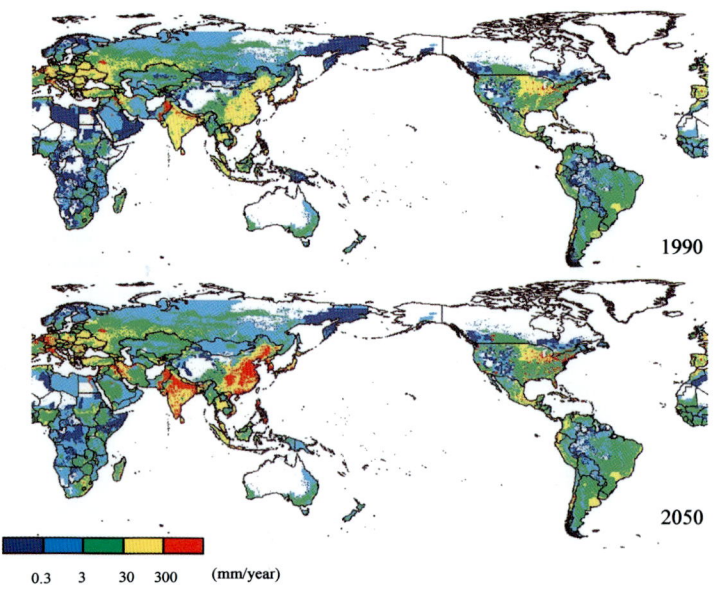

1990

2050

0.3   3   30   300   (mm/year)

Color Plate 6. Water demand per unit area in 1990 and 2050 (see Fig. 11 of Chap. 3)

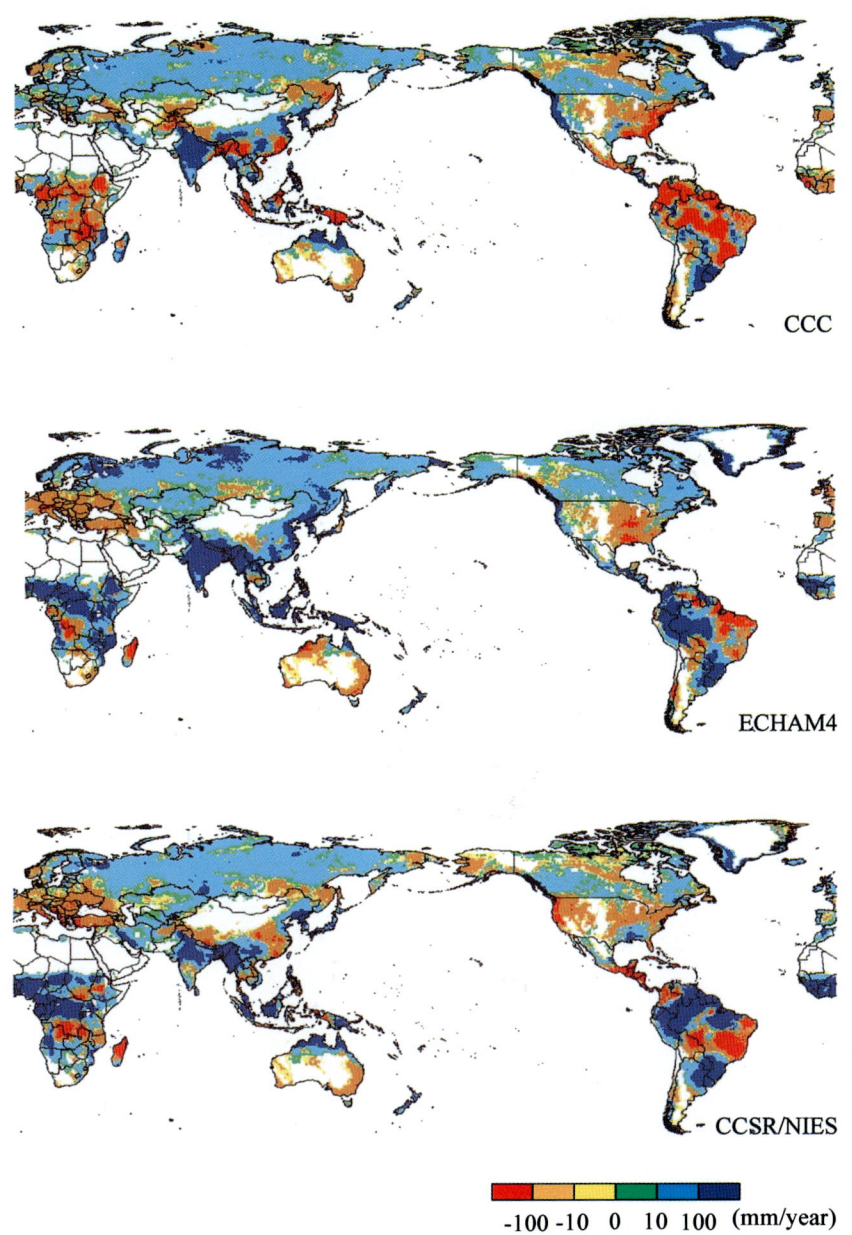

Color Plate 7. Changes in runoff calculated based on the results of the transient experiments of the CCC, ECHAM4, and CCSR/NIES climate models (mean runoff for the 10 years from 2050 to 2059 minus mean runoff for the 10 years from 1980 to 1989) (see Fig. 12 of Chap. 3)

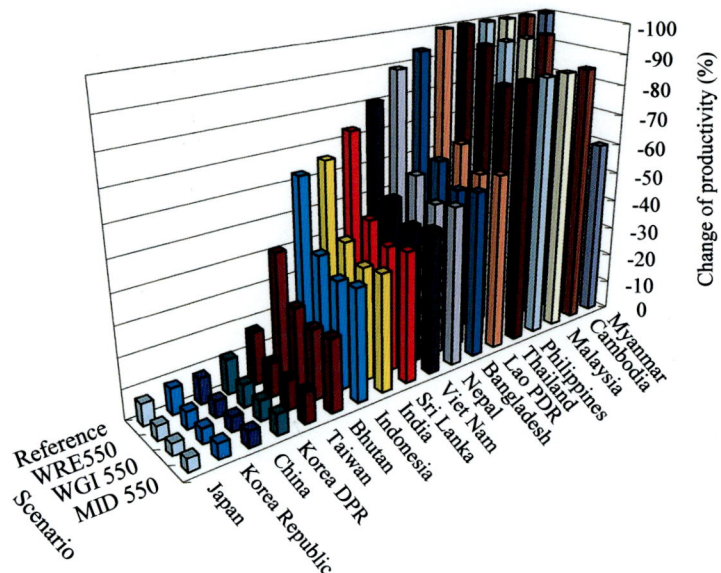

Color Plate 8. Change in winter wheat productivity from 1990 to 2100 (see Fig. 16 of Chap. 4)

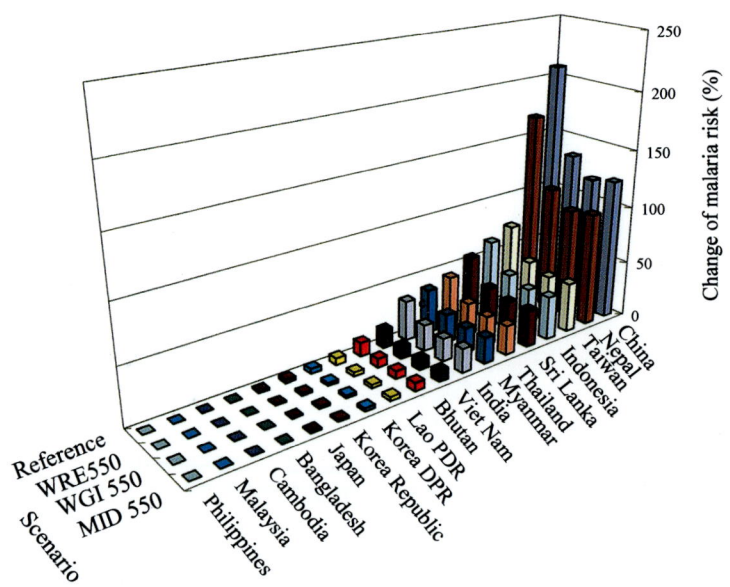

Color Plate 9. Change in population living in area at high malaria risk from 1990 to 2100 (see Fig. 18 of Chap. 4)

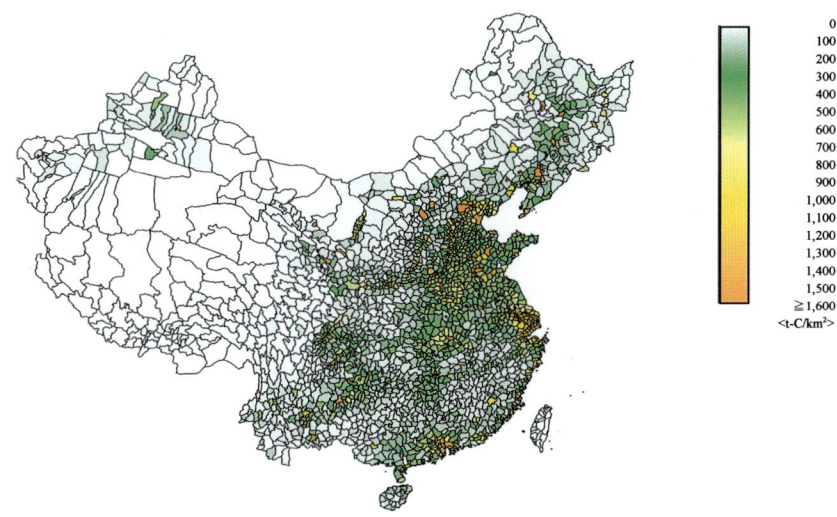

Color Plate 10. CO_2 emission intensity in China for 2010 (see Fig. 5 of Chap. 5)

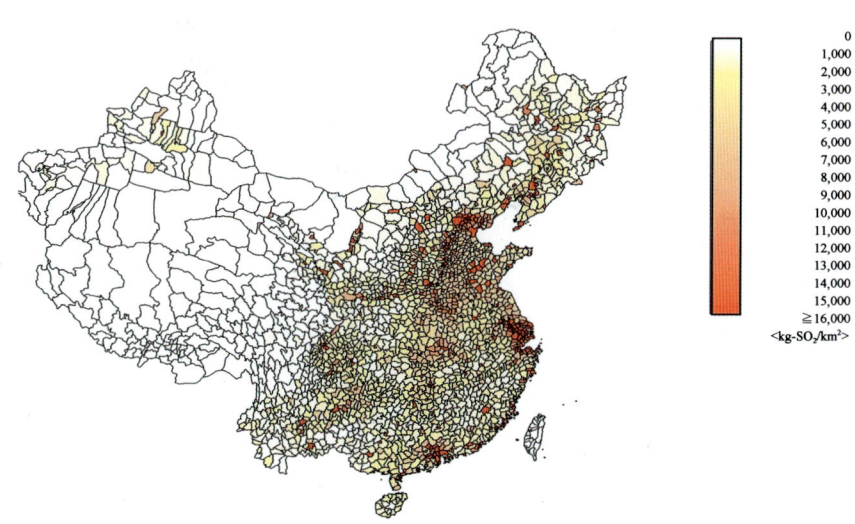

Color Plate 11. SO_2 emission intensity in China for 2010   (see Fig. 6 of Chap. 5)

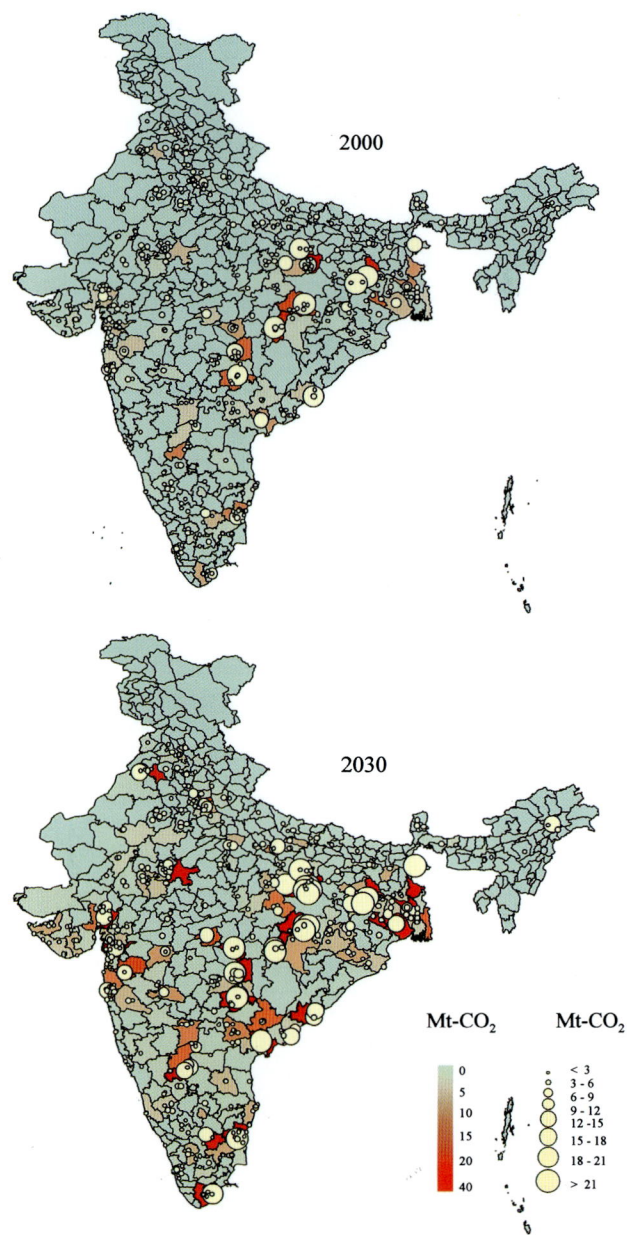

Color Plate 12. Regional distribution of CO_2 emissions in India for 2000 and 2030 in reference scenario (see Fig. 4 of Chap. 6)
Note: Circles show emissions from large point sources.

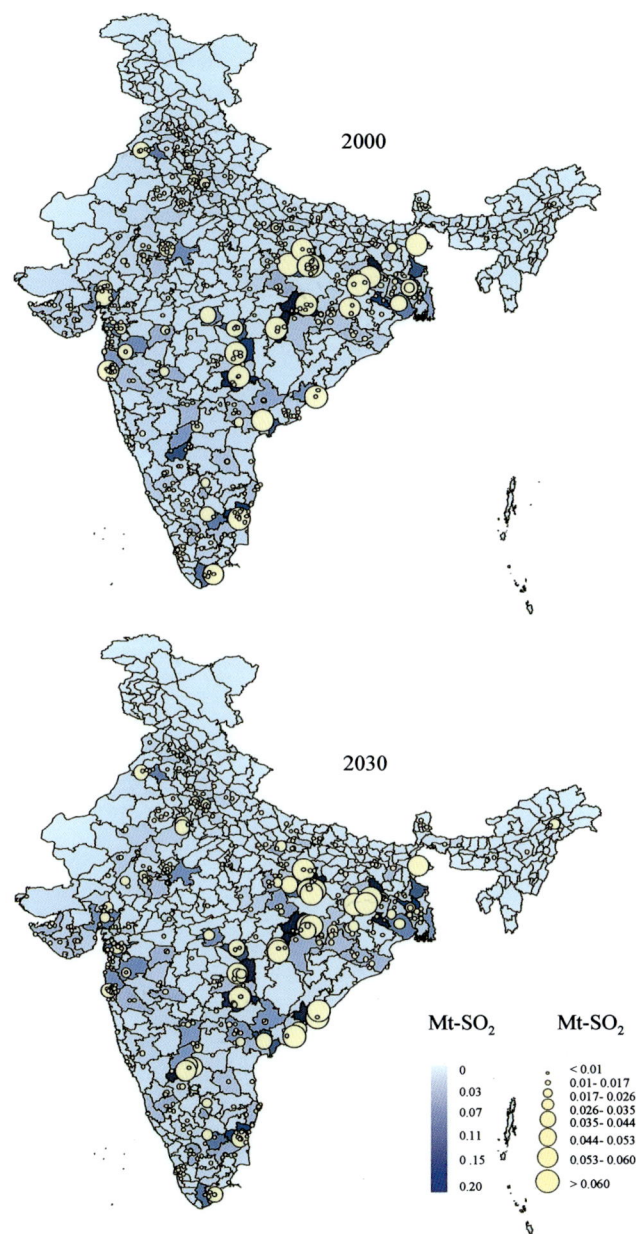

Color Plate 13. Regional distribution of SO₂ emissions in India for 2000 and 2030 in reference scenario (see Fig. 5 of Chap. 6)
Note: Circles show emissions from large point sources.

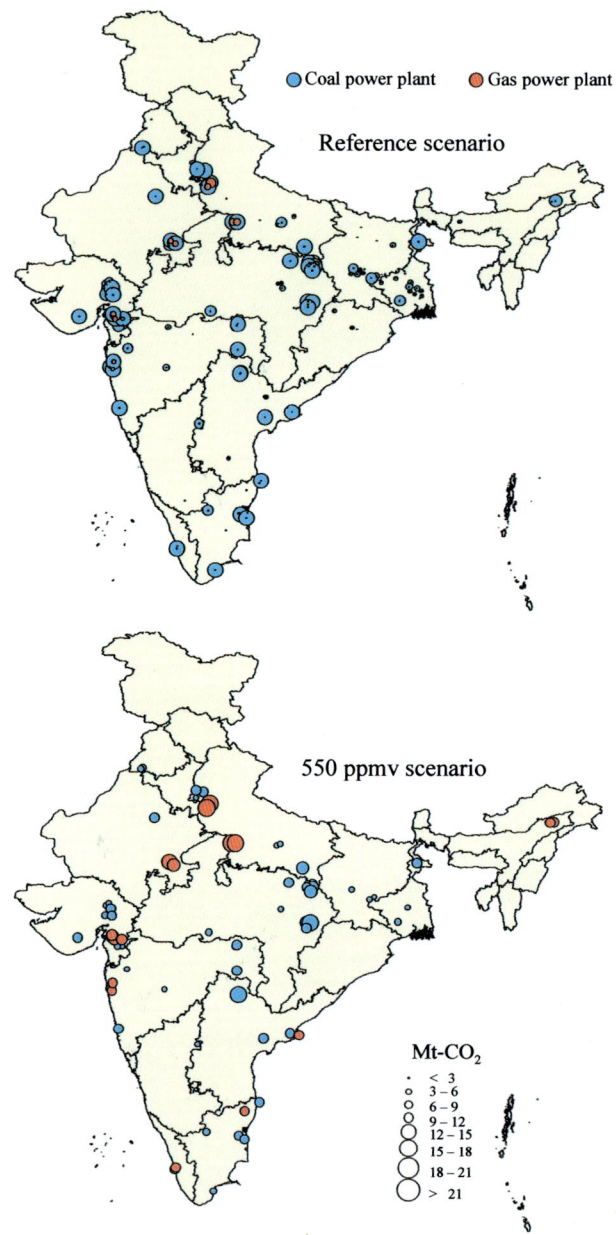

Color Plate 14. Comparison of CO_2 emissions from large power plants in reference and 550 ppmv stabilization scenarios in 2030 (see Fig. 7 of Chap. 6)

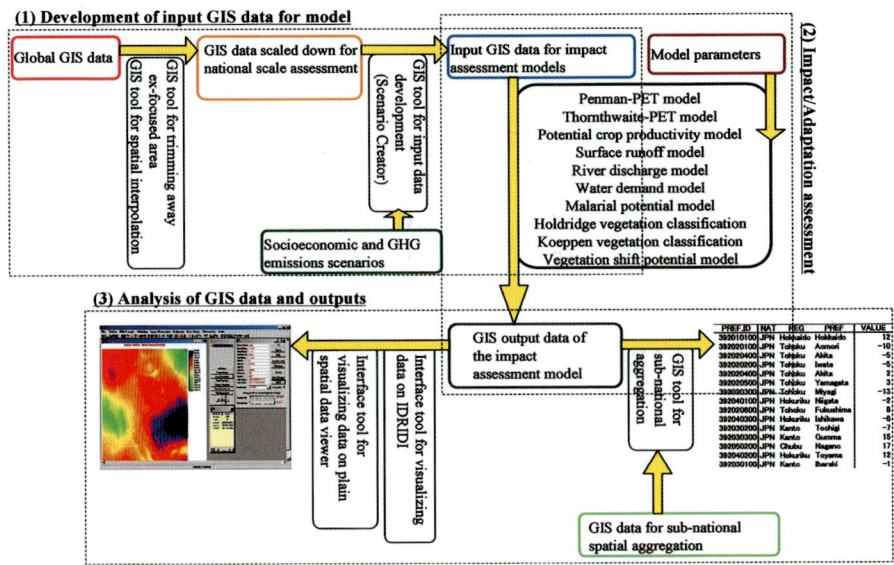

Color Plate 15. Framework of AIM/Impact [Country] (see Fig. 1 of Chap. 12)

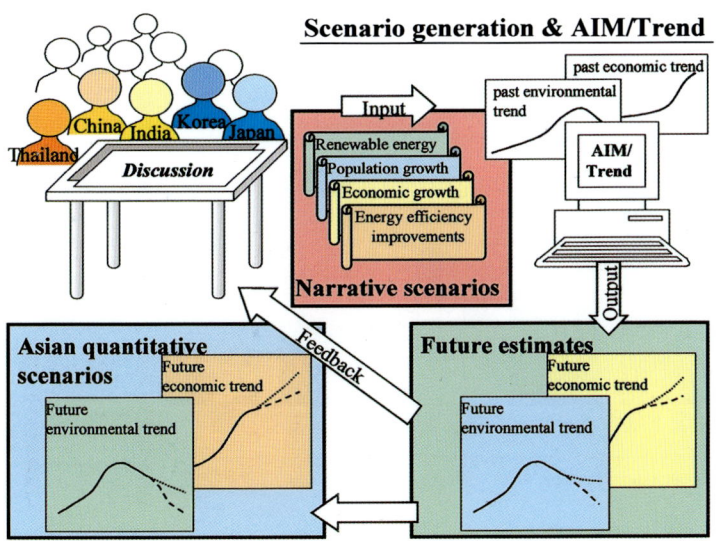

Color Plate 16. Concept of AIM/Trend model (see Fig. 1 of Chap. 13)

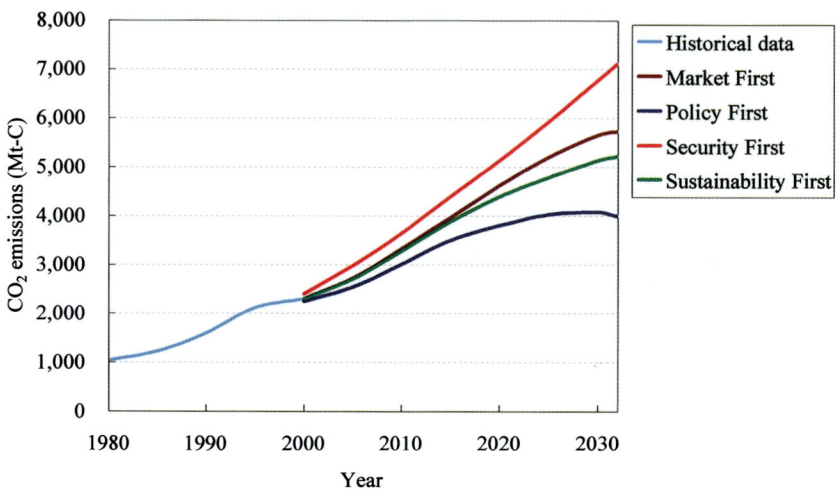

Color Plate 17. Energy related CO_2 emissions in the Asia-Pacific region (see Fig. 9 of Chap. 13)

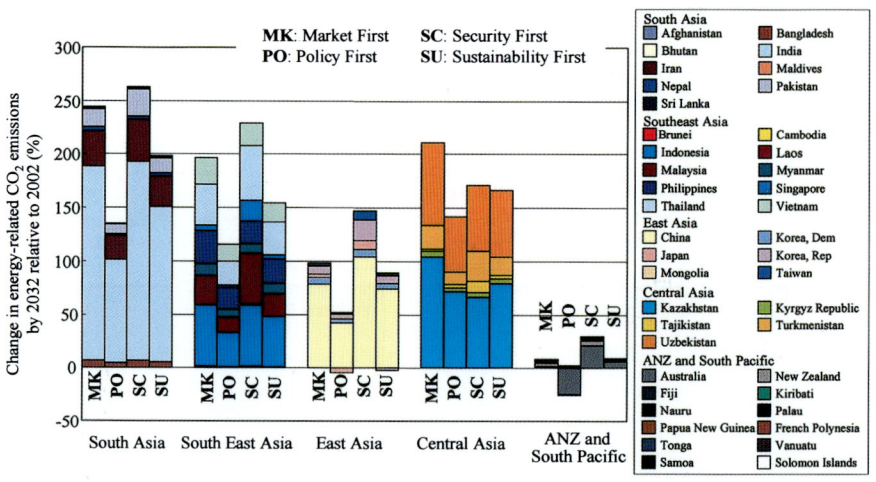

Color Plate 18. Change in energy related CO_2 emissions in sub-regions of the Asia-Pacific region (see Fig. 10 of Chap. 13)

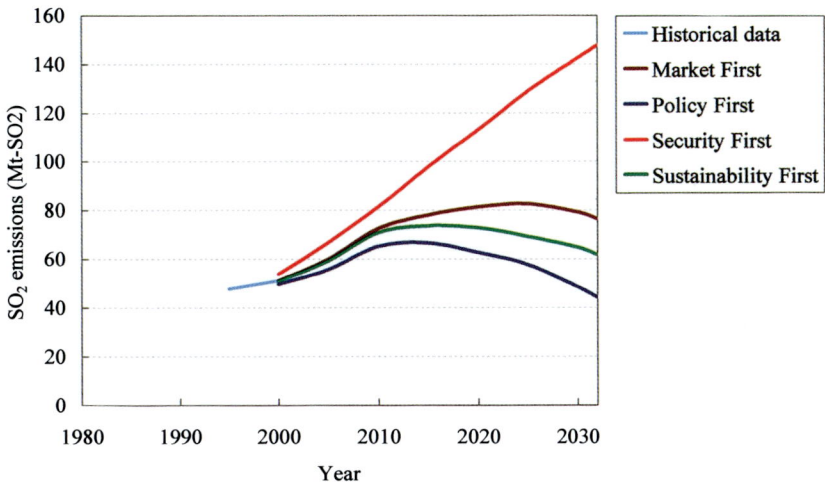

Color Plate 19. Energy related SO$_2$ emissions in the Asia-Pacific region (see Fig. 11 of Chap. 13)

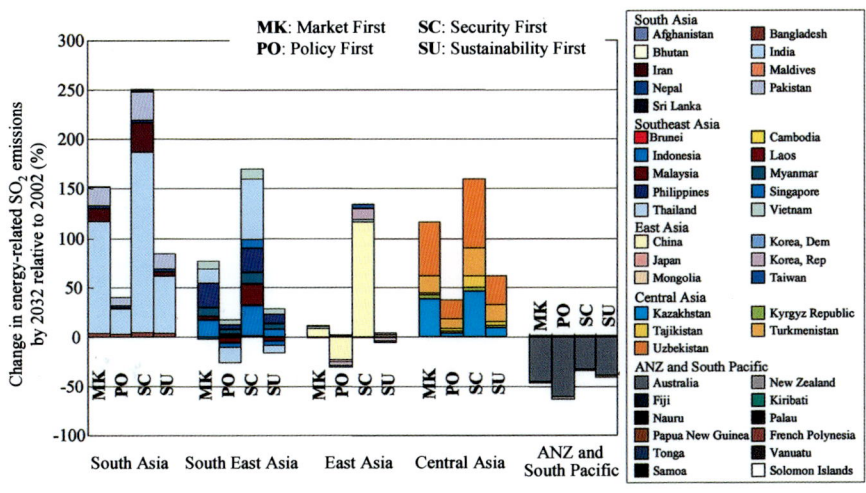

Color Plate 20. Change in energy related SO$_2$ emissions in sub-regions of the Asia-Pacific region (see Fig. 12 of Chap. 13)

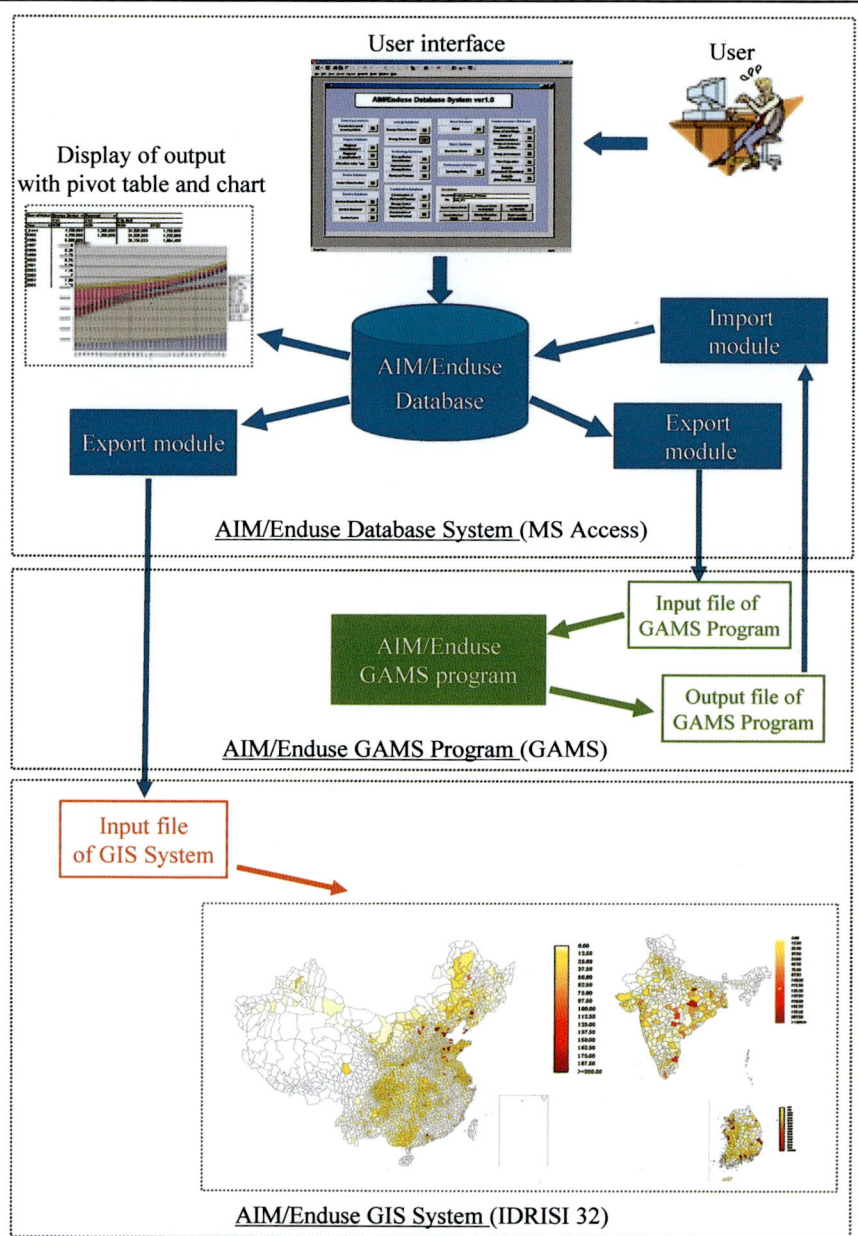

MS Access: Database management software designed by Microsoft Corporation
GAMS: General algebraic modeling system designed by GAMS Development Corporation
IDRISI 32: Geographical information system and image processing software designed
         by Clark Labs, Clark University

Color Plate 21. AIM/Enduse database system, optimization system and GIS
(see Fig. 1.4 of Part IV)

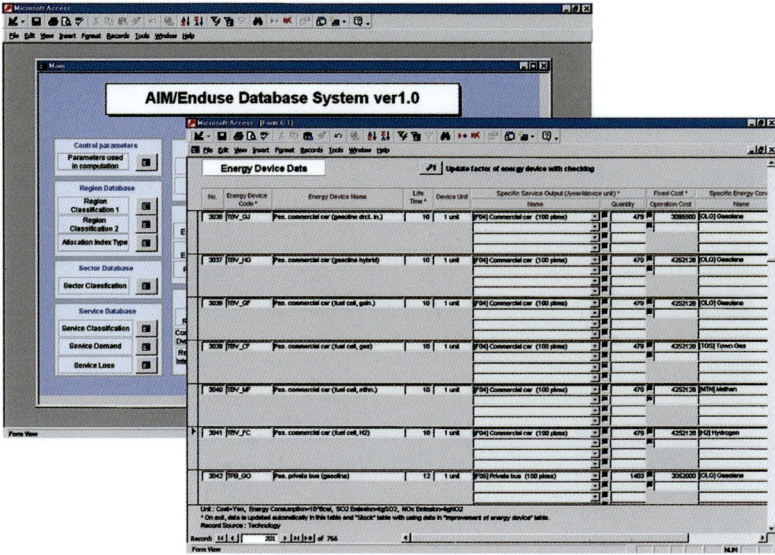

Color Plate 22. User interface for technology table (see Appendix C of Part IV)

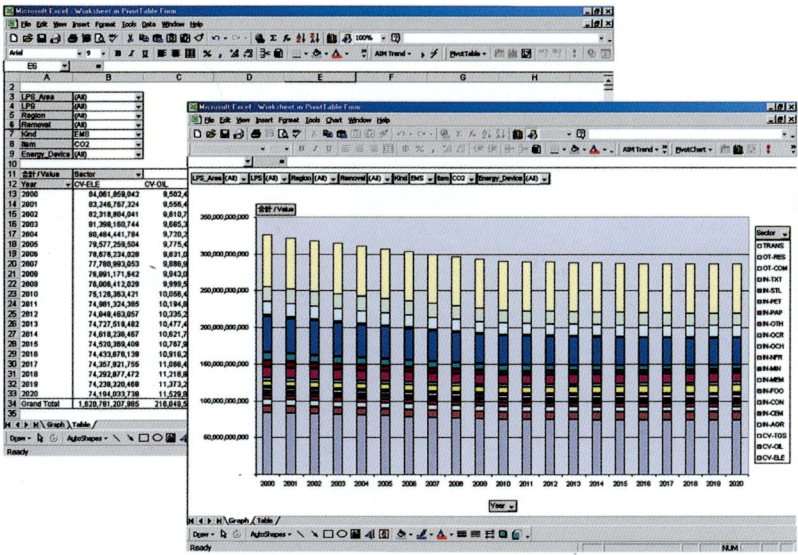

Color Plate 23. Display of output with pivot table and chart (see Appendix C of Part IV)

Color Plate 24. The Seventh International AIM Workshop, 15-17 March 2002, Tsukuba, Japan

Color Plate 25. Photos from the AIM Training Workshop, 5-6 September 2002, Tsukuba, Japan

I. Introduction and Overview

1. AIM Modeling: Overview and Major Findings

Mikiko Kainuma[1], Yuzuru Matsuoka[2], and Tsuneyuki Morita[1]

Summary. The Asia-Pacific Integrated Model (AIM) is a set of computer simulation models for assessing policy options on sustainable development particularly in the Asia-Pacific region. It started as a tool to evaluate policy options to mitigate climate change and its impacts, and extended its function to analyze other environmental issues such as air pollution control, water resources management, land use management, and environmental industry encouragement. More than 20 models have been developed so far, and they are classified into emission models, climate models and impact models from the viewpoint of climate policy assessment. The outline of these models is explained in this chapter. These models have been used as single models or in combinations depending on the policy needs, and they have contributed not only to the governments in the Asian regions, but also to international organizations such as IPCC, UNEP, Eco Asia, ESCAP, and OECD. Previous assessment based on AIM could clarify many important knowledge related to mitigation policies of climate change at global, regional, and country levels. These findings which are summarized in this chapter, have been or are expected to be reflected to the climate policies as well as sustainable development policies.

1.1 Introduction

It is predicted that global climate change will have significant impacts on the society and economy of the Asia-Pacific region, and that the adoption of measures to tackle global climate change will force the region to carry a very large economic burden. Also, if the Asian-Pacific region fails to adopt such countermeasures, it has been estimated that its greenhouse gas emissions will increase to over one-half of total global emissions by 2100. In order to respond to such serious and long-term threats, it is essential to establish communication and evaluation tools for policy makers and scientists in the region. Integrated Assessment Model (IAM) provides a convenient framework for combining knowledge from a wide range of disciplines, and is one of the most effective tools to increase the interaction among these groups.

The Asia-Pacific Integrated Model (AIM) is one of the most frequently used models in the world. The distinguished feature of AIM is that it involves Asian country teams from China, India, Korea, Indonesia, Thailand, Malaysia, Vietnam, Japan and so on, has very detailed description on technologies, and uses information from a detailed geographic information system to evaluate and present the

---

[1] National Institute for Environmental Studies, Tsukuba 305-8506, Japan
[2] Kyoto University, Kyoto 606-8501, Japan

distribution of impacts at the local and global levels. Besides preparing country models for detailed evaluation at the state and national level, we have also developed global models to analyze international economic relationships and climate impacts for evaluating policy options from global viewpoint. Although AIM model has been developed primarily to help respond to climate change problems, it has been extended to be applied to other closely related environmental problems such as air pollution, waste management, and water resources problems.

This chapter is organized as follows: Section 1.2 reviews the global environment issues determining the integrated assessment modeling approach, future directions of integrated assessment model development, and characteristics of AIM. Section 1.3 describes the structure of AIM model and main characteristics of its component models while main findings from global and country assessments are summarized in Section 1.4.

1.2 Integrated Assessment Approach and AIM

1.2.1 Global environmental issues and integrated assessment

When one considers policy making in the context of global environment, the starting point must be the characteristics of global environmental issues themselves. These determine the nature of the assessment approach and also its implementation.

a. Global environmental science is a very broad field, incorporating many distinct scientific disciplines, often with diverse theoretical frameworks and analytical approaches.
b. It also touches, directly and indirectly, upon many social and economic issues. Therefore global environmental issues can be characterized by involving the simultaneous evolution of research and policy making processes.

The importance of these two characteristics cannot be over-emphasized. The need to link wide scientific frontiers/interests to equally wide, and often distinctly differing, political interests and opinions, represents one of the most problematic characteristics of global environmental issues.

c. Many serious environmental problems are found in developing countries. Here the interface between scientific concerns (methodological and practical) and political concerns (economic and social) is characterized by loose linkages.

There are many epistemological gaps which need to be bridged between the science and policy fields, and also between different scientific disciplines. First, policy-makers are intelligent, but often lack the scientific training necessary to be able to interpret and implement detailed research results. Second, scientists have few incentives to link their research results to the policy-making process, and to integrate their own results with those from other scientific fields.

Integrated Assessment (IA) has a vital role to play in bridging these gaps. It is a necessary tool which greatly facilitates the optimal development of institutional and research linkages, projects and policy recommendations.

IA represents the "best available synthesis of current scientific, technical, economic, and sociopolitical knowledge" (IPCC 1995). It differs from disciplinary research as IA aims specifically to inform knowledge and policy decision making using a breadth of knowledge sources from a variety of disciplines.

1.2.2 Integrated assessment models

An Integrated Assessment Model (IAM) is a large-scale computer simulation model to assimilate many different factors and disciplinary inputs. As such it represents a core tool for Integrated Assessment approaches.

Though the first trial of model development for IA can be observed in the beginning of 1970s (Meadows *et al.* 1972; Mesarovic *et al.* 1974), formal IAMs emerged after the late 1970s. Nordhaus (1979), Haefele *et al.* (1981) and Edmonds *et al.* (1985) developed energy-economy integrated models for climate change assessment, and Alcamo *et al.* (1990) developed the first IAM to extend fully from emission to impact for acid rain assessment. In 1990s, IAM studies for climate change assessment have rapidly expanded, and there now exists more than twenty IA models for climate change alone.

IPCC (1996) reviewed twenty-two IAMs for climate policy assessment including AIM. A selection of seven representative models from twenty-two with their characteristics is presented in Table 1. All of the seven models have the components of GHG emissions, climate change, and impacts caused by climate change, which are integrated for the purpose of future scenario assessments as well as climate policy assessments.

IAM studies continue to be developed and the direction of these developments can be classified into the following three. First, modeling targets and phenomena are becoming wider and more detailed than before in order to respond to widened audiences, new policy needs, and new scientific knowledge. Very detailed climate change scenarios were prepared for IAMs (Hulme *et al.* 2002), special dynamic models of land use and land degradation are trying to be integrated with emission and impact models (Hootsman *et al.* 2001; Groeneveld *et al.* 2000), technology factors are trying to be introduced as endogenous variables in IAMs (Weyant and Olavson 1999), and pluralism in value system are trying to be operated and reflected to future projection in IAMs (Janssen *et al.* 1998; van Asselt *et al.* 2002). Furthermore, institutional factors such as governmental regulations, international regime and cultural systems have been proposed to be incorporated with IAMs.

The second direction is to apply IAMs to participatory IA process where stakeholders including policy makers and scientists communicate with each other to recognize the priority of information and decisions. The typical examples of the participatory IAM application are ULYSSES (Urban Lifestyles, Sustainability and Integrated Environmental Assessment) project (van Asselt *et al.* 2001), VISIONS (Integrated Visions for a Sustainable Europe) project (Rotmans *et al.* 2001), and COOL (Climate Options for the Long Term) project (Berk *et al.* 1999). These

Table 1. Seven representative IAMs

model name	model type	model components			reference
		emission	climate	impact	
AIM (Asia-Pacific Integrated Model	Large-scale policy assessment model	complex	complex	complex	---
DICE (Dynamic Integrated Climate & Economy model)	Policy optimization model	simple	simple	simple	Noldhaus (1994)
IIASA (International Institute for Applied System Analysis	Large-scale policy assessment model	complex	simple	complex	WEC/IIASA (1995)
IMAGE2.0 (Integrated Model to Assess the Greenhouse Effect)	Large-scale policy assessment model	complex	complex	complex	Alcamo (1994)
MERGE (Model for Evaluating Regional & Global Effects of GHG Reduction Policies)	Policy optimization model	complex	simple	simple	Manne *et al.* (1993)
MiniCAM (Mini Global Change Assessment Model)	Large-scale policy assessment model	complex	simple	simple	Edmonds *et al.* (1994)
TARGETS (Tool to Assess Regional &Global Environmental & Health Targets for Sustainability)	Strategic policy assessment model	simple	simple	complex	Rotmans *et al.* (1995)

projects supported European scenario development process of long term environmental change and policy introductions with policy makers and citizens.

The third direction of IAM development is to apply IAMs to regional and local assessment rather than global scale assessment. Lorenzoni *et al.* (2000) downscaled the IPCC global emission scenarios (IPCC 2000) into British impact assessment by means of their IAM, Green *et al.* (2000) estimated co-benefit of air pollution abatement policies and global warming mitigation policies for several countries, Amann *et al.* (2001) tries to extend an IAM on acid rain (RAINS) to multi-pollution and multi-effect model including air pollutants of SO_2, NO_x, VOC, NH_3 and to integrate it with country models (NIAM: National Integrated Assessment Model) to assess environmental impact in a national level (Johansson *et al.* 2001).

Such new developments of IAMs increase the audience and users, and IAMs are been adopting in many important processes of environmental policy making as core tool to link science frontier to their processes.

1.2.3 Characteristics of AIM

Our research project of IAM modeling, the Asia-Pacific Integrated Model (AIM), started in 1991, relatively a little later timing than those of other representative IAMs. However, our project has established a unique position among these IAMs.

First, AIM is only one IAM focusing on Asia, and the AIM project has been specifically developed using a collaborative approach of the Asian region, based on an international collaboration program with participation of governments and researchers of the Asian countries. The AIM project has strongly supported Asian developing countries to get their own IA tool, and these countries have already applied own IAMs to their actual policy making processes.

Second, AIM has several unique structures of model integration in addition to the emission-climate-impact integration which is commonly observed in representative IAM models.

1. integration between economic modules and environmental modules in order to analyze tradeoffs between Asian rapid economic growth and its environmental conservation, and to assess sustainable development policies in Asia;
2. linkage between top-down modules and bottom-up modules for comprehensive and consistent assessment of various policy options including top-down macro-economic policies and bottom-up technological options, which are both essential to solve the Asian complicated environmental issues;
3. linkage between short/middle-term simulation modules and long-tem simulation modules in order to integrate urgent Asian policies against current pollution and nature destruction issues with long-term policies for climate change mitigation;
4. linkage between country/local modules and world modules in order to design effective policies for regional collaboration in Asia, that include technology transfer and assistance in environmental investment desired by most of developing countries in the region.

These structures were prepared to respond to actual requests from policy making processes. These unique structures enable us easily to advance our studies toward the new direction of world IAM community written above, and AIM has already started evolving in the new directions.

Third, AIM has been used in the actual policy making process in the Asian region. Several Asian countries have adopted this model as an in-house model for policy assessment, and Environmental Ministers' Congress for Asia and Pacific (Eco Asia) adopted AIM as core tool for policy design. AIM also contributed to other international activities: AIM was selected as reference model in the Special Report on Emission Scenarios (SRES) and in Third Assessment Report (TAR) both of Intergovernmental Panel on Climate Change (IPCC) and also in the Global Environment Outlook (GEO) of United Nations Environmental Program (UNEP). AIM simulation results were used by many other international organizations including OECD, ESCAP, ADB, UNU, and WWF. Recently, the AIM modeling team started to prepare new set of models for Millennium Ecosystem Assessment (MA) and Asia-Pacific Environmental Innovation Strategy Project (APEIS).

1.3 Structure of the AIM Model

1.3.1 AIM for climate policy assessment

The roadmap of AIM models is shown in Fig. 1 and Color Plate 1. These figures also show the correspondence between these models and chapters covering their descriptions.

The original AIM is an integrated 'top-down and bottom-up' model and comprises three main models - the greenhouse gas emission model (**AIM/Emission**), the global climate change model (**AIM/Climate**) and the climate change impact model (**AIM/Impact**) (Morita *et al.* 1993; Matsuoka *et al.* 2000).

Several emission models including both top-down and bottom-up models have been developed to estimate greenhouse gas emissions and other related gases. One is a bottom-up energy model which we call **AIM/Enduse** model. It focuses on the end-use technology selection in energy consumption as well as energy production. It calculates future demand of energy services for several sectors, determines the optimal set of technologies for the demand by total cost minimization based on exogenous energy price, and then, it estimates future energy consumptions as well as future emissions of GHGs. AIM/Enduse was recently evolved to AIM/Local model which is linked to local emission inventories in order to simulate multi-gas emissions from large point source. AIM/Enduse was developed also for more comprehensive bottom-up model (**AIM/Bottom-up**) by adding industrial process module and lifestyle change module to the energy endues model.

For AIM/Emission, three kinds of top-down models were developed, those are, AIM/Energy-Economics, AIM/CGE (Energy), and AIM/Land-Equilibrium. **AIM/Energy-Economics** is a partial equilibrium model to analyze long-term en-

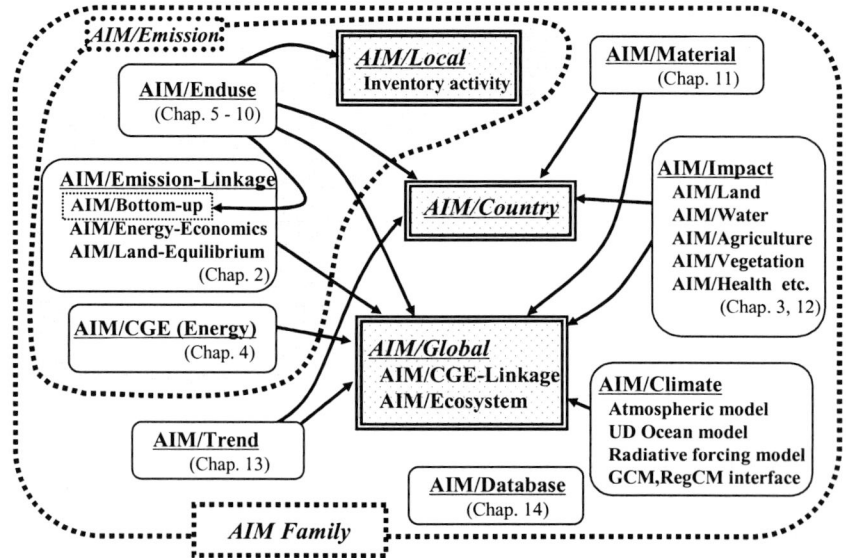

Fig. 1. Roadmap of AIM family (see color plates)

ergy demand and supply. It was mainly used for GHG emission forecast over next one century. **AIM/CGE (Energy)** is a general equilibrium model focused on energy sectors to analyze the relationship between emissions and international trade. It was frequently used for the assessment of Kyoto Mechanism such as emission trade and Clean Development Mechanism. The third top-down model, **AIM/Land-Equilibrium** is a general equilibrium model focused on agriculture and forestry sector in order to analyze land use change based on international agricultural market. This model was applied to quantification of GHG emissions caused by land use change.

For the purpose of quantifying comprehensive emission scenarios for IPCC, a specific linkage model, **AIM/Emission-Linkage** was developed. Three models were combined in this model, those were, AIM/Bottom-up, AIM/Energy-Economics, and AIM/Land-Equilibrium, and then, comprehensive GHGs emissions as well as related gas emissions such as SO_2, NO_x were simulated based on the model.

The estimated GHG emissions and related gas emissions are input into the **AIM/Climate**. Except for CO_2, GHGs emitted into the atmosphere are gradually transformed by chemical reactions, which are calculated within the AIM/Climate. For short-life chemicals such as ozone and OH radicals, pseudo-equilibrium state is assumed, and the oxidation and photochemical reactions of CH_4 and other molecules are represented by simple kinetic equations. The absorption of CO_2 and heat to ocean is calculated using an upwelling-diffusion (UD) model with the oceans divided into a surface mixed layer and an intermediate layer, which extends down to about the 1000 meters. Carbon cycle between atmosphere and terrestrial ecosystem is also reproduced by simple model which was validated by dynamic vegetation simulations.

Global averaged surface temperature changes are calculated by a simple radiative forcing model which is linked to an energy balance/upwelling-diffusion ocean model. Regional distribution of climate parameters is estimated based on the experiments of GCMs and regional climate models. AIM/Climate includes specific interface between simple module of AIM/Climate and GCMs and regional climate models in order to interpolate regional climate distribution.

The **AIM/Impact** treats mainly the impact on water supply, vegetation change, primary production industries such as agriculture and forest products, and on human health such as malaria spread. It can also be used to assess higher-order impacts on the regional economy. For these impact assessments, environmental and socio-economic data of the region have been gathered and filed in a geographical information system (GIS). The resolution of the data is from half degree to five minutes in latitude.

AIM/Database for AIM/Emission as well as AIM impact has been developed to support the development and utilization of AIM models. As mentioned above, AIM model has been developed as an international collaboration program, and the database has also been developed by the collaborative program.

1.3.2 Other models of AIM Family

Over the years of developing AIM models, a variety of new models, which are interrelated and interconnected, have been added to the family shown in Fig. 1 and Color Plate 1. These new models are AIM/Trend, AIM/Material, AIM/CGE-Linkage, AIM/Ecosystems, AIM/Country, and AIM/Global.

   AIM/Trend is developed to project the basic trend of economy, energy and environment in Asia-Pacific region. It covers as wide a range of countries in Asia-Pacific region (42 countries), and it estimates environmental conditions through 2032. It uses simple method (econometric) and develops several scenarios for capacity building. **AIM/Material** intends to estimate economic and environmental effects of environmental investment. It assesses the effects of policy integration for comprehensive environmental problems including solid, water and air pollution. It achieves consistency of material flow along with economic activities. **AIM/CGE-Linkage** integrates original global CGE model with Material, Energy-economics and Land-equilibrium modules to assess global economic development taking into account the international markets. **AIM/Ecosystem** is an extended version of AIM/Impact and intends to evaluate not only the impact of climate change, but also other important environmental issues especially ecosystem assessment. It has several modules to estimate surface runoff, river discharge, potential crops productivity, vegetation classification, and health impacts. **AIM/Country** and **AIM/Global** models integrate various modules at country and global levels respectively in order to assess the country and global sustainable development.

1.4 Major Findings

1.4.1 Findings from global assessment

Major findings from global assessment written in Part II can be summarized as follows:

1. Without the Kyoto target, the global temperature would increase by more than 2°C in 2100. In such a case, severe impacts can be predicted on water resource supply, agriculture productions, vegetation and human health (Chapter 2, 3, 4).
2. Their impacts would be significant and with serious negative damages in the low latitude regions, especially developing countries in tropical and sub-tropical zones, while climate change could cause positive effects in the high latitude regions (Chapter 3).
3. Future development of the Asia-Pacific region has significant influence on global emission scenarios. The Developing Asia-Pacific becomes dominant in climate change issue (Chapter 2).
4. Different development paths require different technology/policy measures and show different costs of mitigation to stabilize atmospheric CO_2 concentrations at the same level. No single type of measure will be sufficient for the timely development, adoption and diffusion of mitigation options for CO_2 stabilization

(Chapter 2).

5. It is predicted that ratification of the Kyoto Protocol may cause economic impacts. However there are several ways including technological development and international cooperation to mitigate the economic impacts and a possibility of promoting the growth of economies (Chapter 4).

1.4.2 Findings from regional and country assessment

The followings are also major findings from regional and country assessment written in Part II and III:

1. It could be possible for the Developing Asia-Pacific to continue high economic growth while maintaining GHG emissions at a low level. Technological progress and technology transfer should be emphasized to maintain low GHG emissions in the Developing Asia-Pacific's economic development. The market mechanism is an efficient way to achieve the diffusion of advanced technologies (Chapter 2, 5, 6, 7, 8, 9, 13).

2. It is important for the Developing Asia-Pacific to introduce sophisticated measures to control GHG emissions before 2030. Robust policy options should be designed to respond to very wide range of alternative development path (Chapter 2, 5). In China, sophisticated policies should be designed at the sectoral level, especially in transport, commerce and the chemical industry (Chapter 5).

3. New environmental policies are required that are designed for the early stages of development in China and India in order to integrate strategies for both the global environment and local environment including pollution control and waste management. Although under a SO_2 mitigation policy regime, the SO_2 and CO_2 trajectories get decoupled in India, extents of SO_2 mitigation and CO_2 mitigation are strongly correlated under a CO_2 mitigation policy regime (Chapter 5, 7, 11, 13).

4. Long-term adaptation investment for climate change could create co-benefit in other policy fields. Chinese current flood damage could be drastically reduced by the adaptation investment in flood prevention infrastructure to mitigate climate change impact (Chapter 12).

5. The LPS would continue to be responsible for considerable part of the carbon emissions. Power sector is the predominant emission source for CO_2 and SO_2. Operational improvements (like heat rate reduction, excess air control etc.), better maintenance, reducing transmission and distribution losses in the power sector would go a long way in emissions mitigation in India and China (Chapter 5, 6).

6. Energy savings or low CO_2 emitting devices could be difficult to introduce into the market in every sector by 2020 without any climate policy measures in Korea. The marginally higher cost of new low CO_2 emitting devices is too large for them to penetrate the Korean market (Chapter 8).

7. Japanese cost to reach Kyoto target depends on Japan's future development pattern and climate policy design. The development toward "Recycle-based Soci-

ety (B1)" could reduce cost and lead economic growth. Japanese cost could also be reduced by increased environmental investment, environmental industry encouragement, technology improvement, integration with waste management policy, shift to green consumption, and introduction of Kyoto mechanism (Chapter 4, 10, 11).

8. International collaboration for capacity building and knowledge transfer on IA and IAM is essential for global participation in climate change mitigation (Chapter 1, 5, 6, 7, 8, 9, 10, 11, 12, 13, 14).

References

Alcomo J, Shaw R and Hordijk L (eds) (1990) The RAINS model of acidification: Science and strategies in Europe, Kluwer, Dordrecht, The Netherlands

Alcamo J (ed) (1994) IMAGE 2.0 - Integrated model of global climate change. Kluwer Academic Publishers, Dordecht

Amann M, Johansson A, Lukewille D *et al.* (2001) An integrated assessment model for fine particulate matter in Europe. Water, Air, and Soil Pollution 130:223-228

Berk MM, Hordijk L *et al.* (1999) Climate options for the long term (COOL). Interim report, NOP, Bilthoven, the Netherlands

Edmonds J and Reilly J (1985) Global energy: Assessing the future, Oxford University Press, New York

Edmons J, Wise M, MacCracken (1994) Advanced energy technologies and climate change: an analysis using the global change assessment model (GCAM). Presentation to the Air and Waste Management Meeting, 6 April, Tempre AZ, Air and Waste Management Association, Pittsburgh

Green C (ed) (2000) Developing country case-studies: integrated strategies for air pollution and greenhouse gas mitigation. Progress Report for the International Co-Control Benefits Analysis Program, The National Renewable Energy Laboratory, The Office of Atmosphere Programs of the US Environmental Protection Agency

Groenveld RA, van Ierland EC (2000) Economic modelling approaches to land use and cover change. NRP Report No.410200045, Wageningen University

Haefele W, Anderer J, McDonald A and Nakicenovic N (1981) Energy in a finite world, Ballinger, Cambridge, MA

Hootsman RM, Bouwman AF, Leemans R, Kreileman GJ (2001) Modeling land degradation in IMAGE 2. RIVM report 481508009, National Institute of Public Health and the Environment, The Netherlands

Hulme M, Lu X, Turnpenny J, Mitchell T *et al.* (2002) Climate change scenarios for the United Kingdom. The UKCIPO2 Scientific Report, Tyndall Centre, Department for Environment, Food and Rural Affairs, UK

IPCC (1996) Climate change 1995: economic and social dimensions of climate change. Bruce JP, Lee H, Haites E (eds), Cambridge University Press, Cambridge

IPCC (2000) Special report on emissions scenarios. Nakicenovic N *et al.* (eds), Cambridge University Press, Cambridge

Jannsen M, de Vries B (1998) The battle of perspectives: a multi-agent model with adaptive responses to climate change. Ecological Economics 26:43-65

Johansson M, Alveteg A, Amann M *et al.* (2001) Integrated assessment modeling of air pollution in four European countries. Water, Air, and Soil Pollution 130:175-186

Lorenzoni I, Jordan A, O'Riordan T, *et al.* (2000) A co-evolutionary approach to climate change impact assessment: Part II. a scenario-based case study in East Anglia (UK). Global Environmental Change 10:145-155

Manne AS and Richels RG (1993) MERGE - A model for evaluation regional and global effects of GHG reduction policies. Energy Policy, 23(1):17-34

Matsuoka Y (2000) Extrapolation of carbon dioxide emission scenarios to meet long-term atmospheric stabilization targets. Environmental Economics and Policy Studies, 3: 255-265

Meadows DH, Meadows DL, Randers J and Behrens WW (1972) The limits to growth, Universe Books, New York

Mesarovic MD and Pestel E (1974) Mankind at the turning point: The second report to the Club of Rome, Dutton, New York

Morita T, Matsuoka Y, *et al.* (1993) AIM - Asian Pacific integrated model for evaluating policy options to reduce GHG emissions and global warming impacts. In "Global warming issue in Asia", Asian Institute of Technology, pp.254-273

Nordhaus WD (1979) The efficient use of energy resources, Yale University Press, New Haven, CT

Nordhaus WD (1994) Managing the global commons: the economics of climate change. The MIT Press

Rotmans J (1995) TARGETS in transition. RIVM report, Bilthoven, The Netherlands

Rotmans J, van Asselt M *et al.* (2001) Integrated visions for a sustainable Europe: changing mental maps:VISIONS final report. ICIS, Maastricht/Utrecht

van Asselt MA, Middelkoop H *et al.* (2001) Integrated water management strategies for the Rhine and Meuse basins in a changing environment. Final Report of the NRP project 0/958273/01, ICIS, Maastricht/Utrecht

van Asselt MA, Rotmans J (2002) Uncertainty in integrated assessment modeling, from positivism to pluralism. Climate Change 54:75-105

WEC and IIASA (1995) Global energy perspectives to 2050 and beyond. World Energy Council, London

Weyant JP, Olavson T (1999) Issues in modeling induced technological change in energy, environmental, and climate policy. Environmental Modeling and Assessment 4:67-85

II. Global Modeling and Application

2. Long-term Scenarios based on AIM Model

Tsuneyuki Morita[1], Kejun Jiang[2], Toshihiko Masui[1], Yuzuru Matsuoka[3], and Ashish Rana[1]

Summary. In order to respond to climate change, it is essential to describe possible future trajectories of greenhouse gas (GHG) emissions in terms of both nonintervention and intervention strategies. This chapter analyzes long-term GHG emissions scenarios according to alternative development paths for the world and major regions, based on the nonintervention emissions scenarios quantified by the Asian-Pacific Integrated Model (AIM). AIM has been revised and applied to the quantification of story lines for scenarios of socioeconomic development, and GHG emissions from energy use, land use changes, and industrial production processes are simulated. A wide range of mitigation policies have been adopted as responses to climate change. The results show that to achieve stabilization at a different GHG concentration level, it is essential to have a policy package to reach the target concentration level, rather than a single policy. Energy efficiency improvements and renewable energy introduction make a key contribution to the reduction of GHG emissions as a result of such a policy package. The mitigation costs could be small without a significant reduction in economic growth. The developing world could substantially reduce GHG emissions compared with nonintervention scenarios with sufficient knowledge transfer from the developed countries.

2.1 Introduction

The Asia-Pacific region, which covers Asia plus the Oceania region, excluding the Middle East, has half the world's population and is experiencing high economic growth, making it a major growth center in the global economy. Many countries in the region share problems that arise from rapid industrialization, population growth, and the increasing concentration of people in cities. With energy consumption increasing rapidly, the region is a major and growing driver of climate change. On the other hand, the region will suffer significant damage from climate change in terms of its impact on water resources, agriculture, ecosystems, and natural disasters. The Asia-Pacific region has been emerging as a dominant force for human responses to climate change.

To enable the Asia-Pacific region to respond to the climate change issue, the first step is to forecast the regional scenarios for greenhouse gas emissions in rela-

---

[1] National Institute for Environmental Studies, Tsukuba 305-8506, Japan
[2] Energy Research Institute, State Development Planning Commission, Guohong Building, A-11 Muxidi Beili, Xicheng District, Beijing 100038, China
[3] Kyoto University, Kyoto 606-8501, Japan

tion to world emissions. Future emissions scenarios are mostly dependent on regional development patterns, and this region has a wide range of options for its development path. This means that future GHG emissions could be diverse, depending on the future development path. The recognition of such diverse non-intervention scenarios is very important in assessing the policy options for responding to climate change, since any reduction in the level of GHG emissions depends not only on the target climatic stabilization level, but also on the rate of increase in GHG emissions in a baseline non-intervention scenario.

It is known that a lot of emissions scenarios have already been quantified or published, including ones for the Asia-Pacific region. The most popular scenarios are the IS92 scenarios published by the IPCC in 1992 (IPCC 1992), and the number of other quantified scenarios comes to more than 150 for China, 30 for India, and 20 for South Asia (Morita *et al.* 1994; Morita and Lee 1998). However, none of these scenarios was explicitly analyzed from the viewpoint of future alternative paths for development in the Asia-Pacific region. Only some scenarios clarified the relationship between development patterns and emissions at the global level (Lashof and Tirpak 1990; WEC 1993). The IPCC activities for the Special Report on Emissions Scenarios (SRES) (Nakicenovic *et al.* 2000) gave us an opportunity to analyze the Asia-Pacific emissions scenarios with explicit consideration of the regional paths for future development.

This chapter presents some results from the AIM model simulations for emissions scenarios for the Developing Asia-Pacific countries and the world with and without climate change policies. In the next section, SRES scenarios are briefly introduced. This is followed by a description of the model. Following this, the results, findings and conclusions are given.

2.2 SRES Scenarios

Each scenario links one of four "story lines" with one particular quantitative model interpretation. All the scenarios based on a specific story line constitute a scenario "family." The following paragraph describes four story lines, driving forces of the SRES scenarios and their relationships. Each story line represents the playing out of different social, economic, technological and environmental developments (or paradigms), which may be viewed positively by some people and negatively by others. Possible "surprise" and "disaster" scenarios were excluded.

The main characteristics of the four SRES story lines and scenario families are:

- The A1 story line and scenario family describes a future world of very rapid economic growth, low population growth and the rapid introduction of new and more efficient technologies. Major underlying themes are convergence among the regions, capacity building and increased cultural and social interaction, with a substantial reduction in regional disparities in per capita income.
  Scenarios in the A1 family were categorized into four groups according to their technological emphasis—on coal (A1C), oil and gas (A1G), non-fossil fuel en-

ergy sources (A1T) or a balanced mix of all three (A1B). The last group, a balanced mix, is simply noted in this chapter as "A1".

- The A2 story line and scenario family describes a very heterogeneous world. The underlying theme is self-reliance and preservation of local identities. Fertility patterns across regions converge very slowly, resulting in high population growth. Economic development is primarily regionally oriented, and per capita economic growth and technological change are more fragmented and slow compared to other story lines.

- The B1 story line and scenario family describes a convergent world with rapid change in economic structures toward a service and information economy, reduction in materials intensity and the introduction of clean and resource-efficient technologies. The emphasis is on global solutions for economic, social and environmental sustainability, including improved equity, but without additional climate initiatives.

- The B2 story line and scenario family describes a world in which the emphasis is on local solutions to economic, social, and environmental sustainability. It is a world with less rapid, and more diverse technological change, but with a strong emphasis on community initiatives and social innovation to find local and regional solutions. While policies are also oriented towards environmental protection and social equity, they are focused on the local and regional levels.

2.3 Model Description

In order to quantify GHG emissions from various sources, a new linkage module of the integrated assessment model was developed and comprehensive storylines of development were established. Future projections were made using the integrated assessment model for energy use, energy production, industrial processes, land-use changes, agricultural production, livestock, etc. from 1990 to 2100 according to the storylines. These projections were finally converted to the GHG emissions scenarios.

A model framework called the AIM/Emission-Linkage model was developed for this emissions scenario study. It links several models to calibrate the data and perform scenario quantification. An important point to note is that the development pattern of the developing Asia-Pacific region should be analyzed in relation to the global regime since international issues will strongly influence the region's future environment, economy, and energy activities. Scenarios for the developing Asia-Pacific region should also be closely related to scenarios for other regions. Hence, the model framework adopted was a global model divided into key regions.

Major emission sources including energy activities, industries, land use, agriculture, and forests can be simulated in the model framework. The structure of the AIM/Emission-Linkage model comprises two kinds of top-down models – energy and land-use – and a set of bottom-up models as shown in Fig. 1. The GHG emissions from energy consumption and energy production are simulated by the en-

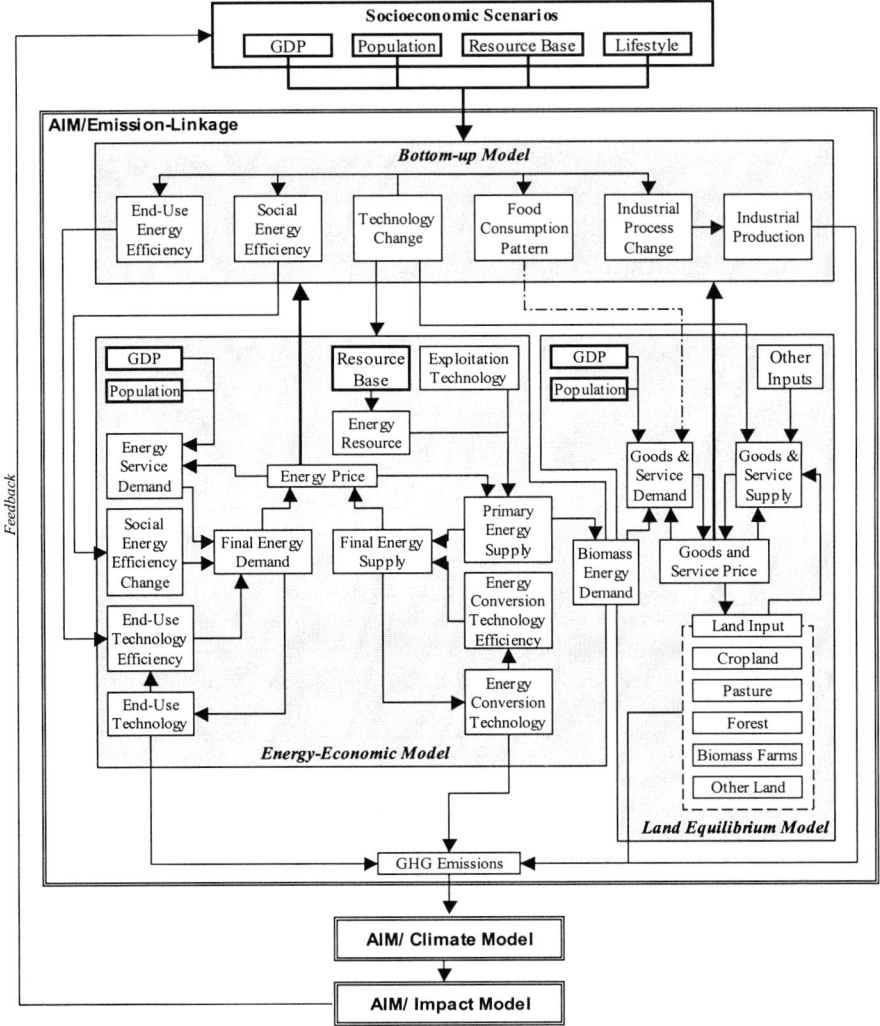

Fig. 1. Outline of the AIM/Emission-Linkage model

ergy model. GHG emissions from land use are derived from the land use model, while GHGs from other emission sources are calculated by simplified industry process models that describe the relationship between GDP per capita and industrial product outputs.

The energy sector top-down module was developed based on the revised Edmonds-Reilly-Barns (ERB) model (Edmonds *et al.* 1983; Edmonds *et al.* 1995), which is widely used for the analysis of emissions. The top-down model for the energy sector provides a consistent, conditional representation of economic, demographic, technical, and policy factors as they affect energy use and production. It is a macroeconomic partial-equilibrium model that deals with energy activities and forecasts energy demand over the long term. It uses the gross domestic product (GDP) and population as future development drivers, combined with other energy-related parameters to forecast energy demand based on the supply and demand balance. Three end use sectors—industrial, residential and transportation— and one energy conversion sector— the power generation sector—are specified in the model. Energy efficiency is described by both technology efficiency and social efficiency improvements. A number of technologies in these four sectors are listed in the model to present different possibilities of technological progress. A link between the bottom-up energy model and the top-down energy model has been developed. A detailed energy use analysis for the developing Asia-Pacific region from the bottom-up model drives the energy use pathway before 2030, while a simplified linkage is presented for other regions in the model. The linked AIM/Enduse model and the energy top-down model comprise the energy model in the model framework.

The top-down land use model is based on the Global Trade Analysis Project (GTAP), which was established in 1992 (Hertel 1997). This model is an applied general-equilibrium model that divides the world into multiple regions. For the sake of this analysis, the land uses for agriculture, livestock, and forests are considered, and the biomass energy demand is considered exogenously. It is designed to explicitly model agriculture and land use, endogenously determine emissions resulting from land use changes, and explore the use of biomass as an element of a strategy for anthropogenic carbon emissions.

The bottom-up models were prepared using the original AIM bottom-up components, which can reproduce detailed processes of energy consumption, industrial production, land use changes, and waste management as well as technology development and social demand changes. The AIM/Enduse model is part of the Asian-Pacific Integrated Model (AIM), which was developed by the National Institute for Environmental Studies (NIES) and Kyoto University (AIM Project Team 1996; Hibino et al. 1996). It is a bottom-up, energy technology model. Based on detailed descriptions of energy services and technologies, it calculates the total energy consumption and production in a bottom-up manner. This model has been used to analyze several key countries in the Asian region, including China, India, Indonesia, and Japan. AIM/Enduse models for key Asian developing countries have been constructed, and the results of analyses using this model have been reported (Jiang et al. 1998; Hu et al. 1996). Among the advantages of bottom-up models, the most important is that their results can be interpreted clearly because they are based on detailed descriptions of changes in human activities and technologies.

The AIM/Emission-Linkage model combines these various components to calculate future GHG emissions in a relatively wide ranging analysis. For the purpose of the model, the world is divided into nine regions: USA, Western Europe, OECD countries and Canada, Pacific OECD, Eastern Europe and the former Soviet Union, Centrally Planned Asia and China, South and East Asia, the Middle East, Africa, and Central and South America. The model has a time horizon extending from 1990 to 2100. The time steps are in units of 5 years up to 2030, followed by time steps at 2050, 2075, and 2100. The GHGs covered in the nonintervention emissions scenarios are CO_2, N_2O, CO, NO_x, and CH_4. Since SO_2 has a strong influence on climate change and is an important pollutant in local areas (Gan 1998; Qi *et al.* 1995), it is also included. CO_2 emissions are analyzed in the intervention scenarios.

2.4 Emissions Scenarios without Climate Change Policies

2.4.1 Assumptions

The data has been compiled from several sources. Based on the descriptions of development patterns (story lines), the quantified key scenario drivers for each scenario used in our model for the Developing Asia-Pacific and the world are listed in Table 1.

The population data resources include estimations by the United Nations and IIASA. There are three patterns for population growth: low (for scenarios A1 and B1), medium (for scenario B2), and high (for scenario A2). The high and low

Table 1. Key scenario drivers assumed for the Developing Asia-Pacific countries and the World

	A1	A2	B1	B2
Developing Asia-Pacific population	4.2 billion in 2050; 2.9 billion in 2100	5.8 billion in 2050; 7.3 billion in 2100	4.2 billion in 2050; 2.9 billion in 2100	4.7 billion in 2050; 5.0 billion in 2100
Developing Asia-Pacific GDP growth rate	6.4% from 1990 to 2050, 4.6% from 1990 to 2100	3.9% from 1990 to 2050, 3.4% from 1990 to 2100	5.6% from 1990 to 2050, 4.0% from 1990 to 2100	5.7% from 1990 to 2050, 3.8% from 1990 to 2100
World population	9 billion in 2050; 7 billion in 2100	15 billion in 2100, Higher growth in non-OECD countries	9 billion in 2050; 7 billion in 2100	11.7 billion in 2100
World GDP	$550 trillion in 2100, High growth in non-OECD countries	$250 trillion in 2100	$350 trillion in 2100	$250 trillion in 2100
GDP/capita trends	OECD: more than $100,000 by 2100; Non-Annex I: >$70,000 by 2100, $14,000 by 2040	Lower growth in non-OECD countries. Disparity rises.	Annex-I: more than $90,000 by 2100; Non-Annex-I: >$30,000 by 2100; Global $40,000	Disparity remains; GDP/capita of OECD is 7 times that of non-OECD (now 13 times).
AEEI	1.2%-1.6%	0.8%-1.0%	1.6%-1.8%	1.0%-1.2%
International trade	High trade Low trade cost	Low trade across regions, high trade within regions; High trade cost	High trade Low trade cost	Low trade across regions High trade cost
Urbanization	Rapid increase	Increase in non-OECD Decrease in OECD	Increase	Decrease

population assumptions were adopted from the research output of IIASA, while the medium population assumptions were taken from the medium case of the World Bank population forecast.

Energy resources in the model include not only conventional energy such as coal, crude oil and natural gas, but also unconventional sources of oil and gas. The energy resources in the simulation are ultimately the recoverable reserves. The energy reserves available for exploitation are determined by progress in energy production technologies, which are described in the story lines. For example, in the A1 scenario, available energy reserves are taken to be plentiful by assuming high levels of improvements in the efficiency of energy exploitation technologies that will make unconventional forms of energy available. Due to the large volume of energy supply, several sub-scenarios were defined according to this possible pathway of energy supply. The A1C scenario describes large amounts of available coal reserves. Oil and gas supply is high in the A1G scenario, while the A1T scenario assumes progress in high technology for renewable energy. A1B is a balanced scenario taking elements from these A1 scenarios. Energy resource availability in Scenario A2 mainly relies on energy resources distribution among the regions. The low technological progress assumption for energy production in scenario A2 means that energy exploitation mainly relies on conventional energy. Due to concern for environment, energy resource pressure is not so great for the B1 and B2 scenarios.

The grades of energy resources used in the model differ on the basis of exploitation costs. When combined with the level of improvements in exploitation technology efficiency (expressed as the rate of improvement in the marginal cost of producing energy), the graded energy resource exploitation cost determines the primary energy production cost (price). Table 2 shows the total for the energy resources assumed in the model.

Improvements in energy efficiency mainly rely on the parameters of social efficiency improvements and technological efficiency improvements based on the energy market. Social efficiency improvements were determined according to factors such as changes in the economic structure, the trend toward dematerialization, patterns in people's lifestyles, transportation, etc. Technological efficiency improvements were set up according to the introduction of different technologies. For example, on the demand side, efficiency was assumed to be relatively low in the A1

Table 2. Assumed total energy resources

	Conventional oil	Conventional gas	Coal	Unconventional oil	Unconventional gas
OECD	1,271	2,186	56,808	12,709	73,062
CIS & EE	1606	2,679	62,439	451	36,628
Developing Asia-Pacific	912	657	20,385	556	9,379
ROW	7,325	3,449	3,368	4,403	45,026
World	11,114	8,971	143,000	18,119	164,095

Unit: Exa (10^{18}) joule.

scenarios since low energy prices provide very little incentive to improve end use energy efficiencies and high income levels will encourage people to pursue comfortable and convenient lifestyles (especially in the household, services, and transportation sectors). This will result in the consumption of much more energy. Efficient technologies are not fully introduced into the end use side, dematerialization processes in the industrial sector are not well promoted, lifestyles become energy intensive, and there is greater use of private motor vehicles in developing countries as per capita GDP increases. Thus, the final energy use in scenario A1 is much higher than in the other scenarios (A2, B1, B2), while the difference in per capita final energy use between Annex 1 countries and non-Annex 1 countries in 2100 is small. In determining energy efficiency in the Developing Asia-Pacific, attention was paid to the processes of social and economic restructuring that have been underway since the 1980s. This is a key issue for short- and medium-term analyses of the Developing Asia-Pacific countries.

2.4.2 Quantified scenarios

From the parameters used for the inputs, outputs were obtained from the model on energy use and GHG emissions.

Global primary energy use will keep increasing up to 2100, except in scenario B1, in which primary energy starts to decrease after 2075. The range for primary energy use by the end of the next century is quite large and is 6.7 times that in 1990 for A1C, while it is only 2.3 times for B1. The primary energy intensity has a similar range, which is from 2.41 GJ/MUS$ for B1 to 7.7 GJ/MUS$ for A2, while it was 16.5 GJ/MUS$ in 1990.

With regard to primary energy demand in the Developing Asia-Pacific, all of the scenarios show increases to support the demand for economic development at least up to 2075, and then a decrease is found for scenario B1. The highest primary energy demand in 2100 (scenario A1C) is 10.5 times that in 1990, while the lowest (scenario B1) is 3.7 times. The growth rate for primary energy use in the Developing Asia-Pacific is higher than the global average. Per capita primary energy use ranges from 75 to 274 GJ, compared with 181 to 403 GJ for OECD countries. A significant catch-up effect is seen in per capita primary energy use although it is still lower than that of the OECD countries. The primary energy mixes also show highly significant changes during the next century. However, the changes are quite different for each scenario, in accordance with the conditions described in the story lines. Figure 2 shows the primary energy mix for scenario A1B, in which renewable energy becomes dominant.

GHG emissions follow the pattern of energy use in the case of CO_2 and SO_2 emissions. Due to the introduction of renewable energy, CO_2 emissions reach a peak between 2030 and 2050 (Fig. 3), except for scenario A2 for the world and A2 for the Developing Asia-Pacific in which coal use will continue and renewable energy use will be limited. Per capita CO_2 emissions in the Developing Asia-Pacific rise from 0.56 tons of carbon (t-C) in 1990 to range from 0.97 to 1.9 t-C in 2030, and 0.84 to 4.8 t-C in 2100, while the corresponding data for the OECD countries are 3.34 t-C in 1990, 2.55 to 4.71 t-C in 2030, and 1.13 to 5.73 t-C in 2100. All the scenarios show that per capita CO_2 emissions do not reach the level of the OECD countries in 2100.

By the year 2100, the range of most GHG emissions scenarios expands significantly for both the world and the Developing Asia-Pacific. For example, CO_2 emissions in these scenarios for the world in 2100 range from 6.0 Giga tons of carbon (Gt-C) per year to 36.8 Gt-C per year, a factor of more than 6 (Fig. 4). They are 2.4 Gt-C, 13.9 Gt-C and 5 Gt-C for the respective scenarios for the Developing Asia-Pacific. Consequently, these results highlight how future emissions estimates can vary according to different development pathways.

Global SO_2 emissions reach a peak between 2020 and 2050, and then start to decrease (Fig. 5). The emissions fall below the 1990 level by 2050 except for scenario A1C. The global SO_2 emissions trajectory follows that of the developing countries. The Kuznets curve was introduced to calculate SO_2 and NO_x emissions. Based on the historical data, per capita GDP is a key factor in controlling the level of SO_2 emissions. Since the per capita GDP in the Developing Asia-Pacific is rapidly increasing, all SO_2 emissions decrease after 2030 (Fig. 6). Emissions are

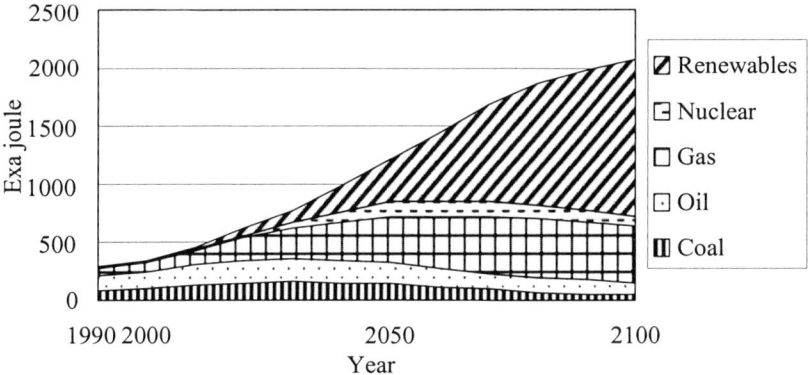

Fig. 2. Global primary energy in A1B scenario

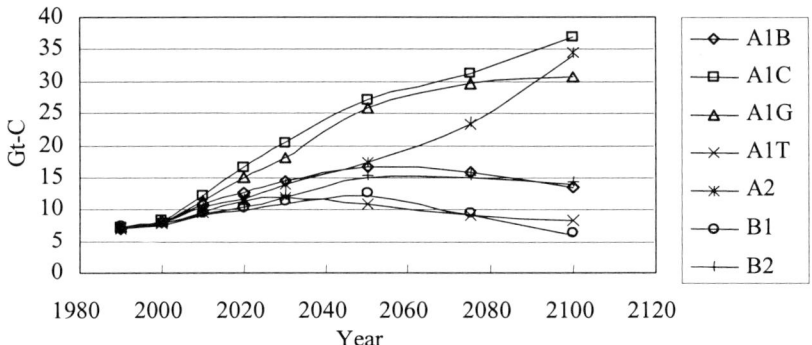

Fig. 3. Global CO_2 emissions

higher in scenario A1C due to the large amount of coal used. The range of SO_2 emissions in 2100 is relatively smaller than that of other gases. This is the result of domestic environmental policies to control pollutant emissions rather than climate change policies.

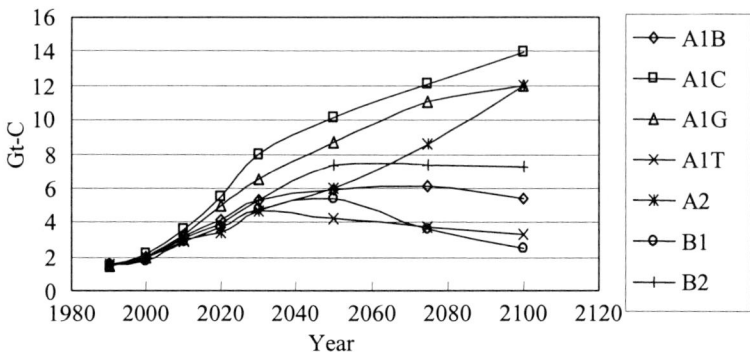

Fig. 4. CO_2 emissions in Developing Asia-Pacific

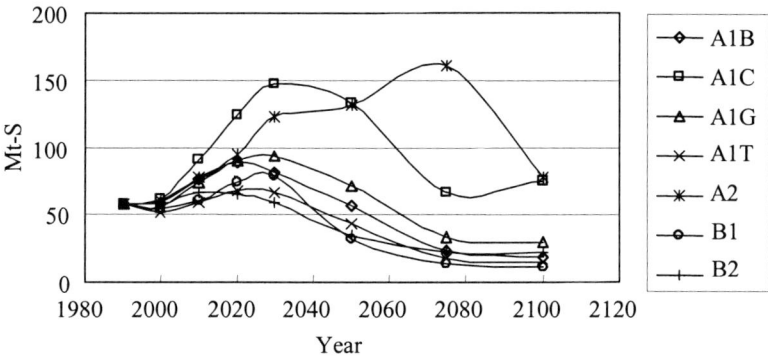

Fig. 5. Global SO_2 emissions

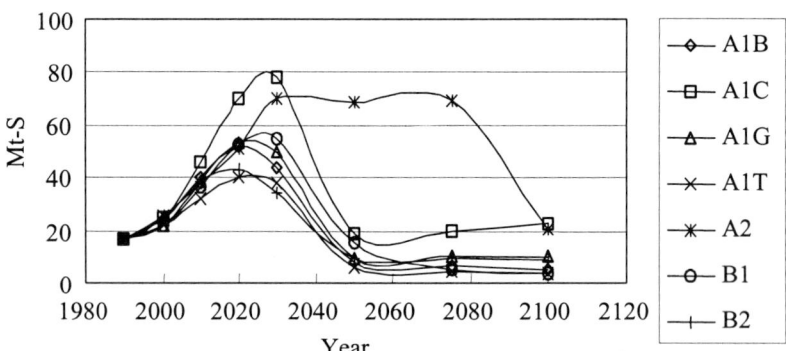

Fig. 6. SO_2 emissions in Developing Asia-Pacific

It is interesting to look at the A1 scenario emissions family. This scenario family has a quite wide diversity. Scenarios A1B and A1T have low emissions, indicating that it is possible for the Developing Asia-Pacific to maintain a high economic growth rate with low emissions. Energy end use technology improvements and large-scale renewable energy recovery will play key roles in the trajectory. Scenarios A1C and A1G have high emissions due to the huge fossil fuel consumption. Their emissions are among the highest of all the groups. If the world follows either of these paths, new technologies for emissions control and new policies will have to be introduced. The same applies to NO_x emissions.

2.5 Emissions Scenarios with Climate Policies

Scenarios in the previous section (SRES scenarios) do not include any explicit mitigation or stabilization policies or measures. As such, they include scenarios ranging from rapidly increasing to decreasing emissions over the next one hundred years. New scenarios based on the wide range of SRES scenarios were quantified as a set of mitigation (policy intervention) scenarios for stabilizing atmospheric GHG concentrations. Therefore, the policy/technology measures assumed in these scenarios are strongly affected by baseline emission trajectories of SRES scenarios as well as by their socio-economic assumptions. They describe mitigation measures and policies (the additional climate initiatives) that would have to be undertaken, in each SRES scenario "world," to achieve stabilization at different levels (450, 550, 650, and 750 ppmv). As a result, the analysis and comparison of scenarios with climate policies (AIM stabilization scenarios) can supply very systematic data to clarify the relationship between the relative contribution of the development path and climatic policy/technology measures. Knowledge of these relationships can in turn enable us to assess robust policy/technology options for different future development paths.

2.5.1 Policy package design for stabilizing global climate

The AIM stabilization scenarios were simulated to quantify the various pathways to reach the desired target for global GHG concentrations by the end of the 21st century. A policy package was designed for this quantification based on the diverging baseline scenarios.
The policy package used in the AIM stabilization scenarios is as follows:

- Improved transportation efficiency. Higher transportation technology efficiency and the introduction of advanced transport technologies, such as electric vehicles and fuel cell vehicles, are included.
- Social efficiency gains. Efficiency improvements from industrial structure changes and lifestyle changes are considered.
- Improved power generation efficiency. More advanced power generation technologies are introduced.

- Improved end use efficiency. Higher end use technology efficiency improvements are adopted.
- Nuclear power progress. Advanced nuclear power generation technologies such as the fast breeder reactor (FBR) are emphasized.
- Incentives for natural gas use.
- Carbon tax. A carbon tax is levied on the basis of carbon emissions.
- Renewable energy incentives. Solar, wind, geothermal, and ocean energy will be well developed.
- Synthesized fuel production.
- Commercial biomass: early introduction, larger share. Commercial biomass will involve low cost technology to bring it to the market.
- Preference for forests.

Population and GDP growth are not designed to be reduced for mitigation, although there will be some reduction in the GDP due to the introduction of the above policies.

All these policies are incorporated in the AIM stabilization scenario analysis based on the merits of each baseline scenario. In the A1B baseline scenario, successful economic development, social prosperity, human equity, etc., are the key factors. Consequently end use technology efficiency improvements and social efficiency improvements are emphasized in the A1B stabilization scenario analysis. Intergenerational equity is considered in the A1 mitigation scenarios to avoid major pressure on CO_2 emissions reductions after 2050. In the A2 scenario failed economic development results in inequity and little improvement in technological efficiency. Hence, technological efficiency improvements, commercial renewable energy utilization, and nuclear technology incentives are adopted in the A2 stabilization scenario simulation. A neutral policy level was maintained for the stabilization scenario analysis of the B2 world, since the B2 baseline scenario already includes an understanding of the importance of human welfare and inequity, as well as environmental solutions. There is no major pressure for policies in the AIM B1 stabilization scenario for a 550 ppmv stabilization level analysis.

In the A1B stabilization scenario family, much stricter policies are required for the 450 ppmv stabilization analysis. A wider range of policies has to be introduced, and strong policies have to be considered in order to attain the large reduction in CO_2 emissions. Early reduction is essential to avoid substantial pressure on social development and technological progress in the latter part of the 21st century. Investment in technology R&D will contribute to CO_2 emissions reductions over the next several decades. High carbon tax rates must also be adopted at an early stage even in the developing countries.

By examining all the policies adopted in the AIM stabilization scenario analyses, some policies such as carbon taxes, end use efficiency improvements, and renewable incentives are seen in all the stabilization scenario analyses. All these policies could be regarded as robust policies.

The quantified policies in this study are shown in Table 3 based on the model parameters.

Table 3. Policy option package for stabilization at 550 ppmv

Policy options	A1B	A2	B1	B2
Transport efficiency improvements	Vehicle fuel use efficiency improvement rate will be 0.14% higher than the BaU case for all regions, starting from 2000	Vehicle fuel use efficiency improvement rate will be 0.14% higher than the BaU case for all regions, starting from 2000	Vehicle fuel use efficiency improvement rate will be 0.1% higher than the baseline case for all regions, starting from 2000	Vehicle fuel use efficiency improvement rate will be 0.1% higher than the baseline case for all regions, starting from 2000
Other end use technology efficiency improvements	-	-	0.1% higher efficiency improvements	0.15% higher efficiency improvements
Power generation efficiency	0.13% higher efficiency improvement	0.15% higher efficiency improvement	0.1% higher efficiency improvement	0.1% higher efficiency improvement
Social efficiency improvement	0.3% higher energy efficiency improvement, additional 0.2% higher energy efficiency improvement in developing countries from 2030 to 2050	0.3% higher energy efficiency improvement, additional 0.2% higher energy efficiency improvement in developing countries from 2030 to 2050	0.1% higher energy efficiency improvement, starting from 2000; additional 0.1% higher energy efficiency improvement in developing countries from 2030 to 2075 (efficiency improvement rate will be 0.1%, 0.2%, 0.1% higher in 2000, 2030 to 2075, 2100 respectively in developing countries)	0.2% higher energy efficiency improvement, starting from 2000; additional 0.2% higher energy efficiency improvement in developing countries from 2030 to 2070

2.5.2 Quantified stabilization scenarios

This section presents the quantified results from AIM/Emission-Linkage for the stabilization scenarios.

Among the same target concentration level stabilization scenarios—for example, the 550 ppmv stabilization group—there is no significant difference in CO_2 emissions trajectories (Fig. 7). Rather, the CO_2 emissions reductions differ due to the different baseline emissions trajectories. They show that CO_2 emissions will increase first then start to decrease in the second half of the 21st century.

Table 3. Policy option package for stabilization at 550 ppmv (continued)

Policy options	A1B	A2	B1	B2
Carbon tax	US$50/t-C Annex 1 countries start from 2000, non-Annex 1 countries start from 2030	US$80/t-C Carbon tax starts from 2000.	US$15/t-C Annex 1 countries start from 2000, non-Annex 1 countries start from 2030	US$60/t-C Annex 1 countries start from 2000, non-Annex 1 countries start from 2030
Nuclear power incentives	-	0.5% higher marginal production cost improvement rate	-	0.2% higher marginal production cost improvement rate
Natural gas incentives	-	0.4% higher marginal production cost improvement rate	0.1% higher marginal production cost improvement rate	0.2% higher marginal production cost improvement rate
Syn-oil	0.1% higher marginal production cost improvement rate	0.15% higher marginal production cost improvement rate	0.1% higher marginal production cost improvement rate	0.15% higher marginal production cost improvement rate
Syn-gas	0.1% higher marginal production cost improvement rate	0.16% higher marginal production cost improvement rate	0.1% higher marginal production cost improvement rate	0.16% higher marginal production cost improvement rate
Biomass incentives	0.1% higher marginal production cost improvement rate	0.2% higher marginal production cost improvement rate	0.1% higher marginal production cost improvement rate	0.2% higher marginal production cost improvement rate
Solar energy	-	0.4% higher marginal production cost improvement rate (3.5 cents/kWh)	-	0.1% higher marginal production cost improvement rate

To achieve CO_2 stabilization at a given level, CO_2 abatement is mainly achieved through a mix of technological progress in the energy end use sector and supply sector, structural changes in the economy with a trend toward dematerialization and lifestyle changes. End use technology efficiency improvements and lifestyle changes are favored mitigation measures in the A1B baseline scenarios. In order to avoid possible damage from climate change to prevent a greater welfare loss, people may invest more in end use technology R&D to attain higher efficiency improvements, and give up their energy-intensive consumption patterns. Advanced energy end use technology could be introduced to save energy, espe-

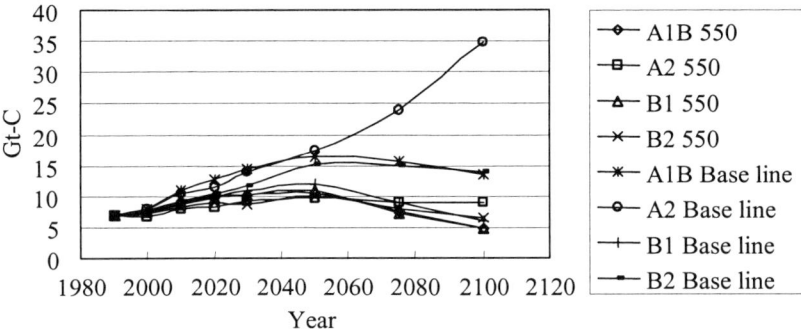

Fig. 7. Global CO_2 emissions in stabilization scenarios

cially fossil fuels. In the A1B world, in order to reach the 450 ppmv stabilization level, early action to reduce GHG emissions becomes essential due to the large reduction needed. If the reduction of GHG emissions is delayed, there will be critical pressure for reductions in the latter half of the 21st century, which may cause social and economic losses. In the A2 baseline scenario, due to the energy resource limitations in the baseline scenarios, CO_2 abatement is mainly through progress towards zero carbon technologies such as renewable energy utilization technologies, nuclear power generation technology, etc. Fossil fuel use could be reduced due to the increase in renewable energy and nuclear energy production, when the cost of such technologies decreases as a result of the large demand for them. End use technology efficiency improvements are also a key countermeasure for CO_2 abatement. The results show that in the A2 world, early GHG emissions reduction is also essential. In the B1 baseline scenario, there is relatively little pressure for the CO_2 emissions reduction to reach the 550 ppmv stabilization level, so the target could be reached by price incentive policies, such as a carbon tax. In the B2 baseline scenario, progress in both energy end use technology and energy supply technology is emphasized.

Technological progress is thus a key issue for CO_2 emissions abatement in the AIM mitigation emissions scenarios. This is because these scenarios embrace the perspective of induced technical change; i.e., an additional environmental constraint accelerates the rates of technological change already implicit in the scenario baseline.

Examining the policies used for emissions reductions in this study, it is seen that some of them are not necessarily adopted in response to climate change, especially in developing countries. For example, technological efficiency improvements in both energy production and energy end use, social efficiency changes, and low carbon technology incentives (nuclear and renewable energy, etc.) have been widely adopted in pursuit of sustainable development, as has been the case in China.

As a result, primary energy will decrease with energy efficiency improvements and the introduction of energy price incentive policies, and the primary energy mix will tend to shift to low carbon energy sources such as natural gas, renewable energy, nuclear energy, etc. (Fig. 8).

Cost analyses were simulated by the AIM/Emission-Linkage model. Table 4 shows the GDP loss for each mitigation scenario and different target level in 2050 and 2100. The results reveal that the GDP loss ranges from 0.1% to 5.9% across the scenarios. Obviously the costs rely on the target level and baseline emissions trajectory. The largest loss occurs in the A1B-450 scenario, at 5.9%.

Applying the designed robust policies to different scenarios results in different CO_2 emissions levels (Fig. 9). Some commonly used policies in the AIM mitigation scenarios could be recommended as essential countermeasures in response to climate change, while they also have benefits unrelated to the climate change concept. Policies such as technological progress in end use and energy supply, social efficiency improvements, renewable energy incentives and carbon taxes can be regarded as robust policies.

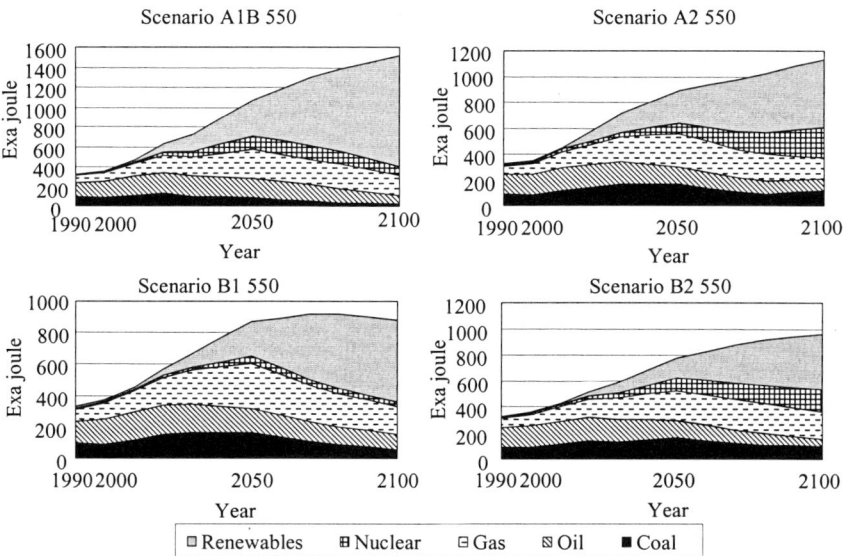

Fig. 8. Global primary energy in AIM stabilization scenarios

Table 4. GDP loss for each scenario at different target levels

	2050	2100
A1B-550	1.0%	2.0%
A1B-650	0.6%	1.0%
A1B-450	3.2%	5.9%
A2-550	1.3%	3.2%
B1-550	0.3%	0.1%
B2-550	0.9%	1.2%

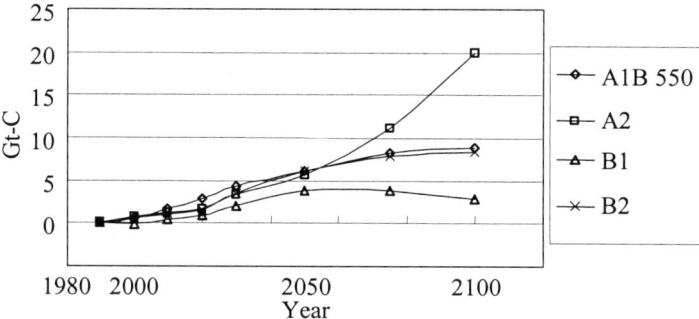

Fig. 9. CO_2 emission reductions with robust policies

2.6 Findings

In this chapter, a model framework to analyze long-term emissions scenarios for the global and the Developing Asia-Pacific was described. Four development patterns were simulated to generate scenarios. Several key findings have been obtained from the results, as follows:

1. The large range of CO_2 emissions in 2100 was simulated for the global and the Developing Asia-Pacific countries.
2. The trend in the Developing Asia-Pacific emissions is similar to that for the world, and the Asia-Pacific future would affect the global future significantly.
3. All the scenarios show that CO_2 emissions in 2100 will be above the level in 1990 for the Developing Asia-Pacific, while some scenarios present the possibility that they will be below 1990 CO_2 emissions at the global level.
4. The growth rate of GHG emissions is higher in the Developing Asia-Pacific countries than the global rate.
5. Half of the scenarios present a decrease in CO_2 emissions after 2050.
6. Technological progress will contribute substantially to low CO_2 emissions.
7. Per capita CO_2 emissions of the Developing Asia-Pacific countries will be below the level of the OECD countries over the next 100 years.
8. SO_2 and NO_x emissions of the Developing Asia-Pacific countries will decrease rapidly after 2020.
9. The high economic growth scenarios (scenarios A1) give a wider range of CO_2 emissions trajectories than for the low economic growth scenarios.
10. The global market and global governance are especially important key factors for CO_2 emissions scenarios.

Furthermore, a set of mitigation scenarios was simulated by the AIM/Emission-Linkage model based on the nonintervention emissions scenarios. Key findings from the results of this modeling are as follows:

1. The targeted stabilization levels could be reached through the adoption of various policies. All the mitigation scenarios from AIM show a trend toward various stabilization levels.
2. Wide-ranging policy packages are needed, rather than a single policy, in order to mitigate the socioeconomic effects of responses to climate change.
3. In the A1 and A2 world views as well as for 450 ppmv stabilization, an early reduction in GHG levels is essential to avoid serious pressure on social development and technological progress in the second half of the 21st century.
4. Integration between global climate policies and domestic environmental policies could effectively reduce GHG levels in the developing regions for the next two or three decades.
5. Technological progress and lower energy consumption play a very important role in stabilization.
6. Knowledge transfer to developing countries is a key issue that should be emphasized to motivate developing countries to participate in early CO_2 emissions reductions.
7. Technological efficiency improvements for both energy use technology and energy supply technology, social efficiency improvements, renewable energy incentives and the introduction of energy price incentives, such as a carbon tax, can be regarded as robust policies.
8. Robust technology/policy measures include efficiency improvements in end use technologies and social systems, as well as the introduction of renewable energy.

2.7 Conclusions

Based on the above findings, the following points form our conclusions from the Asian viewpoint:

1. The future development of the Asia-Pacific region has a significant influence on global emissions scenarios. The Developing Asia-Pacific countries will achieve a dominant position in climate change issues.
2. It is possible for the Developing Asia-Pacific to continue high economic growth while maintaining GHG emissions at a low level.
3. Technological progress and technology transfer should be emphasized to maintain low GHG emissions in the economic development of the Developing Asia-Pacific.
4. It is important for the Developing Asia-Pacific to introduce sophisticated measures to control GHG emissions before 2030.
5. Robust policy options should be designed to respond to a very wide range of alternative development paths.

From the global common viewpoint, the major conclusions can be summarized as follows:

1. Different development paths require different technology/policy measures and involve different costs of mitigation to stabilize atmospheric CO_2 concentrations at the same level.
2. Secondly, no single type of measure will be sufficient for the timely development, adoption and diffusion of mitigation options for CO_2 stabilization. Policy integration across an array of technologies, sectors and regions is the key to the successful promotion of climate change policies.
3. The level of technology/policy measures in the beginning of the 21st century will be significantly affected by the choice of the development path over the next one hundred years.
4. Several robust policy options across the different worlds are identified for stabilization. Technological efficiency improvements for both energy use technologies and energy supply technologies, social efficiency improvements, renewable energy incentives and the introduction of energy price incentives, such as a carbon tax, can be regarded as robust policies.
5. Large and continuous energy efficiency improvements and afforestation are common features of mitigation scenarios in all the different SRES worlds. The introduction of low-carbon energy is also a common feature of all scenarios, especially the introduction of biomass energy over the next one hundred years, as well as the introduction of natural gas in the first half of the 21st century.

References

AIM Project Team (1996) A guide to the AIM/Enduse model. AIM Interim Paper, IP-95-05, Tsukuba, Japan

China statistical yearbook (1985-1996) China Statistical Publishing House

Edmonds J, Reilly J (1983) A long-term global energy-economic model of carbon dioxide release from fossil fuel use. Energy Economics 5: 75-88

Edmonds J, Wise M, Barns D (1995) Carbon coalitions: the cost and effectiveness of energy agreement to alter trajectories of atmospheric carbon dioxide emissions. Energy Policy 23: 309-335

Edmonds J, Wise M, Sands R, Brown R, Kheshgi H (1996) Agriculture, land use, and commercial biomass energy: a preliminary integrated analysis of the potential role of biomass energy for reducing future greenhouse related emissions. Pacific Northwest National Laboratory, Washington D.C., U.S.A.

Gan L (1998) Energy development and environmental constraints in China. Energy Policy 26: 119–128

Hertel T (1997) Global Trade Analysis. Cambridge University Press

Hibino G, Kainuma M, Matsuoka Y, Morita T (1996) Two-level mathematical programming for analyzing subsidy options to reduce greenhouse-gas emissions. Working Paper, IIASA, Laxenburg

Hu X, Jiang K, Liu J (1996) Application of AIM/Emission model in P.R. China and preliminary analysis on simulated results. AIM Interim Paper, IP-96-02, Tsukuba, Japan

Intergovernmental Panel on Climate Change (IPCC) (1992) Climate change 1992: the supplemental report to the IPCC scientific assessment. Cambridge University Press, Cambridge

Jiang K, Hu X, Matsuoka Y, Morita T (1998) Energy technology changes and CO_2 emission scenarios in China. Environment Economics and Policy Studies 1: 141-160

Jiang K, Masui T, Morita T, and Matsuoka Y (2000) Long-term GHG Emission Scenarios for Asia-Pacific and the World .Technological Forecasting and Social Change, 63, 207-229.

Jiang K, Morita T, Masui T, Matsuoka Y (2000) Global Long-term Greenhouse Gas Mitigation Scenarios based on AIM. Environmental Economics and Policy Studies, 3, 2, 239-254.

Kram T, Morita T, Riahi K, Roehrl RA, Rooijen SV, Sankovski A, and Vries BD (2000) Global and Regional Greenhouse Gas Emissions Scenarios. Technological Forecasting and Social Change, 63, 335-371.

Lashof DA, Tirpak D (1990) Policy options for stabilizing global climate. 21P-2003 U.S. Environmental Protection Agency, Washington D.C.

Morita T, Lee H (1998) Appendix to emissions scenarios database and review of scenarios. Mitigation and Adaptation Strategies for Global Change 3(2-4): 121-131

Morita T, Matsuoka Y, Penna I, Kainuma M (1994) Global carbon dioxide emissions scenarios and their basic assumptions, 1994 survey. CGER, Tsukuba

Morita T., Nakicenovic N., Robinson J (2000) Overview of Mitigation Scenarios for Global Climate Stabilization Based on New IPCC Emission Scenarios (SRES). Environmental Economics and Policy Studies, 3, 2, 65-88.

Morita T. Robinson J (2001) Greenhouse Gas Emission Mitigation Scenarios and Implications. In Climate Change 2001: Mitigation, Cambridge University Press, 115-166.

Nakicenovic *et al.* (2000) Special report on emissions scenarios. Cambridge University Press, Cambridge

Nakicenovic, N., Victor N, Morita T (1998) Emission Scenarios Database and Review of Scenarios. Mitigation and Adaptation Strategies for Global Change, 2(2-4), 95-131.

Qi L, Hao L, Lu M (1995) SO_2 emission scenarios of eastern china. Water, Air and Soil Pollution 85: 1873-1878

World Energy Council (WEC) (1993) Energy for tomorrow's world. World Energy Council, St. Martin's Press, New York

3. Potential Impacts of Global Climate Change

Hideo Harasawa[1], Yuzuru Matsuoka[2], Kiyoshi Takahashi[1], Yasuaki Hijioka[1],
Yoko Shimada[3], Yosuke Munesue[4], and Murari Lal[5]

Summary. AIM/Impact model, an integrated assessment model of climate change impacts, has been developed in order to evaluate future climate change impacts and to support decision making on the global/Asia scale. AIM/Impact model consists of sub-models for evaluating impacts on major vulnerable sectors (water, agriculture, ecosystem, human health) and linkages among them. In this chapter, the general framework of AIM/Impact and examples of model outputs are introduced with a brief description of the sub-models.

3.1 Introduction

There is concern that anticipated climate change will cause significant negative damage on ecosystems and various sectors of human life. The degree of climate change and its damage depends on the pattern of future greenhouse gases (GHGs) emissions. The spatial and temporal distribution of climate change impact will be unequal, since the degree of climate change varies spatially and the adaptive capacity in relation to climate change is quite different among those affected according to their physical, economic, and social environments. The direct physical impacts of climate change on each sector may be interrelated and cause higher-order impacts. In order to evaluate alternative policies on GHGs mitigation, the consequent impacts, including higher-order effects, need to be assessed, while analysis on adaptation for the mitigation of future impacts is also important.

The AIM/Impact model, an integrated assessment model of climate change impacts, has been developed in order to evaluate future climate change impacts considering these complicated interrelationships and to support decision making on the global/Asia scale. The AIM/Impact model consists of sub-models for evaluating impacts on major vulnerable sectors (water, agriculture, ecosystem, human health) and linkages among them.

In this chapter, the general framework of AIM/Impact is presented first. Secondly, examples of model outputs are provided with a brief description of the sub-models.

---

[1] National Institute for Environmental Studies, Tsukuba 305-8506, Japan
[2] Kyoto University, Kyoto 606-8501, Japan
[3] Kobe Institute of Health, Kobe 650-0046, Japan
[4] Tokyo Institute of Technology, Tokyo 152-8552, Japan
[5] Indian Institute of Technology, New Delhi 110016, India

3.2 Framework of AIM/Impact Model

Figure 1 shows the linkages between the sub-models developed for the AIM/Impact study. Some sub-models have already been developed and sectoral impacts of climate change are estimated for various future climate scenarios projected using General Circulation Models (GCMs).

The FOOD sub-model consists of a productivity model for 12 crops and an agricultural trade model. The potential productivity changes caused by climate change are estimated using a 5° x 5° spatial resolution. Then, based on the estimated changes in crop productivity, the agricultural trade model calculates the allocation of the production of, and demand for, crops and other commodities that maximize social welfare.

The HEALTH sub-model examines the impact of malaria infection. It evaluates the suitability of climatic factors for the malaria mosquito to reproduce, and estimates the extent of different possible levels of malaria infection.

The VEG sub-model estimates the impact of climate change on several forest and other vegetation types. The model simulates forest collapse in regions where the rate of climate change is too high for the existing vegetation patterns to continue. The VEG sub-model also determines the value of human services provided by forests that are lost due to climate change. Work to modify the model so that it simulates dynamic changes to the vegetation is continuing.

The HYDRO sub-model uses information on climate, soil and terrain to simulate surface runoff and river discharges. The WATER sub-model estimates the future water demand at the national level and assigns that demand to each grid block, so creating a spatial distribution of water demand. Sub-models of the AIM/Emission model and of the AIM/Climate model such as the ENERGY and the CLIMATE ones are also interrelated with the sub-models of AIM/Impact in various ways. For example, the CLIMATE sub-model provides future climate scenarios for the sub-models of AIM/Impact with processing of the spatial GCM projections and observed climatology.

The sub-models that have been developed and will be developed are now related in a complex way. This complexity did not exist during the initial development stage of the project and the reasons for this are as follows:

1. It was necessary and efficient to consider long-term climate change problems simultaneously with other short-term environmental problems to develop realistic policies, especially for emissions abatement and impact adaptation. Initially, using individual sub-models with future climate scenarios was sufficient for estimating the damage caused solely by climate change in each region for each sector. However, the problems have become more complicated and recent policy requirements meant that the sub-models developed in the earlier stages had to be revised.
2. Since the initial stage, the significance of the feedback effect caused by changes in vegetation patterns and other sectors affected by climate change has been recognized. Nevertheless, due to the limited computer resources and

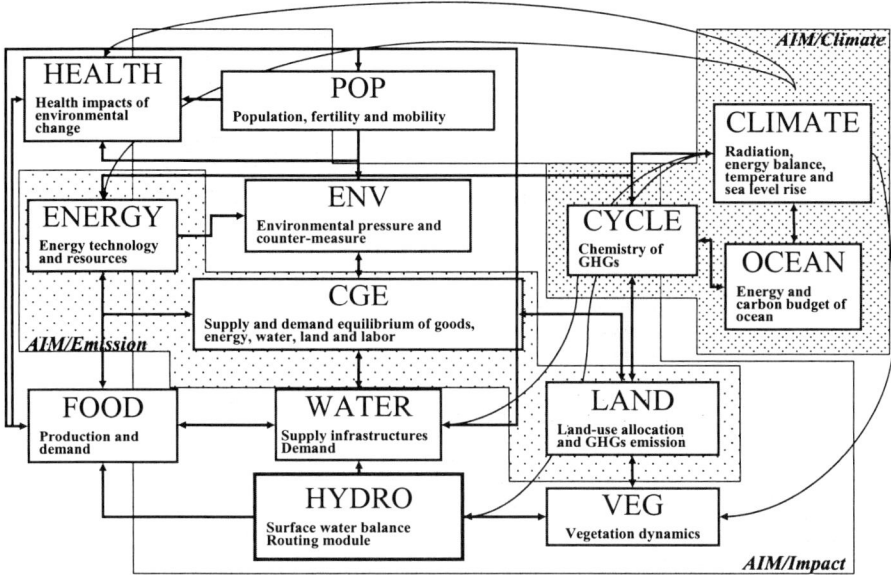

Fig. 1. Framework of AIM/Impact model (see color plates)

the complexity of the feedback processes, this issue had not received much detailed attention when reassessing the model. As climate models are refined and computer technology improves, it is becoming easier and more feasible to consider the feedback processes for carbon and other GHGs within GCM simulations. The importance of studies on the impact of climate change on vegetation is growing.

3. In order to investigate efficient strategies for mitigating and adapting to the impacts of climate change, it is necessary to estimate and compare the monetary value of the damage caused by climate change, the monetary value of the damage alleviated by various adaptation strategies, and the cost of adaptation strategies. A framework for this economic assessment, mainly relying on the CGE model, is required to quantitatively assess the various impacts of climate change in financial terms.

3.3 Agriculture

The productivity of agricultural land will be greatly influenced by future environmental changes. For example, climate-induced changes are expected to have profound impacts on potential crop yields, and so influence the distribution of cropping patterns in the Asia-Pacific area.

In order to assess the impact of climate change on agriculture, studies are necessary using a framework that can evaluate related direct and indirect effects.

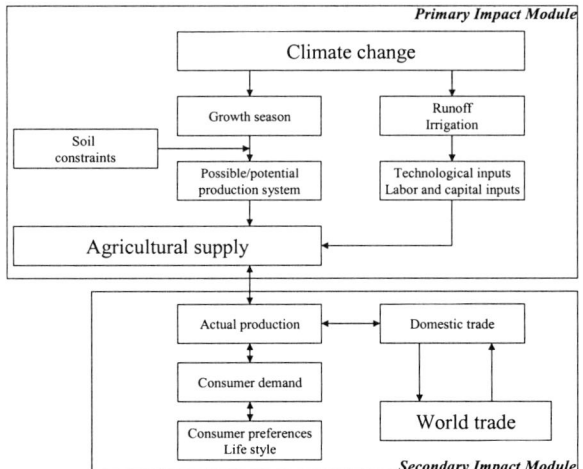

Fig. 2. Framework of agricultural impact model

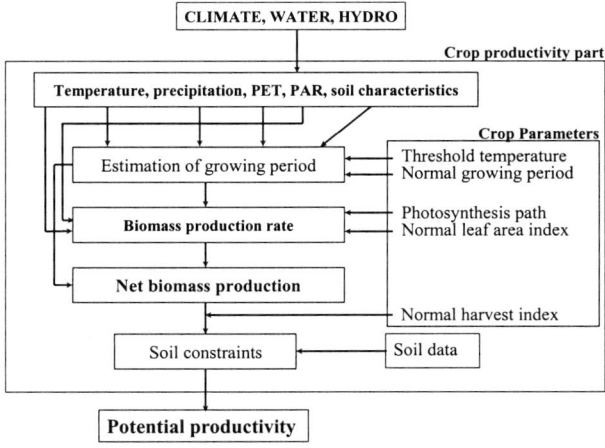

Fig. 3. Model used to estimate potential crop productivity

Figure 2 shows the framework of the agricultural impact study (Takahashi *et al.* 1997). The two basic assumptions for this study are: (1) climatic change will directly affect land and water resources (the primary effects); and (2) changes in land and water resources will affect economic activities (the secondary effects).

A potential crop productivity model was developed and linked to regional climate and environmental soil data to illustrate the first assumption. The method used for estimating potential crop productivity is based on that used by the Food and Agriculture Organization (FAO 1978-1981). The number of days suitable for crop cultivation (the growing period) are counted using climate data, then the crop growth during the growing period is simulated using the growth parameters for

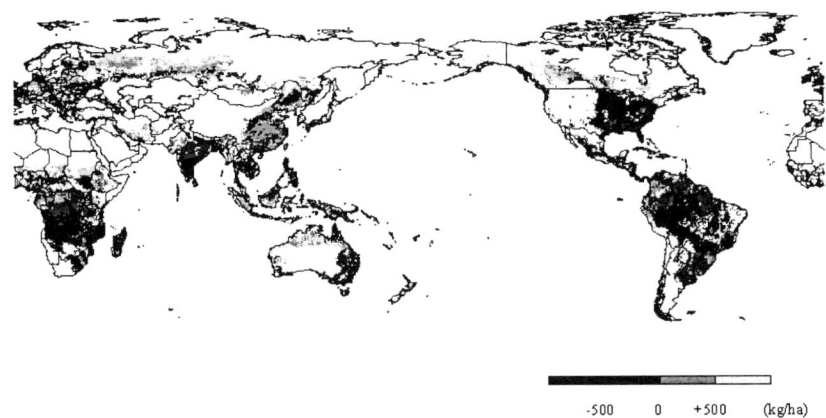

-500   0   +500   (kg/ha)

Fig. 4. Change in the potential productivity of rice from 1990 to 2050 under the climatic conditions projected using the CCSR/NIES GCM (see color plates)

each crop. Figure 3 shows the framework for estimating potential crop productivity. This model requires data for: daily mean temperatures; mean daytime temperatures; precipitation; potential evapotranspiration (PET); photosynthetically active radiation; and soil characteristics. The physical and chemical properties of soil such as soil texture and soil slope were used to estimate the suitability of land areas for agriculture. The calculations used data for these properties on a 5-minute resolution grid to account for their high spatial variability. Figure 4 shows the change in the potential productivity of rice from 1990 to 2050 under the climate conditions projected with CCSR/NIES GCM. In general, the productivity of rice crops will increase in areas of high-latitude and decrease in areas of low-latitude.

An agricultural trade model based on the GTAP general equilibrium model developed at Purdue University (Hertel 1997) was used to assess the impact of climate change on the economy through changes in crop productivity within each region (Takahashi *et al.* 1999). Regions and production sectors treated in the model are shown in Tables 1 and 2. In the agricultural trade model, potential changes in the productivity of rice, winter wheat, tropical maize and other crops - as estimated in the potential crop production model - are taken as the changes in the Hicks-neutral technology parameters for their respective production sectors.

Table 3 shows the changes in producer prices, agricultural production and social welfare in some regions under changed climatic conditions (IS92a emissions for 2100). Comparing the changes in social welfare per capita, India is found to be the country likely to suffer the most damage. This reflects the significant decline in the productivity of wheat in India and the comparatively large share of agricultural products purchased using private funds. Whilst a rich country can use trade to mitigate the economic damage caused by decreases in crop productivity, a poor country cannot adapt in the same way, as a result, its social welfare deteriorates.

Table 1. The thirty countries and regions in the agricultural trade model

Country	Code
Australia	AUS
New Zealand	NZL
Japan	JPN
Korea, Republic	KOR
Indonesia	IDN
Malaysia	MYS
Philippines	PHL
Singapore	SGP
Thailand	THA
China	CHN
Hong Kong	HKG
Taiwan (China)	TWN
India	IDI
Other South Asia	RAS
Canada	CAN
U.S.A.	USA
Mexico	MEX
Central America and Caribbean	CAM
Argentina	ARG
Brazil	BRA
Chile	CHL
Other South America	RSM
EU	E_U
Austria, Finland, Sweden	EU3
European Free Trade Area	EFT
Central European Associates	CEA
Former USSR	FSU
Middle East and North Africa	MEA
South Africa	SSA
Rest of the world	ROW

Table 2. The eight production sectors in the agricultural trade model

Sectors	Description
Paddy	Paddy
Wheat	Wheat
Other grains	Other grains
Other crops	Non-grain crops
Livestock	Wool
	Other livestock
Other agricultural products	Processed rice
	Meat products
	Milk products
	Other food products
	Beverages and tobacco
	Textiles
	Leather
Manufacture	Apparel
	Lumber
	Pulp paper
	Petroleum and coal
	Chemicals, rubbers and plastics
	Nonmetallic minerals
	Primary ferrous metals
	Nonferrous metals
	Fabricated metal products
	Transport industries
	Machinery and equipment
	Other manufacturing
Services	Electricity, water and gas
	Construction
	Trade and transport
	Other private services
	Other governmental services
	Ownership of dwellings

Table 3. Changes in producer prices, agricultural production and social welfare in some regions under changed climate (IS92a emissions, the year 2100 compared to 1990)

	JPN	CHN	IDI	CAN	USA	E_U
Producer price change (%)						
Rice	-0.01	-1.58	17.96	-40.16	-0.06	-4.93
Wheat	4.91	8.47	125.11	-13.10	4.76	8.92
Other grains	1.81	0.79	1.80	-43.59	-1.46	-3.36
Other crops	-0.01	-0.28	1.90	2.76	-0.10	-0.05
Livestock	-0.19	-0.09	2.84	-1.22	-0.59	-0.04
Other agricultural products	-0.15	-0.01	0.30	-0.35	-0.07	0.04
Manufacture	0.03	-0.12	-1.10	0.61	0.03	-0.02
Services	0.03	-0.16	-0.93	0.69	0.02	-0.02
Production change (%)						
Rice	0.11	-0.25	-1.76	105.99	0.23	2.03
Wheat	-6.60	-3.97	-7.64	115.07	2.87	-3.64
Other grains	-15.56	-1.39	-1.33	89.41	-4.04	-6.50
Other crops	0.11	-0.07	-4.25	-2.26	0.25	-0.03
Livestock	0.09	-0.24	-2.27	0.94	0.03	-0.22
Other agricultural products	0.11	-0.27	-4.73	0.69	0.04	-0.22
Manufacture	-0.01	0.31	-0.37	-1.62	0.03	0.05
Services	0.00	0.00	-2.62	-0.02	0.01	0.01
Consumer price index (%)	0.001	0.001	6.047	0.513	0.017	-0.010
Income change per capita (%)	0.026	-0.236	-0.617	0.833	0.026	-0.009
Social welfare change (%)	0.022	-0.219	-4.892	0.343	0.009	0.003

3.4 Vegetation

Figure 5 shows the process used to account for the economic impact of forest loss caused by climate change (Munesue *et al.* 2000). The impacts of climate change on several forest types and other vegetation types are projected. These vegetation types are based on the vegetation climate-zones classified according to the Holdridge scheme (Holdridge 1947) under current and future climatic conditions and an assumed maximum rate for vegetation transfer that follows the shift in suitable climatic conditions due to climate change. The model simulates forest collapse in regions where the rate of climate change is too great to allow the existing vegetation types to shift. The black cells (cells are classified to a forest type under the present climate and to a non-forest type under the future climate) and gray cells (cells are classified to a forest type under the present climate and to a different forest type under the future climate, but the shift cannot occur smoothly) in Fig. 6 indicate forests that collapse under situations of low, medium or high rates of climate change. The global mean temperature increases between 1990 and 2100 under these scenarios are 0.85°C, 2.08°C, and 3.52°C. The maximum rate of forest transfer is assumed to be 1 km/yr in this model trial. The forest area will shrink by 2%, 5% and 9% of the total land area, and the replacement of forest types at risk of shrinking will be 3%, 7% and 12% of the total land area. The most damaged forests are in the northern parts of Eurasia and North America.

The economic value of the loss due to forest collapse caused by climate change is estimated using values from the National Center for Ecological Analysis and Synthesis (Costanza *et al.* 1997), and comes to US$49.7 billion, $US126.4 billion, and $US219.5 billion at 1994 dollars respectively under the three scenarios.

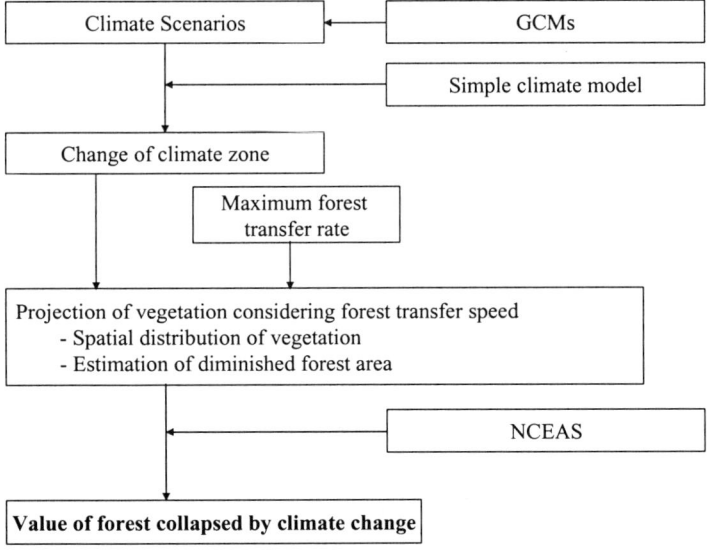

Fig. 5. Process of estimating the economic cost of forest loss from climate change

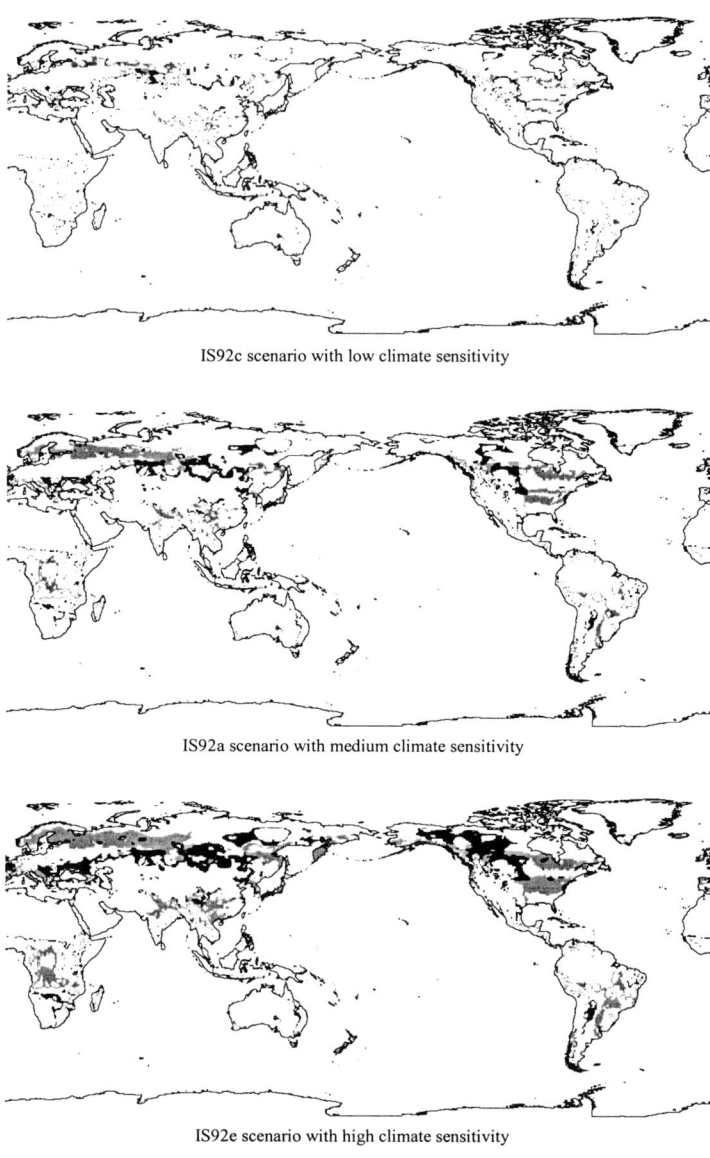

IS92c scenario with low climate sensitivity

IS92a scenario with medium climate sensitivity

IS92e scenario with high climate sensitivity

■ Cells are classified to a forest type under the present climate and to a non-forest type under the future climate (diminishment of forest).

▓ Cells are classified to a forest type under the present climate and to a different forest type under the future climate, but the shift cannot occur smoothly (replacement of the forest type with the risk of diminishment).

Fig. 6. Forest collapse under low, mid, high-climate change scenarios (IS92c emissions with low climate sensitivity, IS92a emissions with medium climate sensitivity, IS92e emissions with high climate sensitivity) (see color plates)

3.5 Health

Air and water pollution, as well as solid and hazardous wastes, impact directly on human health. Global climate changes will also affect human health in many ways. For example, global warming will change the vegetation distribution and increase ground temperatures. This will allow the habitat of the Anopheles mosquito, which is the malaria vector, to expand. In addition, the time taken for the development of the malaria protozoan will shorten and its reproductive potential will increase. As a result, the risk of malaria outbreaks on a global scale is predicted to increase. A sub-model of the AIM/Impact model was developed to quantitatively estimate this risk (Matsuoka *et al.* 1995).

Figure 7 shows the assessment framework used in this study. The two major components of the framework are the relationship between sporogony and temperature, and the eco-climatic index model which shows the climatic response of vectors. These components are supported by the soil moisture sub-model and outputs from the equilibrium experiments of GCMs. The primary climatic variables of this framework are surface temperature and precipitation distributed spatially and temporally for both the current situation and that expected under $2xCO_2$ climate conditions.

Figure 8 shows the potential for malaria infection calculated under current and expected future climate conditions. The yellow indicates areas where the risk of malaria is mesoendemic and the red highlights areas where it is hyperendemic (see color plates).

Table 4 shows the predicted expansion of the population living under endemic malaria when the global mean temperature increases by 2.0°C. The risk of malaria in China, Indonesia and India increases due to this change. It is concluded that the population expected to be at risk of malaria infection in the Asia-Pacific region will increase by 30%.

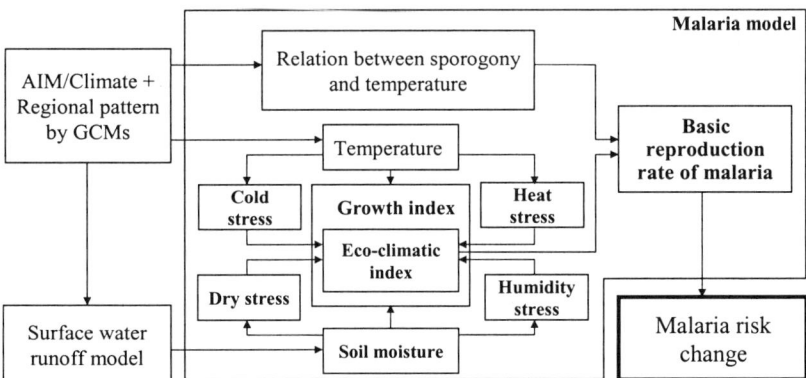

Fig. 7. Framework of malaria impact model

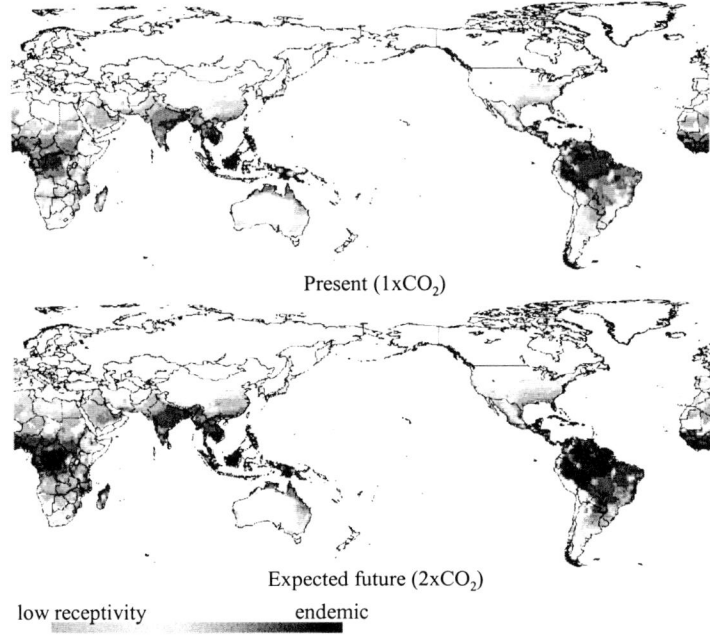

Present (1xCO$_2$)

Expected future (2xCO$_2$)

low receptivity   endemic

Fig. 8. Expansion of the area affected by malaria (Upper: current climate, Lower: 2xCO$_2$ climate change, GCM: GFDL q-flux equilibrium experiment) (see color plates)

Table 4. Population at risk of malaria (millions)

Country	Reported values		Present climate		Future climate	
	Country Population	Malarious Area	Low Risk	High Risk	Low Risk	High Risk
Afghanistan	18.14	10.39	3.65	0.00	18.14	0.00
Bangladesh	98.66	2.29	64.16	64.16	64.16	64.16
Bhutan	1.42	0.18	0.79	0.73	0.79	0.79
Cambodia	7.28	2.36	5.76	5.76	5.76	5.76
China	1059.52	975.82	646.93	54.60	807.30	132.33
India	750.90	728.33	721.40	522.31	731.45	648.04
Indonesia	163.39	155.63	122.00	74.76	138.17	107.62
Iran	44.21	34.83	12.92	0.00	29.76	0.00
Iraq	15.90	15.90	9.39	0.00	15.90	0.00
Laos	4.12	3.30	3.79	3.64	3.79	3.79
Malaysia	15.56	15.44	13.14	12.16	13.14	12.16
Myanmar	37.15	33.22	30.93	23.25	36.06	29.39
Nepal	16.63	10.76	14.09	4.76	14.95	10.22
Oman	1.24	1.24	0.92	0.00	0.48	0.00
Pakistan	96.18	96.18	91.89	0.00	95.86	0.00
Philippines	54.38	16.82	30.55	30.55	30.55	30.55
Saudi Arabia	11.54	4.51	8.68	0.00	7.97	0.00
Sri Lanka	15.84	11.51	12.67	9.17	12.67	12.67
Syria	10.27	6.38	1.91	0.00	10.27	0.00
Thailand	51.30	46.09	44.61	43.98	44.61	43.98
Turkey	49.27	49.27	14.78	0.00	49.27	0.00
U.A.E.	1.33	1.33	1.07	0.00	1.07	0.00
Viet Nam	59.61	44.03	45.96	41.37	45.96	45.96
Yemen	6.85	3.01	1.08	0.00	2.38	0.00
Papua New guinea	3.33	3.33	2.97	2.14	3.15	2.81
Solomon	0.27	0.27	0.08	0.00	0.00	0.00
Vanuatu	0.14	-	0	0.00	0.00	0.00
Total	2594.43	2272.42	1906.12	893.34	2183.61	1150.23

3.6 Water

Figure 9 shows the model used to predict changes in water resources (Takahashi *et al.* 2001). In the surface runoff model, the runoff volume is calculated using grid climate information and surface information, as a result, the volume of water potentially available for use within river basins can be estimated. In the water demand model, the future water demand in each river basin is calculated using estimates of the current water demand and population distribution together with economic development scenarios. Moreover, with the changes in the water available for use relative to water demand, future qualitative changes in the balance between water supply and demand can be understood. An assessment of the balance between supply and demand was carried out for each river basin.

Water demand was estimated for the household, industrial, and agricultural sectors (Fig. 10). For each sector, water demand was first estimated for each country based on population changes and growth in economic activities as well as scenarios for improvements in water use efficiency. Next, the spatial distribution of the water demand was estimated using estimates of population and land use distribution (cropland density). Finally, the estimated spatial values were aggregated to provide a total value for water demand in each river basin.

To estimate the surface runoff in each grid, one of the bucket-type models (Vorosmarty *et al.* 1989) was used with inputs of monthly mean climate data and surface data at a $0.5° \times 0.5°$ spatial resolution. Surface runoff was calculated as the amount of effective precipitation (precipitation minus evapotranspiration, but including consideration of snow coverage during periods of low temperatures) exceeding the maximum amount of water able to be retained in the soil. Evapotranspiration was calculated as a function of the monthly mean PET using the method of Penman (FAO 1992) and the level of soil-water saturation.

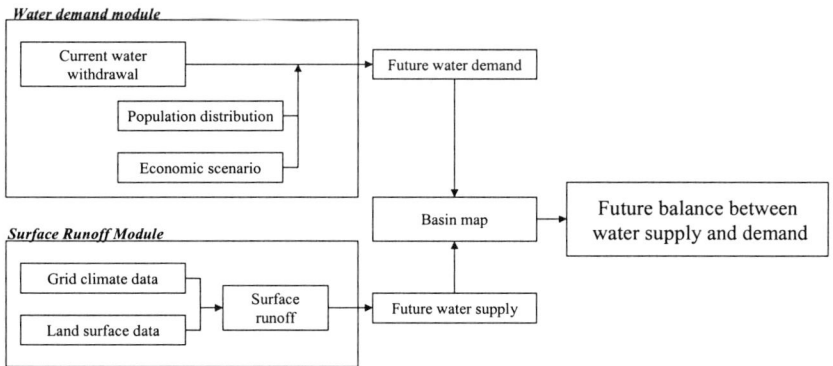

Fig. 9. Model used to estimate water resources

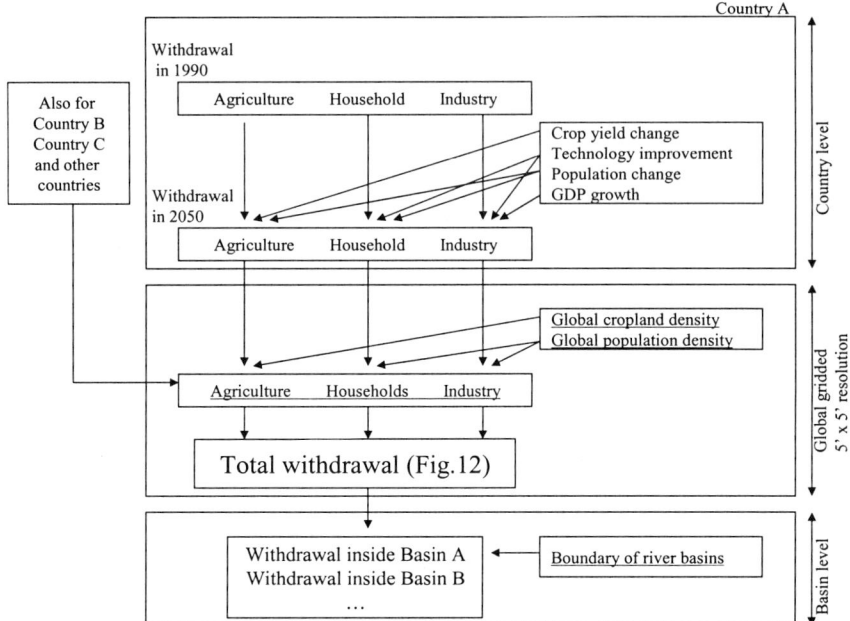

Fig. 10. Model used to estimate water demand

Figure 11 shows the spatial distribution of water demand density (mm/year, a total of three sectors) with a 5° x 5° resolution in 1990 and 2050. Areas with a very high density of water demand exist in various regions, including eastern China, Southeast Asia, India, Japan, eastern USA and Europe. Of these, eastern China and India will have extremely high demand due to the rapid increases in demand during the first half of the 21st century.

When the potential trends for change from the current situation are examined, it is seen that in the industrialized countries where population should be stable, water demand will increase only in the industrial sector. It will not change significantly in the household sector, and should decline in the agricultural sector. In contrast, increases will occur simultaneously in all three sectors in the developing countries, reflecting the trends of both population growth and economic development. The increase in the industrial sector will be the largest.

Figure 12 shows changes in the calculated runoff based on the results of the transient experiments (atmospheric CO_2 concentration is assumed to increase at an annual rate of 1%) of the CCC, ECHAM4, and CCSR/NIES climate models. This is equivalent to [mean runoff for the 10 years from 2050 to 2059] minus [mean runoff for the 10 years from 1980 to 1989]. Thus, it can be seen that some increases in runoff will occur throughout the whole of Siberia. Large increases in runoff in southern China and the Indian subcontinent are also common features. On the other hand, completely opposite change patterns are shown by the climate models for central Africa and northern South America. This shows that the choice

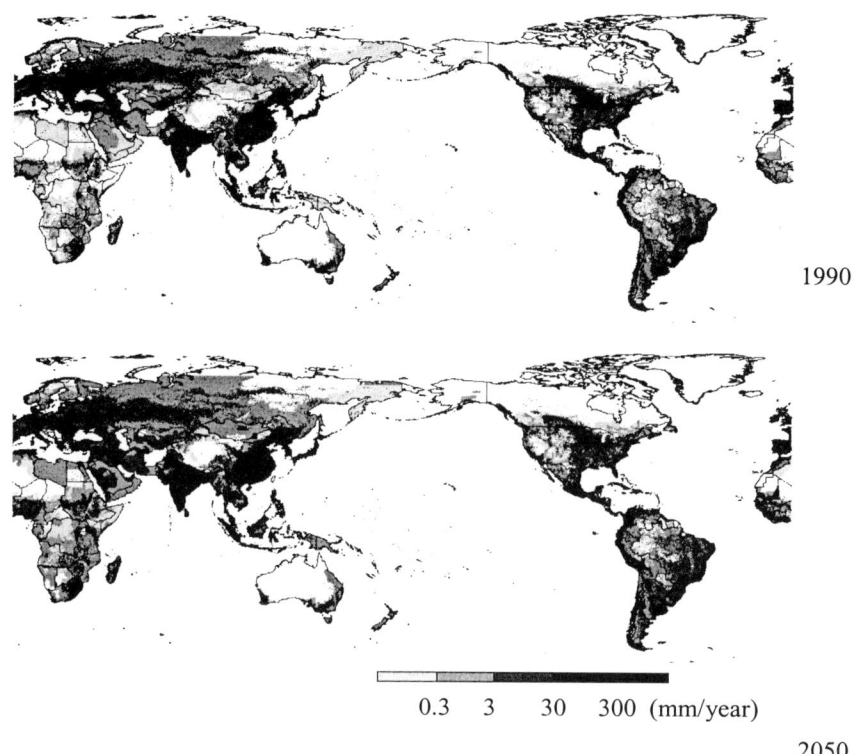

1990

0.3   3   30   300 (mm/year)

2050

Fig. 11. Water demand per unit area in 1990 and 2050 (see color plates)

of climate model significantly influences the spatial distribution of changes in runoff.

As an example of the comparison between water supply and demand, Figure 13 shows the annual changes in the supply and demand balance from 2050 to 2059 for the Ganges and Mekong river basins. For the Ganges river basin, the ratio will not change greatly since the increase in runoff is expected to match the increase in water demand due to population growth and economic development. However, the CCSR/NIES model shows a different behavior for the Ganges river basin compared with the other two models. For the Mekong river basin, the supply - demand ratio will increase with the increasing water demand, but the value of the ratio will not be high compared to other basins where water is scarce.

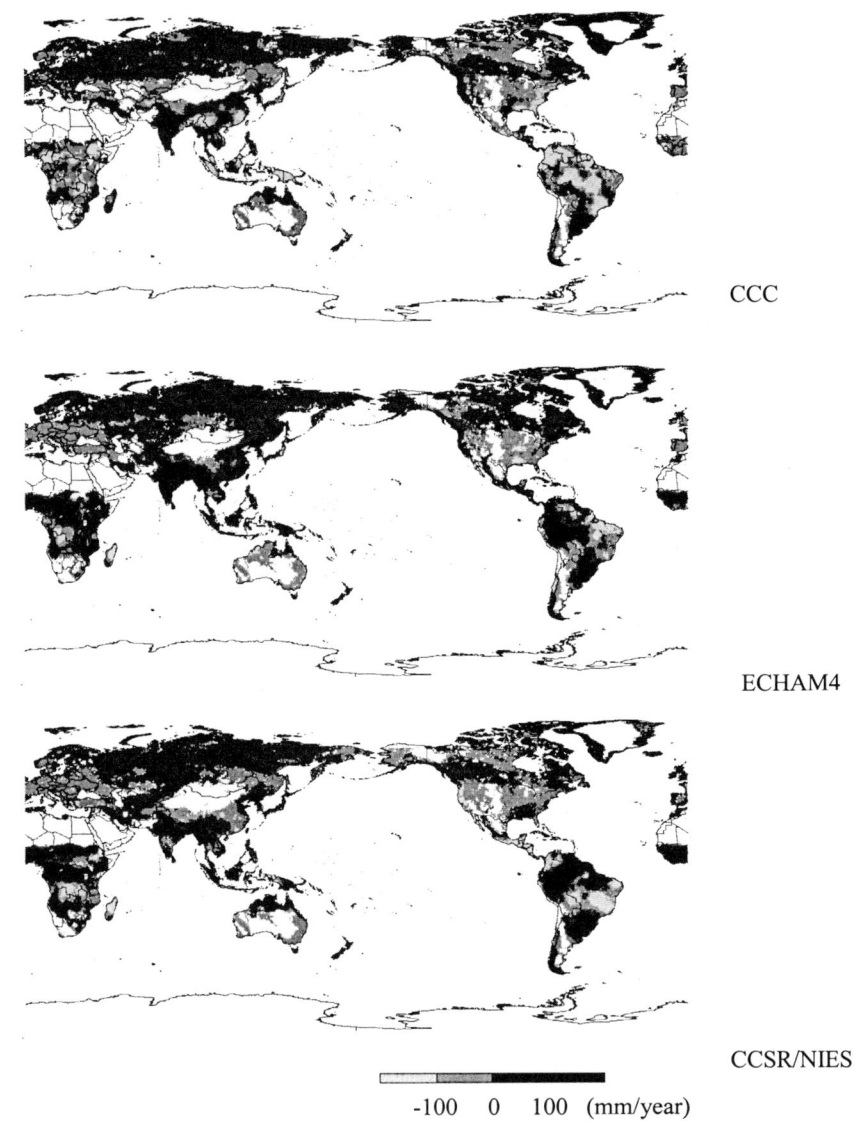

CCC

ECHAM4

CCSR/NIES

-100   0   100   (mm/year)

Fig. 12. Changes in runoff calculated based on the results of the transient experiments of the CCC, ECHAM4, and CCSR/NIES climate models (mean runoff for the 10 years from 2050 to 2059 minus mean runoff for the 10 years from 1980 to 1989) (see color plates)

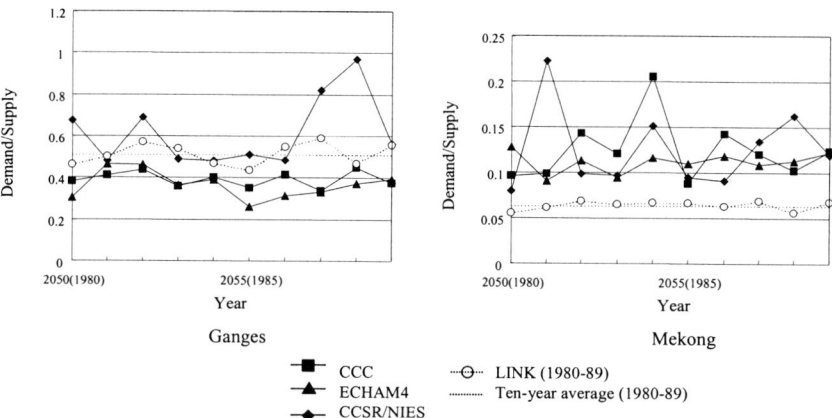

Fig. 13. Annual change in the supply and demand ratio for water from 2050 to 2059 for the Ganges and Mekong river basins

3.7 Climate Change Scenarios used in AIM/Impact

The impacts of climate change have been assessed using climate scenarios based on the latest GCM experiments at each research stage. Currently, the transient experiments with the GCMs listed in Table 5 are mainly used for creating future climate scenarios (Takahashi *et al.* 2001). To compensate for the disadvantages of GCMs, such as the coarse spatial resolution, historically observed data provided by LINK (New *et al.* 1998) is amalgamated with the simulated results of GCMs. The development of climate scenarios based on regional climate models or statistical downscaling methods have also been investigated, although they have not yet been used for practical impact assessment.

Research collaboration with the GCM modeling project inside NIES has been promoted to take advantage of the fact that NIES has both the impact analysis project and the climate modeling project. AIM/Impact not only receives projections of future climatic conditions from GCM experiments, but it also provides the results of the impact modeling to GCM modelers so feedback processes can be depicted.

Table 5. Transient experiments of the GCMs currently used to create climate scenarios

	CCSR	CCCma	CSIRO	GFDL	HADCM2	ECHAM4	NCAR
Institution, country	Tokyo University and National Institute for Environmental Studies, Japan	Canadian Center for Climate Modelling and Analysis	Australia's Commonwealth Scientific and Industrial Research Organization	Geophysical Fluid Dynamics Laboratory, USA	Hadley Centre for Climate Prediction and Research, UK	Deutsches Klimarechenzentrum, Germany	National Centre for Atmospheric Research, USA
Resolution (A-GCM)	5.6°×5.6° 20layer	3.7°×3.7° 10layer	3.2°×5.6° 9layer	4.5°×7.5° 9layer	2.5°×3.75° 19layer	2.8°×2.8° 19layer	4.5°×7.5° 9layer
Resolution (O-GCM)	2.8°×2.8° 17layer	1.8°×1.8° 29layer	3.2°×5.6° 21layer	4.5°×3.75° 12layer	2.5°×3.75° 20layer	2.8°×2.8° 11layer	1°×1° 20layer
CO_2 concentration (control run)	345ppmv	295ppmv	330ppmv	300ppmv	323ppmv	354ppmv	330ppmv
CO_2 concentration (transient run)	1%/yr	1%/yr	0.9%/yr	1%/yr	1%/yr	1%/yr	1%/yr
Simulated period (control run)	1890-2099 210yr	1900-2100 200yr	1881-2100 219yr	1958-2057 100yr	1860-2099 240yr	1860-2099 240yr	1901-2036 136yr
Climate sensitivity	3.5°C	3.5 °C	4.3 °C	3.7 °C	2.5 °C	2.6 °C	4.5 °C
Reference	Emori *et al.*, 1999	Reader and Boer, 1998	Hirst *et al.*	Manabe and Stouffer, 1996	Johns *et al.*, 1997	Roeckner *et al.*, 1996	Meehl, 2000

3.8 Future Direction for Model Improvement

The role of studies on the assessment of climate change impacts on a global scale has been changing in recent years. The main purpose of impact assessments on a global scale was to alert people to the seriousness of climate change and to prompt the promotion of policies to mitigate GHG emissions by identifying the regions that could suffer severe damage as a result.

Nowadays, however, the roles that impact studies on a global scale are being expected to fulfill has become diversified. In order to investigate the feasibility and effectiveness of strategies for dealing with climate change problems in a more realistic way, the impact on the natural environment and human wellbeing as a consequence of choices regarding alternative paths to future socioeconomic development should be assessed by taking into account not only climate change itself, but also the wider contributing factors, such as the socioeconomic background and the state of the natural environment. The framework of analysis to achieve a more comprehensive approach is expected to be developed and applied to the ongoing international activities for comprehensive policy evaluation that focus on wider contexts, such as sustainable development and ecosystem conservation. Moreover, since it is possible for the natural environment and human system altered by climate change to affect the climate system itself through feedback mechanisms, impact assessment is expected to be closely linked with climate systems analysis and to provide the information required for considering the consequences of feedback effects.

Taking into consideration the expected role for impact assessment on a global scale, our best efforts are being directed towards the creation of linkages among sectoral assessment models, as explained in section 3.2, as well as to refine the following aspects of the AIM/Impacts model:

- Development of impact and adaptation assessment methods that consider not only climate change, but also future changes in socioeconomic and other environmental factors.
- Extension of data collection on adaptation strategies and the development of methods of evaluating the cost-benefit ratio of the strategies.
- Closer collaboration with climate system modelers to take into account feedback effects.

Adding to these refinements, in order to analyze adaptation strategies in more realistic and effective ways, top-down analysis on a global scale that considers international trade, technology transfer, climate change, and the bottom-up analysis on a national scale that considers concrete adaptation measures, will be linked by maintaining consistency in the socioeconomic background in addition to the physical and monetary variables.

References

Costanza R, d'Arge R, de Groot R, Farber S, Grasso M, Hannon B, Limburg K, Naeem S, O'Neill RV, Paruelo J, Raskin RG, Sutton P, van den Belt M (1997) The value of the world's ecosystem services and natural capital. Nature 387: 253-260

FAO (1978-1981) Report on the Agro-Ecological Zones Project, Vol.1-4, World Soil Resource Report 48, Food and Agriculture Organization of the United Nations, Rome

FAO (1992) Crop water requirements, Irrigation and Drainage Paper 24, Food and Agriculture Organization of the United Nations, Rome

Hertel TW (1997) Global trade analysis, Cambridge University Press

Holdridge LR (1947) Determination of world plant formations from simple climatic data. Science 105: 367-368

Matsuoka Y, Kai K (1995) An estimation of climatic change effects on malaria. Journal of Global Environmental Engineering 1: 43-47

Munesue Y, Takahashi K (2000) Evaluation of climate chnage impact on vegetation and its economic value. Environmental Science 13(3): 329-337

New M, Hulme M, Jones P (1998) Representing twentieth century space-time climate variability. I: Development of a 1961-1990 mean monthly terrestrial climatology. J. Climate 12: 829-856

Takahashi K, Harasawa H, Matsuoka Y (1997) Climate change impact on global crop production. Journal of Global Environmental Engineering 3: 145-161

Takahashi K, Harasawa H, Matsuoka Y (1999) Impacts of climate change on food production. - An Economic Assessment -. Journal of Global Environmental Engineering 5: 1-9

Takahashi K, Matsuoka Y, Shimada Y, Harasawa H (2001) Assessment of water resource problems under climate change - considering inter-annual variability of climate derived from GCM calculations-. Journal of Global Environmental Engineering 7: 17-30

Takahashi K, Matsuoka Y, Okamura T, Harasawa H (2001) Development of climate change scenarios for impact assessment using results of General Circulation Model simulations. Journal of Global Environmental Engineering 7: 31-45

Vorosmarty CJ, Morre III B, Grace AL, Gildea MP (1989) Continental scale model of water and fluvial transport: an application to South America. Global Biogeochemical Cycle 3: 241-256

4. Cost Analysis of Mitigation Policies

Mikiko Kainuma[1], Yuzuru Matsuoka[2], Tsuneyuki Morita[1], Toshihiko Masui[1], and Kiyoshi Takahashi[1]

Summary. This analysis evaluates the economic and environmental impacts of climate change policies. Firstly, the economic impacts of the Kyoto Protocol are analyzed to assess the short-term effects of climate change policies. It is found that the GDP loss to Japan, the USA, the EU, and Russia will be 0.42%, 0.56%, 0.44%, and 0.25%, respectively, if the Annex B countries ratify the Kyoto Protocol and reduce their emissions without emissions trading and without taking carbon sinks into account. On the other hand, the GDP loss to Japan and the EU will grow if the USA does not ratify the Kyoto Protocol, and these losses will be recovered if the Kyoto mechanisms are adopted. Atmospheric stabilization scenarios are then examined to estimate long-term economic and environmental impacts. It is found that the global mean temperature will increase 1.8-2.8°C by the year 2100 under 550 ppmv scenarios. Impacts on crop productivity in India and the incidence of malaria in China are estimated to become very serious.

4.1 Introduction

The Kyoto Protocol was adopted in 1997 as a first step toward stabilizing the atmospheric concentrations of greenhouse gases. The most important feature of the Protocol is the quantified emissions reduction targets. These would result in a reduction in emissions of greenhouse gases from Annex B countries in the 2008-2012 period to about 5 percent below their 1990 levels. For the purpose of meeting these commitments, the Protocol established the principles of emissions trading, joint implementation and the Clean Development Mechanism (CDM). The Protocol also approved the principle that removal of atmospheric carbon using carbon sinks formed by direct human-induced land use changes and forestry activities could be counted as a means of managing emissions.

In March 2001, President Bush announced that the United States would not ratify the Kyoto Protocol. Since the United States emits the largest amount of CO_2 of any country in the world, the influence of this decision on other industrialized nations could be quite significant.

This paper analyzes the effect of the decision of the USA as well as other important factors related to the reduction of greenhouse gas emissions. Several cases are studied. One case assumes ratification of the Protocol by Annex B countries, including the United States. Another case assumes that it becomes international law without ratification by the United States. Other factors such as price-induced

---

[1] National Institute for Environmental Studies, Tsukuba 305-8506, Japan
[2] Kyoto University, Kyoto 606-8501, Japan

technological change and a movement to boycott the products of non-ratifying countries are also considered.

Even though there are many hurdles to effectively reducing greenhouse gas emissions under the Kyoto target, it is necessary to go further to meet the ultimate goal of the United Nations Framework Convention on Climate Change (UNFCCC); "stabilization of greenhouse gas concentrations in the atmosphere at a level that will prevent dangerous anthropogenic interference with the climate system."

As an example of a stabilization scenario, the target of stabilizing the atmospheric concentration of CO_2 to less than 550 ppmv is considered. The global temperature changes and the rise in sea levels are estimated. Estimated climate changes are used to analyze the climatic impacts on crop production and infectious diseases in the Asia-Pacific region.

4.2 Structure of AIM/CGE (Energy) for Mitigation Analysis

The AIM/CGE (Energy) model is a recursive dynamic equilibrium model of the world economy used to analyze the effects of climate stabilization policies (Kainuma *et al.* 1999). The model divides the world into 21 geopolitical regions. To analyze the impacts of the Kyoto Protocol, Annex B countries are categorized into the following regions: Japan, Australia, New Zealand, the United States of America (USA), Canada, the European Union (EU), and Eastern Europe and the Former Soviet Union (EEFSU). The AIM model focuses in detail on the Asia-Pacific region, which is divided into 10 regions: China, Taiwan, the Republic of Korea, Hong Kong, Singapore, India, Indonesia, Malaysia, the Philippines, and Thailand. Other regions are Latin America, Middle East Asia and North Africa, Sub-Saharan Africa, and the Rest of World (ROW).

Goods are aggregated into seven energy goods and four non-energy goods. The energy goods are coal, crude oil, petroleum and coal products, natural gas, nuclear energy, renewable energy, and electricity. The non-energy goods are aggregated into four categories. The first category includes energy-intensive products; the second includes agriculture, other manufactures and services; the third includes transport industries; and the last is savings.

Figure 1 shows the structure of the AIM/CGE (Energy) model used for the cost analysis. The model has three sectors—the production, household, and government sectors—in each region. CO_2 and other greenhouse gases are emitted by each of these sectors.

The production of electricity and non-energy goods involves the use of fossil fuels and the emission of CO_2 in the production sector. In addition, the use of automobiles and other direct uses of fossil fuels emit CO_2 in the household and government sectors. It is assumed that the household sector has carbon emission rights and distributes them to the other sectors and within the household sector itself. Fossil fuels cannot be used without having carbon emission rights. The price of these carbon rights depends on several factors such as emissions targets and the

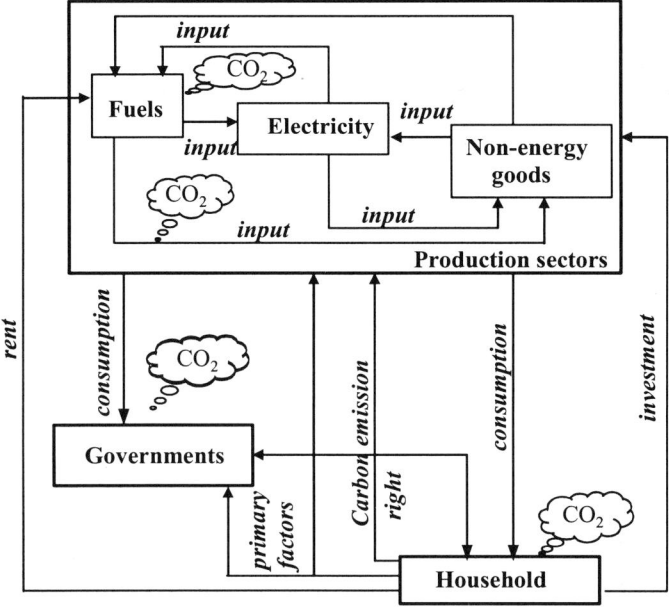

Fig. 1. The structure of AIM/CGE (Energy) for mitigation analysis

method of emissions trading. The household sector also supplies primary factors to the production and government sectors. An agent in the household sector determines consumption and savings. The marginal propensity to save is a calibrated function of a weighted aggregate of regional and global rates of return on fixed capital. Regional investment is calculated using the GDP growth rate and regional and global rates of return. Investment is balanced with savings on a global scale. The model allows for trade in intermediate goods. AIM assumes identical preferences in all countries for foreign versus domestic goods; i.e., the elasticity of substitution is the same for all regions. Domestic and imported goods are not perfect substitutes.

Figure 2 shows the nesting of the production structure in AIM. All industries have a similar production structure. Output is calculated by including primary factors, intermediate goods, and energy. Energy is nested into fossil fuels and electricity, and fossil fuels are in turn nested into fuel goods and carbon emission rights. It is assumed that the elasticity between fuel goods and carbon rights equals zero. Therefore, carbon rights become a constraint on production functions.

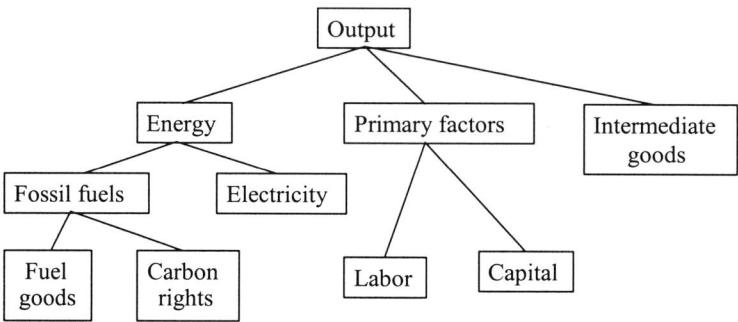

Fig. 2. Nesting of the production structure (elasticity of substitution: energy/primary/intermediates =0.3 for energy intensive products and 0.2-0.5 for other products; electricity/fossil fuel=0.3; fuel/fuel =1; labor/capital=1; fuel/carbon=0)

4.3 Cost of the Kyoto Protocol

4.3.1 Scenario assumptions

The emissions reduction target adopted at COP3, held in Kyoto, is analyzed using the AIM model. The reduction target for each country compared to the 1990 emissions level is as follows: Australia: 0.8%, New Zealand: 0%, FSU: 0%, Japan: -6%, Canada: -6%, USA: -7%, EU: -8%. It is assumed that several policy measures such as a carbon tax and the Kyoto mechanisms are used to meet these targets.

Besides the reference scenario, three sets of scenarios are examined. In the first set, it is assumed that each country should reduce emissions without adopting a flexibility mechanism. In the second set, emissions trading is assumed without any restrictions on the amount of carbon traded. In the third set, CDM is also counted besides emissions trading without restriction of tradable amounts. In each set, the impacts of participation or non-participation of the USA in the Kyoto Protocol are examined. Also examined are price-induced technological change and a movement to boycott goods exported by non-ratifying countries. In the price-induced technological change scenario, it is assumed that technologies shift to energy-saving types as the price of energy rises. In the boycott movement scenario, it is assumed that the price of goods exported by non-ratifying countries is 10% higher than the price in the reference scenario.

In addition to above scenarios, restriction case with one-third of the reduction commitment and the effects of carbon sinks are analyzed. The quantities absorbed by the sinks are assumed to be as follows: EU: 9.84 Mt-C/year, FSU: 19.46 Mt-C/year, Australia: 0 Mt-C/year, Canada: 12.0 Mt-C/year, Japan: 13.0 Mt-C/year, New Zealand: 0.2 Mt-C/year, USA: 28.0 Mt-C/year.

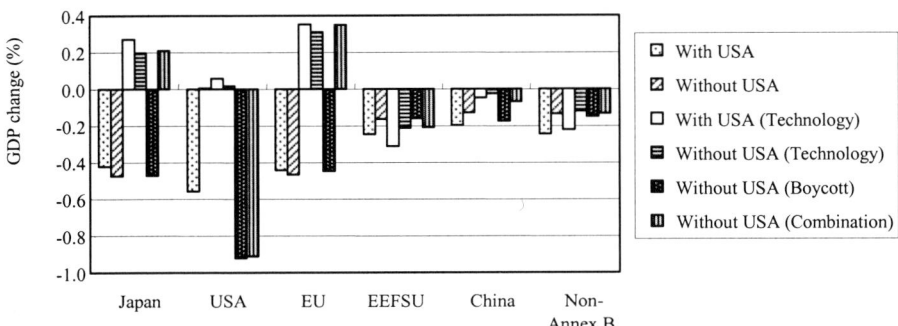

Fig. 3. Percentage change in the GDP in 2010 without emissions trading

4.3.2 GDP changes in 2010

Figure 3 shows the percentage reduction in the GDP compared to the reference scenario without emissions trading. Six scenarios are compared. They are classified according to three factors: ratification and non-ratification by the USA, price-induced technological development and a boycott movement. The combination scenario is one that includes both price-induced technological development and a boycott movement.

The GDP loss for the USA is the highest in the "with USA" scenario, even though the carbon price is the lowest among Japan, the USA, and the EU. The GDP change is negative in every region in the "with USA" scenario, while that for the USA becomes positive in the "without USA" scenario. The GDP changes become positive for Japan, the USA and the EU, while remaining negative for Russia in the "priced-induced technological change with USA" scenario. The absolute value of the impacts decreases in the "priced-induced technological change without USA" compared to the "with USA" scenario. This is because it is assumed that there is a very high potential for technological change if the price of energy increases substantially. The GDP change for the USA is negative and the amount of the loss in the "boycott movement" scenario is larger than the "with USA (ratifying the Kyoto Protocol)" scenario. This means that the economic loss to the USA is greater than if the USA ratifies the Protocol as long as other countries boycott US goods. However, the impacts are also negative for Japan, the EU and Russia. The effect of a boycott movement on the world economy is not good. In the "technological change+boycott movement" scenario, the GDP changes are positive for Japan and the EU, and the negative impact on China is lessened.

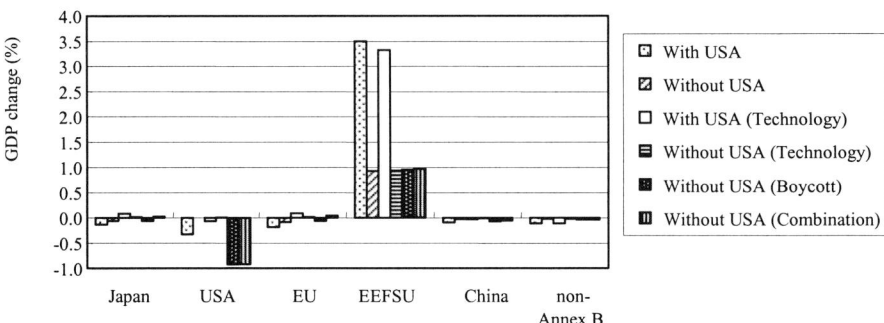

Fig. 4. Percentage change in the GDP in 2010 with emissions trading

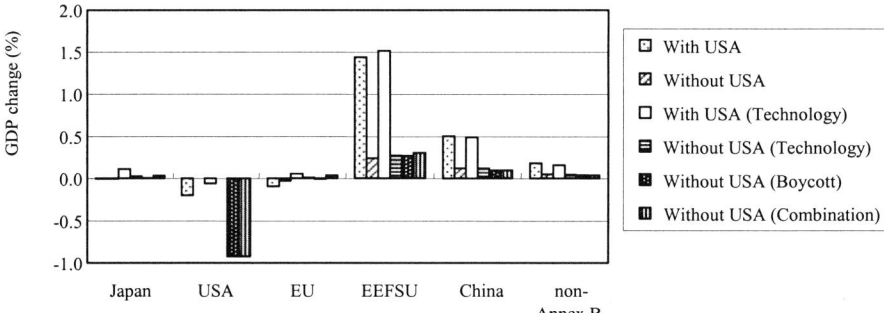

Fig. 5. Percentage change in the GDP in 2010 with emissions trading and CDM

Figure 4 shows the percentage change in the GDP in 2010 in the "emissions trading with no restrictions" scenarios. Again six scenarios are analyzed. In these scenarios, positive impacts for Russia are observed, especially in the "with USA" scenarios and the "price-induced technological change" scenario. Although a negative impact for the USA with the "boycott" scenario is observed, the impact is not large for other regions.

Figure 5 shows the percentage change in the GDP in the "emissions trading and CDM with no restrictions" scenarios. Positive impacts for Russia and non-Annex B countries are observed, especially in the "with USA" and "price-induced technological change" scenarios. Impacts are minimal for Japan, the USA and the EU.

When the assumption is that the amount of tradable carbon is restricted, for example, to one-third of the emissions reduction, the impacts on Annex B countries become greater compared to the "no-restrictions" scenarios. The impacts for Russia are negative in the "without USA" scenarios. The gains for Russia from emissions trading are much less than those corresponding to the "no-restrictions" scenarios.

When carbon sinks are taken into account, the GDP impacts are substantially reduced. The GDP impacts on Japan, the USA, the EU, and Russia in 2010 are -0.28, -0.47, -0.41, and -0.23, respectively, in the "with USA" scenario. These would be -0.33, 0.01, -0.43, and -0.15, respectively, in the "without USA" scenario. Impacts decrease in the emissions trading scenario. These would be -0.04, 0.01, -0.06, and 0.6 for Japan, the USA, the EU and Russia, respectively, in "the emissions trading without USA" scenario.

Table 1 shows the GDP changes in three sets of scenarios. The GDP gain for China is the highest in the "CDM case with USA" scenario. This gain is lowered in the "without USA" scenario.

Table 2 shows the corresponding carbon price. The carbon price for Japan is the highest in all the scenarios. If the CDM is assumed in addition to emissions trading, the carbon price becomes very low. This is especially true in the case of the "without USA" scenarios. If price-induced technological change occurs, the price becomes much lower.

Table 1. Percentage change in the GDP in 2010

	Japan	USA	EU	FSU	China	Non-Annex B
without emissions trading						
With USA	-0.42	-0.56	-0.44	-0.25	-0.20	-0.24
Without USA	-0.48	0.01	-0.47	-0.16	-0.13	-0.14
With USA (technology)	0.27	0.06	0.35	-0.31	-0.05	-0.22
Without USA (technology)	0.20	0.02	0.31	-0.21	-0.03	-0.12
Without USA (boycott)	-0.47	-0.92	-0.45	-0.16	-0.17	-0.15
Without USA (combination)	0.21	-0.91	0.35	-0.21	-0.07	-0.13
with emissions trading						
With USA	-0.14	-0.33	-0.19	3.50	-0.09	-0.11
Without USA	-0.06	0.00	-0.08	0.92	-0.03	-0.02
With USA (technology)	0.08	-0.07	0.08	3.32	-0.04	-0.11
Without USA (technology)	0.01	0.00	0.01	0.94	-0.01	-0.03
Without USA (boycott)	-0.06	-0.92	-0.06	0.96	-0.07	-0.04
Without USA (combination)	0.02	-0.92	0.04	0.97	-0.06	-0.04
with emissions trading and CDM						
With USA	0.00	-0.20	-0.09	1.44	0.51	0.19
Without USA	0.00	0.00	-0.03	0.24	0.12	0.05
With USA (technology)	0.11	-0.06	0.06	1.51	0.49	0.16
Without USA (technology)	0.03	0.00	0.01	0.27	0.12	0.05
Without USA (boycott)	0.00	-0.92	-0.01	0.27	0.10	0.04
Without USA (combination)	0.04	-0.92	0.04	0.31	0.10	0.04

Table 2. Price of carbon in 2010

	Japan	USA	EU	FSU
without emissions trading				
With USA	343	177	256	0
Without USA	337	0	250	0
With USA (technology)	252	145	195	0
Without USA (technology)	247	0	191	0
Without USA (boycott)	355	0	264	0
Without USA (combination)	260	0	200	0
with emissions trading				
With USA	69	69	69	69
Without USA	25	0	25	25
With USA (technology)	61	61	61	61
Without USA (technology)	23	0	23	23
Without USA (boycott)	27	0	27	27
Without USA (combination)	25	0	25	25
with emissions trading and CDM				
With USA	38	38	38	38
Without USA	10	0	10	10
With USA (technology)	35	35	35	35
Without USA (technology)	9	0	9	9
Without USA (boycott)	11	0	11	11
Without USA (combination)	11	0	11	11

Unit: US$/t-C

4.3.3 Carbon leakage

Figure 6 shows CO_2 emissions changes in 2010 under the "no-trading," "emissions trading" and "emissions trading and CDM" scenarios without the USA compared to the emissions of the reference scenario. Carbon leakage in non-Annex B countries is observed in the no-trading scenario. Leakage in non-Annex B countries decreases as the GDP recovers in the Annex B countries in the trading scenario. In the "emissions trading and CDM" scenario, as the amount of allowable emissions of the non-Annex B countries is restricted to the reference scenario, theoretically there is no leakage. The emissions of Russia in the "trading" scenario are higher than those in the "no-trading" scenario by 0.16 Gt-C.

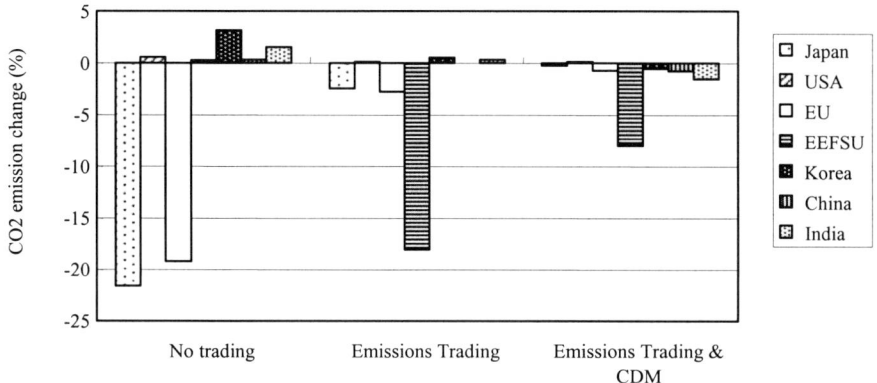

Fig. 6. CO_2 emissions changes in 2010 compared to the reference scenario

4.3.4 Major findings

The change in the GDP in 2010 to achieve the target of the Kyoto Protocol is less than 0.3% in any region if emissions trading is assumed. It can be said that achievement of the target will not have a major influence on any single economy. If it is assumed that there is a movement to boycott the goods of the non-ratifying countries, the GDP loss will grow further. On the other hand, there is a possibility of reducing the economic loss or even expanding the economy by promoting the introduction of energy conservation technologies.

No ratification by the United States lowers the carbon price and decreases the CDM incentive. In this case, total greenhouse gases will increase compared to the reference scenario.

4.4 Long-term Emissions Reduction Scenarios

4.4.1 Long-term mitigation scenarios

The economic impacts of the Kyoto Protocol have been evaluated using the AIM model. Although the Kyoto Protocol is very important as a milestone in climate policy, it is necessary to reduce emissions by more than the reduction target specified by the Protocol in order to achieve the long-term goals of the UNFCCC. Considerable efforts have been devoted to estimating the different stabilization pathways in the post-SRES experiments (IPCC 2001a). Based on these experiments, stabilization of emissions pathways and economic as well as climatic impacts were studied using the AIM model. The emissions model examines several im-

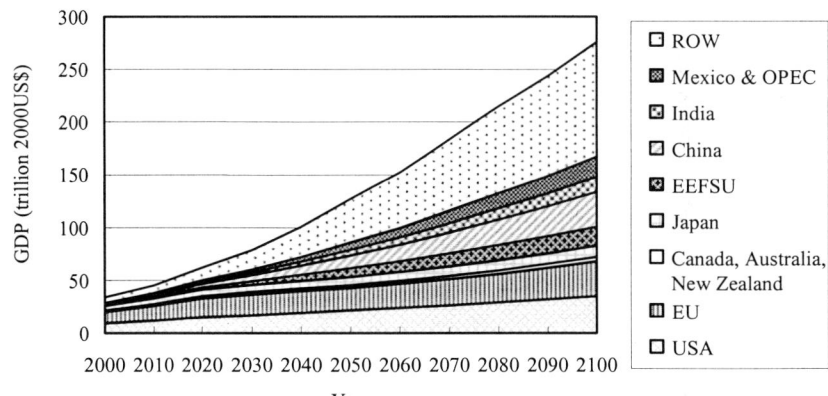

Fig.7. Projection of the world GDP under the reference scenario

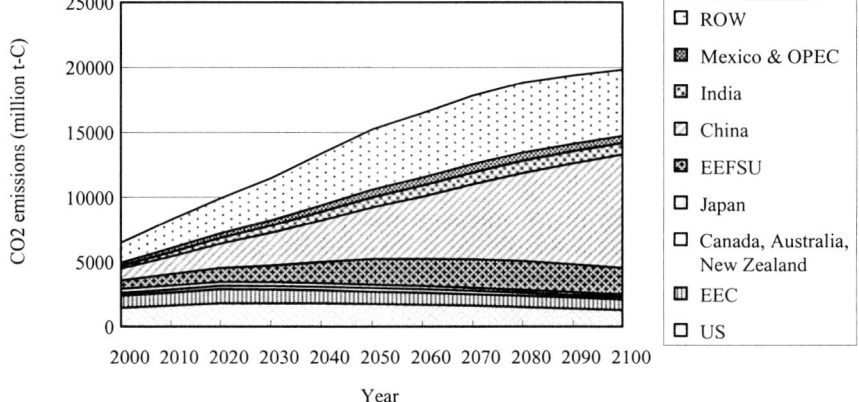

Fig. 8. Projection of CO_2 emissions under the reference scenario

portant variables such as GDP changes, energy consumption, carbon emissions, and marginal costs. Outputs of the emissions model are fed as inputs into the climate model. Estimated climate changes under different scenarios are used to estimate the climatic impacts on food production and infectious diseases focusing on the Asia-Pacific region.

Atmospheric stabilization scenarios were used that limit the atmospheric concentration of CO_2 at 550 ppmv. As a reference scenario the driving forces used by the SRES B2 scenario (IPCC 2000) were taken. It is assumed that carbon emissions can be traded without quantitative limitations on trading cases within the allowable emissions. The global allowable emissions are specified in the 550 ppmv scenarios and they are allocated according to the population.

Figure 7 shows a projection of the world GDP from 2000 through 2100. World growth rates during this period vary from 1.25% to 3.16%, with an average of 2.1%. The highest is that of China, which varies from 1.5% to 6.4%. Figure 8

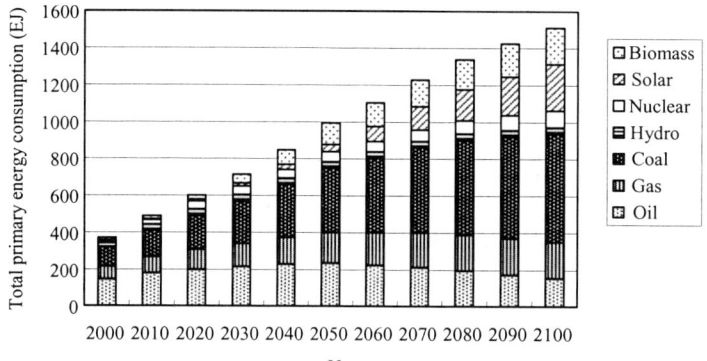

Fig. 9. World energy demand in enduse sectors under the reference scenario

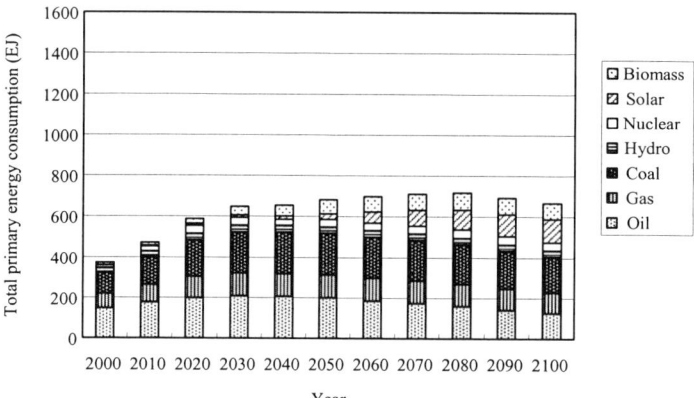

Fig. 10. World energy demand in enduse sectors under the WRE 550 scenario

shows a projection of world CO_2 emissions. It is projected that China will become the top CO_2-emitting country after 2020. The growth rate in world CO_2 emissions will follow a downward curve, whereas that of China will increase. The growth rate in CO_2 is much higher than the growth rate of the GDP in China under this reference scenario. This is because energy efficiency in China is estimated to be lower than that of the developed countries in this case.

The results of the reference scenario are compared with three 550 ppmv scenarios. The three 550 ppmv scenarios examined are WRE 550, WGI 550, and MID 550. The WRE scenario was proposed by Wigley *et al.* (1995) to find the optimal path to 550 ppmv from the economic point of view. WGI 550 is a scenario proposed by IPCC Working Group I (IPCC 1995). It is a path aimed at avoiding an abrupt change in emissions in achieving the 550 ppmv target. The MID 550 scenario is proposed, representing the mean of these two scenarios.

Figure 9 shows world energy demand in enduse sectors in the reference scenario and Fig. 10 shows the results of the WRE 550 scenario. In the reference scenario, the use of solid energy increases from 18% in 2000 to 48% in 2100, and

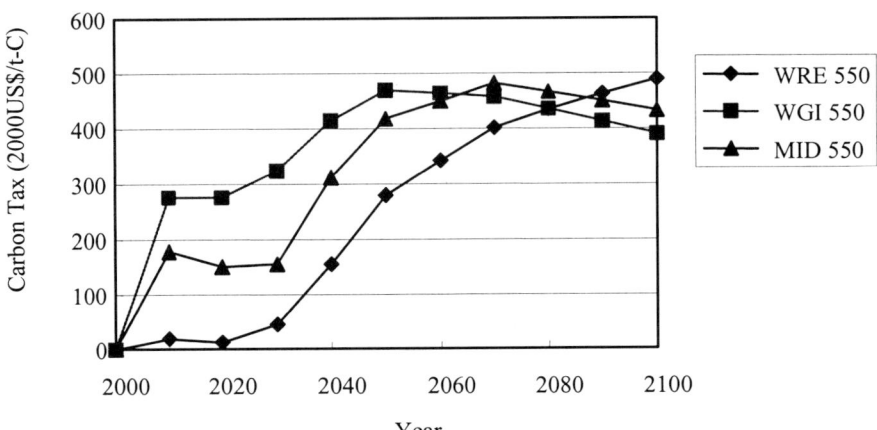

Fig. 11. Projection of marginal costs to reduce emissions

more than half is used in China in 2100. The world final energy demand in the WRE 550 scenario decreases to nearly half that in the reference scenario in 2100. This reduction comes about mainly from cutting coal use. The share of coal in the WRE 550 scenario becomes 29% in 2100. Electricity demand will increase in the policy scenarios. The share of electricity will increase from 17% in 2000 to 29 % in 2100 in WRE 550.

4.4.2 GDP changes until 2100

Figure 11 shows the projections of the marginal costs for reducing emissions. The marginal costs for the WGI 550 scenario are the highest through the year 2060, those of MID 550 become the highest from 2060 through 2080, and then those of WRE 550 become the highest from 2090 onwards. Although the constraint of the WGI 550 scenario is the severest until 2070, the marginal costs become the second highest in 2060. The restructuring of the energy system at an early stage will decrease the marginal costs after 2050.

Figure 12 shows the projections for GDP changes compared to the reference scenario. The GDP loss in the WGI 550 scenario increases until 2050 and then recovers, while that of WRE 550 increases until 2070. The GDP loss in WRE 550 is the highest among the three 550 scenarios in 2100. This means that even if the impact of WRE 550 is minimal for the first three decades, it will become large later. The environmental impacts that will be discussed in Section 4.4.4 show that if the timing of reductions is early, this will reduce the impacts compared to scenarios in which action is taken later.

Figure 13 shows the consumption changes relative to the reference scenario. The values shown are the present discounted values for macroeconomic consumption change with respect to the reference scenario in trillions of 2000 US dollars through 2050. The discount rate is 5%. Consumption in India will grow at the

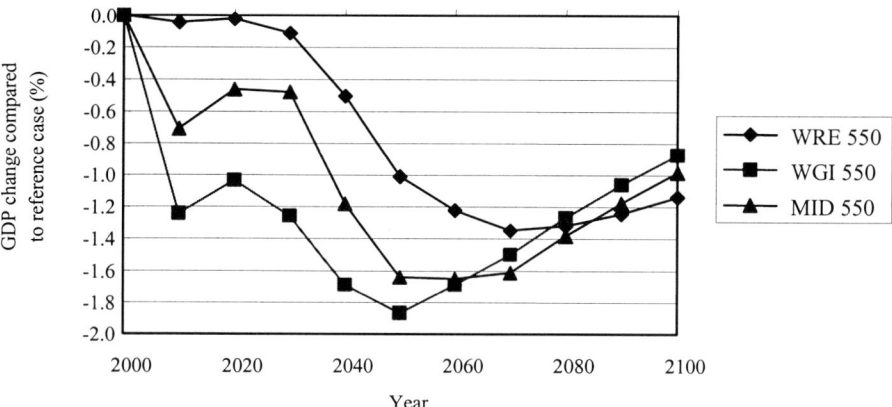

Fig. 12. Projection of GDP changes compared to the reference scenario

highest level, especially in the WGI 550 scenario. Consumption will decline in Annex B countries and China. The greatest decline is in the US, followed by China, under the assumptions of the policy scenarios.

4.4.3 Global climate change

The impact on the global mean temperature is shown in Figure 14. By 2100, the temperature rises by 2.77°C compared with the 1990 value in the reference scenario. The results of the IPCC SRES range from 1.4°C to 5.8°C (IPCC 2001b). As the economic assumptions of the reference scenario are taken from the SRES B2 scenario, it is lower than the average of the SRES range. While the SRES B2

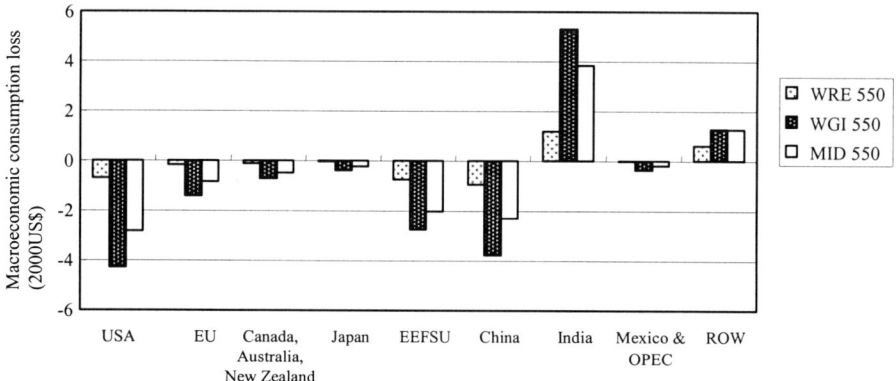

Fig. 13. Present discounted value of macroeconomic consumption loss with respect to the reference scenario (in trillion 2000 US$)

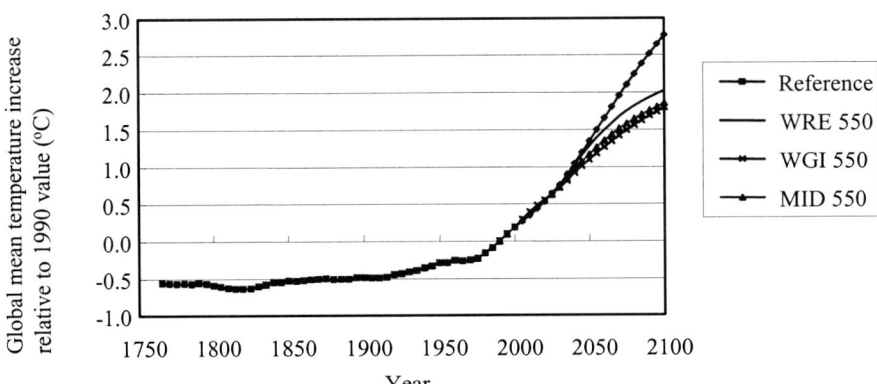

Fig. 14. Temperature increases relative to the 1990 value

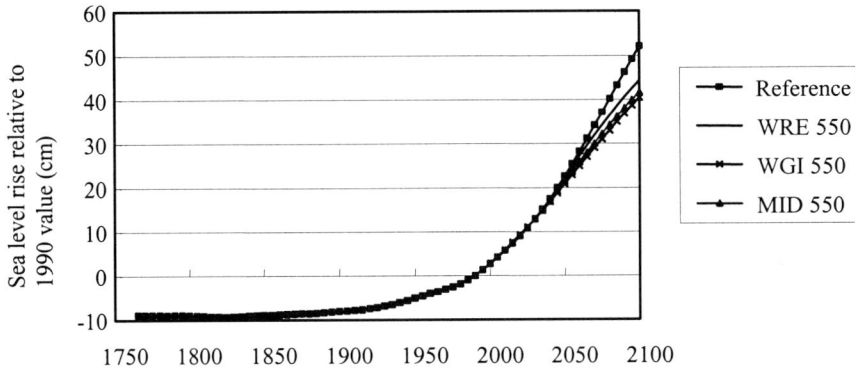

Fig. 15. Rise in the sea level relative to the 1990 value

emissions range from 10.8 Gt-C to 21.8 Gt-C in 2100, the result of the reference scenario is 21.9 Gt-C. This is because the reference scenario uses the assumptions of population and GDP from SRES B2, but does not focus so much on environmental sustainability.

The temperature increase in the WGI 550 scenario is the lowest, at 1.79°C in 2100. The increase in the WRE 550 scenario is 2.02°C. Although the targets of these two scenarios are the same, there is a 0.23°C difference in the temperature increase in 2100. These 550 ppmv scenarios can decrease the temperature by 0.75°C to 0.98°C in 2100 compared to the reference scenario. Although the macroeconomic consumption loss of the WRE 550 scenario is lower than that of the WGI 550 scenario, its impact on climate change is greater.

The globally averaged sea level rise relative to that of 1990 is shown in Figure 15. It ranges from 40 cm to 52 cm. The global sea level in 2100 is projected to rise by 9 cm to 88 cm for the full range of SRES scenarios. The sea level in the WRE 550 scenario rises higher by 3.9 cm than that in the WGI 550 scenario.

4.4.4 Potential impacts in the Asian region

Climate change has direct or potential impacts on water resources, agricultural production, natural ecosystems, and human health, even if socioeconomic interactions are ignored. In the real world, global trade, immigration, and measures for adaptation modify the direct impacts. Hence, there are two stages of the impact study: the direct and indirect stages. In this study, the direct impacts under the reference and 550 ppmv scenarios were considered.

Figure 16 shows the changes in winter wheat productivity in 2100 compared to 1990. The productivity of wheat will decline significantly in Sri Lanka, Malaysia, Korea-PDR, Burma, and other tropical countries.

Figure 17 shows changes in rice productivity in 2100 compared to 1990 under the reference and 550 ppmv scenarios. A slight decrease in rice production is expected in most countries, while a slight increase is expected in Bhutan and Taiwan. The productivity decline in India is projected to be the highest.

Air and water pollution, as well as solid and hazardous wastes, affect human health directly. Global climate change will also affect human health in the future in many ways. For example, global warming will result in increasing temperatures and changing vegetation close to the ground. This will allow expansion of the habitat of the anopheles mosquito, which is the malaria vector. In addition, the development period of the malaria protozoa will shorten and their reproductive potential will increase. As a result, it is expected that the global risk of a higher incidence of malaria will increase.

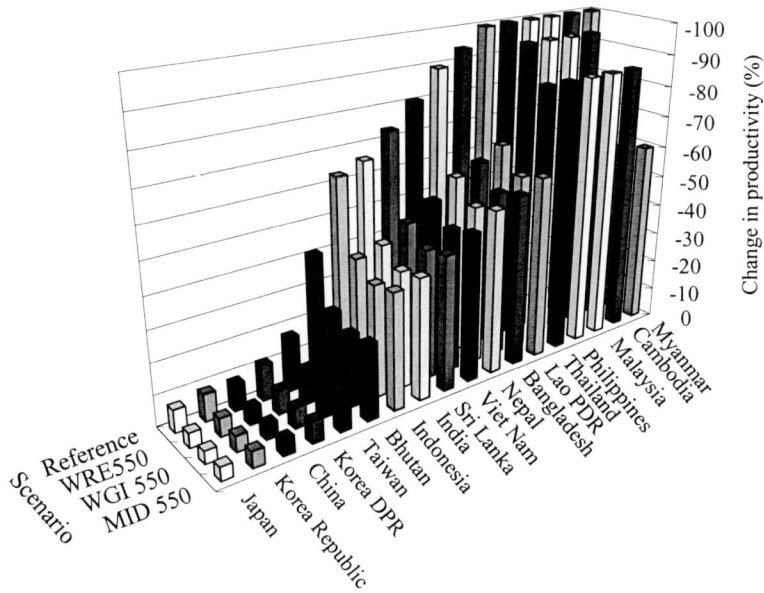

Fig. 16. Change in winter wheat productivity from 1990 to 2100 (see color plates)

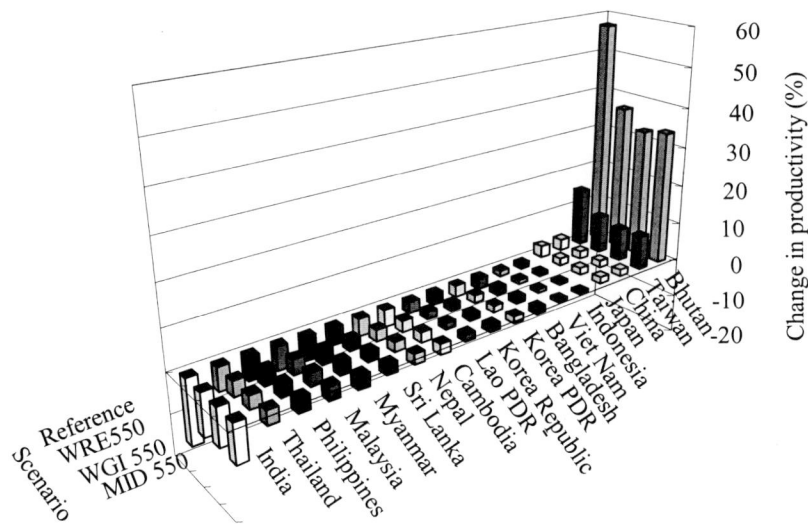

Fig. 17. Change in rice productivity from 1990 to 2100

Figure 18 shows the changes in the populations living where there is endemic malaria. The risk of a higher incidence of malaria in China, Nepal, Taiwan, Indonesia, and Sri Lanka will increase due to the environmental changes. Although China may be little affected by climate change in terms of food productivity, the impact it will experience in terms of malaria risk will be the greatest in the Asian region. Even under the 550 scenarios, the population living in areas at risk will double in 2100 compared to 1990.

4.4.5 Major findings

Several climate change stabilization pathways are examined by the AIM model. An urgent task is to take action to combat global warming as it is an irreversible process and, once it occurs, the probability is very high that it will have multiplier effects that will further expand its impact. It is necessary to consider the many uncertainties involved in human activities such as population growth, economic development, and technological innovation, as well as uncertainties in natural processes to estimate future CO_2 emissions and to make plans for climate stabilization. The scenario approach is a practical means of analyzing the policy options under such uncertainties.

Although the emissions of developing countries were lower than those of developed countries in 2000, it is expected that they will grow much faster than those of developed countries in the future. The estimated impacts on developing countries in the Asian region are serious. The model projects that the early implementation of emissions reductions can minimize the scale of such impacts in the future.

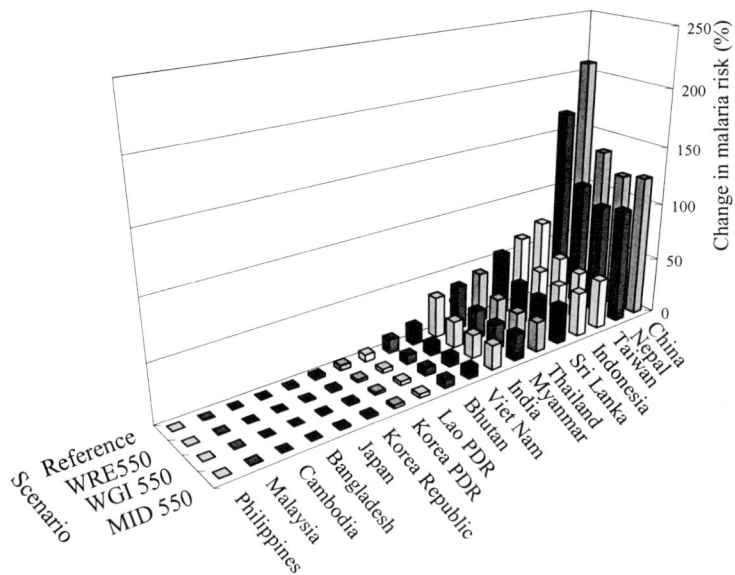

Fig. 18. Change in population living in areas at high malaria risk from 1990 to 2100 (see color plates)

4.5 Concluding Remarks

Without the Kyoto target, the global temperature will increase by more than 2°C in 2100. In such a case, severe impacts of climate change can be predicted on agricultural productions and human health.

It is predicted that ratification of the Kyoto Protocol may have negative economic impacts. However, there are several ways to mitigate these economic impacts as well as possibilities for promoting the growth of economies. For example, if investments are shifted to energy-saving technologies, there is a high likelihood of improving the economy. It is found that the GDP will increase if the price-inducing mechanism is enhanced and the CDM is introduced.

References

IPCC (1995) Climate Change 1995. Cambridge University Press
IPCC (2000) Special Report on Emissions Scenarios. Cambridge University Press
IPCC (2001a) Climate Change 2001, Mitigation. Cambridge University Press
IPCC (2001b) Climate Change 2001, The Scientific Basis. Cambridge University Press
Kainuma M, Matsuoka Y, Morita T (1999) Analysis of post-Kyoto scenarios: the Asian-

Pacific Integrated Model. Special Issue of The Energy Journal, The Costs of the Kyoto Protocol: a Multi-Model Evaluation, pp207-220

Wigley TML, Richels R, Edmonds JA (1995) Economic and environmental choices in the stabilization of CO_2 concentrations: choosing the "right" emissions pathway. Nature, 379: 240-243

III. Regional and Country Modeling

5. Application of AIM/Enduse Model to China

Xiulian Hu[1], Kejun Jiang[1], and Hongwei Yang[1]

Summary. Along with rapid economic growth, as a nation, China has become the second highest consumer of energy in the world and the greatest emitter of CO_2. The amount of future CO_2 emissions in China and the means of reducing these emissions are currently important issues to resolve. Base on sectoral and technological information, this study analyzed future CO_2 emissions scenarios with respect to several cases and assessed the effects of different policy options using the AIM/Country model. It also analyzed CO_2 reduction costs for various sectors. From the results, it was found that CO_2 emissions will increase along with the rapid economic development in China, but it is possible to gradually minimize the growth rate of CO_2 emissions through technological progress directed towards efficient markets and the adoption of policies for CO_2 reduction. Technological progress plays a key role in CO_2 emissions mitigation and local air pollution abatement.

5.1 Introduction

In China, the annual average GDP growth rate was 8.9% in the 1980s following the country's economic reforms and the opening up of its economy. From 1991 to 2001 the average growth rate increased to 9.86%, giving China one of the highest economic growth rates in the world. Rapid economic development has stimulated major social changes in China. The reform of economic mechanisms, changes to government functions, nurturing of the market economy, raising the quality of life, etc., comprise the major social changes in China. The industrial structure is also changing. For example, secondary industries have increased their share of overall economic activity, expressed as a common characteristic of countries in the early phase of industrial development. During the same period, personal income rapidly increased and residential consumption patterns also changed greatly. The period over which these changes will take place will be much shorter in China than was the experience of the developed countries.

Along with rapid economic development, energy production and consumption has increased very quickly. In 2000 the figures were 901 Mtoe (million ton oil equivalent) and 903 Mtoe respectively, with annual growth rates of 5.2% and 4.8% from 1980 to 2000. China is now one of the world's major countries with respect to energy production and consumption. One significant characteristic of energy production and consumption in China is that coal plays a larger role than in most other countries. In 1994, coal accounted for 78% of total energy production

[1] Energy Research Institute, State Development Planning Commission, Guohong Building, A-11 Muxidi Beili, Xicheng District, Beijing 100038, China

and 48% of final energy consumption. This level of energy production and consumption creates serious environment problems and is indicative of low energy efficiency. Since there are limitations on energy resources and little capacity to import energy into China from the international market, it is difficult to change this situation in the near future. Following the increase in energy consumption together with the invariable energy structure, CO_2 emissions in China increased from 0.41 billion tons of carbon in 1980 to 0.65 billion tons of carbon in 1990. The share of CO_2 emissions in China in relation to total world CO_2 emissions has been growing, and this trend can be expected continue in future. Rapid economic growth, with fossil fuels meeting a large part of energy consumption, and the demand for a better life style will drive China into becoming a conspicuous consumer of energy and emitter of CO_2 compared to the rest of the world.

Considering this development trend, a better understanding of future trends in relation to energy demand and CO_2 emissions is critical, and it is also important to assess the effects of various means of reducing CO_2 emissions in China. Many research activities on CO_2 emissions scenarios in China have already been conducted. Some significant scenarios were provided by the IPCC (IS92 and SRES scenarios) (Nakicenovic 2001), IEW, WEC, GREEN, SGM and several other groups (Houghton *et al.* 1995). It is noticeable that the range of variability in the results of these scenarios is very large (Zhou *et al.* 1997). Up to now, most of these research activities have been based on a macroeconomic approach and did not consider the far-reaching technological progress and structural social changes that have been observed in recent years in China. The study described in this report tries to assess the effects of such technological progress and structural social changes as well as possible policies on future energy consumption and CO_2 emissions based on more detailed data, including energy services and processes, energy technologies, and various social changes.

Since 1994, the Energy Research Institute has worked together with the AIM team at NIES to develop an AIM/Country model for China. After several years work, the AIM-China model was developed with the inclusion of 26 sectors involving more than 500 technologies. Energy demand and CO_2 emissions by 2030 were simulated, while some policies that focus on technological progress and collaboration were assessed as to their contribution to controlling CO_2 emissions.

5.2 Input Assumptions and Simulations

Growth in economic development is a key assumption for this research. It is commonly viewed that China can continue to develop very quickly as long as no major social upheaval occurs. Many studies had been conducted to forecast future economic development in China. By synthesizing several forecasts on economic development in China, annual average GDP growth rates of 9% from 1990 to 2000, and 7.5%, 6.5%, and 5.5% from 2000 to 2010, 2010 to 2020, and 2020 to 2030, respectively, were used for the economic development scenario in this research.

Population is another key factor for making forecasts in this research. Major aspects of population growth that needed to be considered include: planned population policy; the decline in fertility accompanying the expected increase in income; the shift in the age of marriage to later in life; the prospect of fewer children due to family working patterns involving both the husband and wife typically having jobs; the desire for fewer children among well-educated people; the increase in the average life span; the decline in the death rate, and so on. It is believed that the growth in the population will follow the pattern of developed countries in the Asian region. Based on these factors, and with reference to the results of some other forecasts on population, the population scenario used in this research was that the population will be 1.28 billion in 2000 and 1.39 billion, 1.47 billion and 1.54 billion in 2010, 2020 and 2030, respectively, with an average natural growth rate of 1.14% from 1990 to 2000 and 0.88%, 0.55% and 0.45% from 2000 to 2010, 2010 to 2020, and 2020 to 2030. The rate of urbanization was taken as 32%, 38%, 43%, and 49% in 2000, 2010, 2020 and 2030, respectively.

According to the current statistics on China's national economy as well as the available data, energy end users in this study are divided into five sectors; the industrial, agricultural, services, residential and transport sectors. Table 1 gives the classification of these sectors and their sub-divisions. Every sector is split into several sub-sectors, or a products or services mode. The industry sector is classified into sub-sectors, and then every sub-sector includes one or more products. For example, the non-ferrous metals sub-sector includes a number of products such as copper, aluminum, zinc and lead. For the transport sector, under every sub-sector, transport modes for passenger transport and freight transport are given. The residential sector is split into urban and rural to match the different development patterns in each. Different technologies related to the demand for services are collected for every sub-sector and product. Energy services and technology selection for each sector or product is determined so that energy consumption and CO_2

Table 1. Classification of energy end user sectors

Sectors	Industrial sector	Agricultural sector	Residential sector	Services sector	Transport
Sub-sectors or products	Iron and steel Non-ferrous metals Building materials Chemical industry Petrochemical industry Paper-making Textile Machinery Power generation Oil refinery	Irrigation Farming work Agricultural products processing Fishery Animal husbandry	Urban energy use Rural energy use Space heating Cooling Lighting Cooking and hot water Household electric appliances	Space heating Cooling Lighting Cooking and hot water Electric appliances	Railway transport Road transport Waterways transport Air transport

emissions can be estimated.

Energy services scenarios for major sectors and products are listed in Table 2. These scenarios reflect the situation of economic development and structural social change in China. The scenarios determine the increase in the level of demand for energy services that will be met by the selected technologies.

Table 3 lists the major technologies used in this model. In AIM-China, these energy use technologies are mainly broken down into three categories:

1. Technologies for services production: they are technologies to satisfy services

Table 2. Energy services forecast by industrial sector (index, 1990=1)

Energy Services	1990	2000	2010	2020	2030
Steel	1	1.8	2.1	2.3	2.6
Cement	1	3.3	5.2	6.0	6.6
Glass	1	2.0	3.5	4.3	4.9
Paper	1	2.1	2.6	3.2	3.7
Bricks	1	1.3	1.8	2.1	2.4
Soda ash	1	2.4	0.5	6.1	7.0
Caustic soda	1	1.9	3.0	3.6	4.1
Fertilizer	1	1.5	1.8	2.1	2.3
Services area	1	1.3	2.0	2.5	3.1
Traffic volume for private cars	1	21.0	118.7	523.3	916.6
Car traffic volume in cities	1	19.8	120.5	534.0	924.7
Car traffic volume in rural areas	1	55.6	68.3	209.7	682.5
Railway passenger transport	1	1.3	1.8	3.5	6.1
Steam locomotives	1	0.0	0.0	0.0	0.0
Diesel locomotives	1	2.0	2.8	4.5	6.2
Electric locomotives	1	2.5	3.6	7.9	16.4
Railway freight transport volume	1	1.5	2.0	2.4	2.6
Road transport	1	2.2	4.0	5.7	7.4
Waterways transport	1	1.3	1.6	1.6	1.6
Air transport	1	4.9	12.9	80.0	182.0
Rural population	1	1.0	1.0	1.0	0.9
Rural households	1	1.2	1.2	1.2	1.2
Rural household land area	1	1.0	0.9	0.9	0.9
Urban population	1	1.4	1.8	2.1	2.5
Urban households	1	1.5	2.0	2.5	3.1
Urban household area	1	1.3	1.6	1.8	2.1
Cooking in urban households	1	0.0	2.0	2.5	3.1
Electric cooking in urban areas	1	0.0	4.7	6.5	8.6
Hot water in urban areas	1	17.5	54.7	111.3	175.2
Space heating in urban areas	1	1.9	2.9	4.1	5.4
Air conditioners in urban areas	1	46.7	141.7	296.7	689.3
Fans in urban areas	1	1.9	2.8	3.7	4.7
Lighting: C in urban areas	1	1.3	1.3	1.6	1.9
Lighting: F in urban areas	1	10.0	25.1	34.3	46.9
Refrigerators in urban areas	1	2.3	4.6	7.8	10.9
Color TVs in urban	1	2.3	4.4	7.0	10.6
Black/white TVs in urban areas	1	1.3	1.7	0.5	0.0
Washing machines in urban areas	1	2.5	4.6	8.4	14.2

supply. These technologies include the renewal of various old technologies and newly installed technologies.

2. Technologies for energy recovery utilization: including various technologies for residual heat, combustible gases and black liquor recovery and their utilization.
3. Technologies for energy conversion: in-plant electric power generators, technologies for thermal energy conversion (e.g. industrial boilers) as well as electric power generation using residual heat and combustible gases, etc.

More than 500 technologies have been collected for the analysis, which covers the major technologies used in every sector defined in Table 1. Some advanced technologies have been taken into account even though they are not currently used in China. Basic data for the technologies include the purchase price, annual rate of

Table 3. Energy services technologies used in this simulation

Classification	Technologies (equipment)
Iron and steel	Coke oven; Sintering machine; Blast furnace; Open hearth furnace (OH); Basic oxygen furnace (BOF); AC-electric arc furnace; DC-electric arc furnace; Ingot casting machine; Continuous casting machine; Continuous casting machine with rolling machine; steel rolling machine; Continuous steel rolling machine; Equipment for coke dry quenching; Equipment for coke wet quenching; Electric power generated with residue pressure on top of the blast furnace (TRT); Equipment for coke oven gas; OH gas and BOF gas recovery; Equipment for co-generation
Non-ferrous metals	Aluminum production using the sintering process; Aluminum production using the combination process; Aluminum production using the Bayer process; Electrolytic aluminum using the upper-insert cell; Electrolytic aluminum using the side-insert cell; Crude copper production with flash furnace; Crude copper production using an electric furnace; Blast furnace; Reverberator furnace; Lead smelting-sintering in a blast furnace; Lead smelting using a closed blast furnace; Zinc smelting using the wet method; Zinc smelting using the vertical pot method
Building materials	Cement: Mechanized shaft kiln; Ordinary shaft kiln; Wet process kiln; Lepol kiln; Ling dry kiln; Rotary kiln with pro-heater; Dry process rotary kiln with pre-calciner; Self-owned electric power generator; Electric power generator with residue heat
	Bricks and tiles: Hoffman kiln; Tunnel kiln
	Lime: Ordinary shaft kiln; Mechanized shaft kiln
	Glass: Floating process; Vertical process; Colburn process; Smelter
Chemical industry	Equipment for synthetic ammonia production: Converter; Gasification furnace; Gas-making furnace; Synthetic column; Shifting equipment for sulphur removal
	Equipment for caustic soda production: Electronic cells using the graphite process; Two-stage effects evaporator; Multi-stage effects evaporator; Equipment for rectification; Ion membrane method
	Calcium Carbine production: Limestone calciner; Closed carbine furnace; Open carbine furnace; Equipment for residual heat recovery
	Soda ash production: Ammonia and salt water preparation; Limestone calcining; Distillation column; Filter
	Fertilizer production: Equipment for organic products production; Equipment for residual heat utilization
Petrochemical industry	Facilities for atmospheric and vacuum distillation; Facilities for rectification; Facilities for catalyzing and cracking; Facilities for cracking with hydrogen adding; Facilities for delayed coking; Facilities for light carbon cracking; Sequential separator; Naphtha cracker; De-ethane separator; Diesel cracker; De-propane cracker; Facilities for residual heat utilization from ethylene
Paper-making	Cooker; Facilities for distillation; Facilities for washing; Facilities for bleaching; Evaporator; Crusher; Facilities for de-water; Facilities for finishing; Facilities for residue heat utilization; Facilities for black liquor recovery; Co-generator; Back pressure electric power generator; Condensing electric power generator

diffusion rate, unit energy consumption, life span, year of entering the market and year of obsolescence, production capacity, etc. Table 4 gives a sample list of technological data for the coke making process in the steel industry.

Prices and emissions factors for energy and other materials inputs are listed in Table 5.

In the analysis of the AIM-China model, particular attention was paid to natural gas supply and traditional biomass supply. According to the possible future for natural gas production and imports, no limitation was placed on natural gas use except in the residential sector. Natural gas use in the residential sector is strongly limited by government policy and investment in infrastructure. Traditional biomass use was limited to maintaining it at the most at the 1994 level among

Table 3. Energy services technologies used in this simulation (continued)

Classification	Technologies (equipment)
Textiles	Cotton weaving process; Chemical fiber process; Wool weaving and textile process; Silk production process; Printing and dyeing process; Garment making; Air conditioners; Lighting; Facilities for space heating
Machinery	Ingot process: Cupola; Electric arc furnace; Fan
	Forging process: Coal-fired pre-heater; Gas-fired pre-heater; Oil-fired pre-heater; Steam hammer; Electric-hydraulic hammer; Pressing machine
	Facilities of heat processing: Coal-fired heat processing furnace; Oil-fired heat processing furnace; Gas-fired heat processing furnace; Electric processing furnace
	Cutting process: Ordinary cutting; High speed cutting
Irrigation	Diesel engine; Electric induct motor
Farming work	Tractor; Other agricultural machines
Agricultural products processing	Diesel engine; Electric induction motor; Processing machines; Coal-fired facilities
Fishery	Diesel engine; Electric induction motor
Animal husbandry	Diesel engine; Electric induction motor; Other machines
Space heating for dwellings	Heat supplying boiler in a thermal power plant; Boiler for district heating; Dispersed boiler; Small coal-fired stove; Electric heater; Brick bed linked to a stove (Chinese KANG)
Cooling in dwellings	Air conditioner; Electric fan
Lighting in dwellings	Incandescent lamp; Fluorescent lamp; Kerosene lamp
Cooking and hot water in dwellings	Gas burner; Bulk coal-fired stove; Briquette-fired stove; Kerosene stove; Electric cooker; Cow dung-fired stove; Firewood-fired stove; Methane-fired stove
Electric Appliances	Television; Cloth washing machine; Refrigerator; Others
Space heating in the services sector	Heat supplying boiler in the thermal power plant; Boiler of district heating; Dispersed boiler; Electric heater
Cooling	System of central air conditioner; Air conditioner; Electric fan
Lighting	Incandescent lamp; Fluorescent lamp
Cooking and hot water	Gas burner; Electric cooker; Hot water pipeline; Coal-fired stove
Electric appliances	Copying machine; Computer; Elevator; Others
Passenger and freight transport	Railways (passenger and freight): Steam locomotive; Internal combustion engine locomotive; Electric locomotive
	Highway (passenger & freight): Public diesel vehicle; Public gasoline vehicle; Private vehicle; Large diesel freight truck; Large gasoline vehicle; Small freight truck
	Waterways (passenger & freight): Ocean-going ship; Coastal ship; Inland ship
	Aviation (passenger & freight): Freight airplane; passenger airplane

rural residents.

To assess the effect of technological progress and the alternative policy options for energy consumption and CO_2 emissions in China, several cases were defined to run the model. The cases defined in Table 6 include the "frozen technology case," the "market case" and the "policy case." They are described in the following.

The frozen technology case, only used for comparison with other cases, can also be called the no technological progress case. It is presumed that the present services production technologies and energy efficiency will remain at the same level as in 1990 without any technological progress. However, this does not mean that energy consumption for this case will increase at the same rate of growth as that for economic development.

In the market case, in a properly functioning market it is assumed that technologies can be selected based on market mechanisms. It will consider technological options after rational assessment of the economic benefits provided by the energy services technologies. This case is designed to emphasize the contribution of market mechanisms to energy use conservation and CO_2 emissions reduction. China is at the stage of economic reform shifting from a planned economy to a market economy and this is expected to be completed in the next 10 years (Zheng *et al.* 1995). The results of the analysis for this case could be used to explain the benefits of market mechanisms, and it is also used as a base line emissions scenario for the medium term.

Table 4. Technological parameters in the coke making process of the steel industry

Name of technologies	Output mass	Input materials	Fixed price (yuan/t[*])	Life span (yr.)	Year of intro-duction	Year of obsoles-cence	Output	Input
							(100Mcal/t[*])	
Small cok-ing oven	Coke	Coal	185	20	1930	2030	36.46	76.00
	Coke	Utility*	185	20	1930	2030	36.46	4.80
	Coke	Electricity	185	20	1930	2030	36.46	0.40
	Coke oven gas	Coal	185	20	1930	2030	5.20	76.00
Large cok-ing oven	Coke	Coal	210	20	1970	2050	36.46	68.00
	Coke	Utility**	210	20	1970	2050	36.46	4.80
	Coke	Electricity	210	20	1970	2050	36.46	0.44
	Coke oven gas	Coal	210	20	1970	2050	5.80	68.00
International Advanced coking oven	Coke	Coal	330	30	1985	2100	36.46	59.89
	Coke	Utility*	330	30	1985	2100	36.46	4.05
	Coke	Electricity	330	30	1985	2100	36.46	0.23
	Coke oven gas	Coal	330	30	1985	2100	7.20	59.89
New ad-vanced cok-ing oven	Coke	Coal	380	30	1995	2100	36.46	54.20
	Coke	Utility*	380	30	1995	2100	36.46	4.26
	Coke	Electricity	380	30	1995	2100	36.46	0.23
	Coke oven gas	Coal	380	30	1995	2100	7.60	54.20

[*] ton steel output
[**] "Utility" used in this table is defined in Appendix A of Part IV.

The policy case is defined in order to analyze the effects of climate policies in reducing CO_2 emissions. As a commonly used method, the policy case was defined here as the levying of a carbon tax of 100 yuan per ton of carbon and returning all the revenues from this carbon tax as subsidies for the diffusion of advanced technologies. The introduction of a carbon tax is assumed to begin from 2000. The introduction of advanced technologies would be promoted by policies that contribute to CO_2 emissions reductions. Analysis of the carbon tax does not mean a pure tax; it can be considered here to be a mixture of comprehensive policies such as an energy tax, government regulations, etc. It is only used here as a means of introducing the carbon tax as a modeling parameter.

The four other cases are selected in order to estimate CO_2 reduction costs. As discussed above, subsidies can be invested in advanced technologies to reduce fixed costs (the price of the technology), thus expanding the introduction of effi-

Table 5. Price and CO_2 emission factors

	Emission factor	Price				
		1990	1994	2000	2010	2030
Coal	10062	2.86	2.90	3.00	3.20	3.50
Coke	12300	8.82	8.89	9.00	9.40	9.80
Crude oil	7811	8.00	9.00	10.00	11.00	12.00
Gasoline	7658	25.00	26.00	26.50	29.00	31.00
Kerosene	7748	25.00	26.00	26.50	29.00	31.00
Diesel	7839	25.00	26.00	26.50	29.00	31.00
LPG	6833	4.29	10.00	14.00	16.00	18.00
NGS	5639	4.29	10.00	12.00	13.00	16.00
Town gas	5835	4.00	6.00	8.00	8.60	9.60
Black liquid	10751	0.00	0.00	0.00	0.00	0.00
Electricity	26449	34.90	35.00	37.00	40.00	43.00
Biomass	0	0.03	0.03	0.04	0.05	0.08
Utility*	14374	4.00	4.16	4.40	4.50	4.80
Steam	14374	4.00	4.16	4.40	4.50	4.80
Water	0	0.30	0.42	0.60	0.80	1.00
Recycled steel	0	1100.00	1120.00	1150.00	1200.00	1300.00

Unit: emission factor: g-C/100Mcal
     price: 1990 yuan/100Mcal for energy and 1990 yuan/ton for materials (water and recycled steel)
* "Utility" used in this table is defined in Appendix A of Part IV.

Table 6. Definition of the cases

	Technology selection	Tax rate (Yuan/t carbon)	Subsidy
Frozen technology			
Market	√		
Policy	√	100	All tax revenues

cient technologies to save energy and reduce CO_2 emissions. By using the linear program sub-module for optimal subsidy assignment, an estimate can be made of the lowest additional cost that would be required at the national level to reduce CO_2 emissions. In the model calculation, four subsidy cases were defined to analyze the reduction of costs. The four cost estimation cases are: a subsidy used as equivalent to the revenue from a 50 yuan per ton carbon tax, a 100 yuan per ton carbon tax, a 500 yuan per ton carbon tax and a 1000 yuan per ton carbon tax.

5.3 Simulation Results

Based on the technologies and services assumption, by using AIM-China, the forecast results for energy demand and CO_2 emissions according to the different cases for China are presented in Fig. 1.

It can be clearly seen from Fig. 1 that CO_2 emissions will increase quickly with economic development in China. Compared with the base year of 1990, CO_2 emissions will be 1.7, 2.5 and 3.8 times the 1990 level in 2000, 2010 and 2030 for the market case, which can be regarded as a possible development case, 1.6, 2.1 and 3.3 times for the policy case, which represents the lowest growth rate of CO_2 emissions. There is little possibility for China to adopt some intervention policies within the near future, so the CO_2 emissions for the market case are very important. All the results show that China will play an important role in world's energy production and consumption, as well as in its CO_2 emissions.

From the CO_2 emissions in the market case, it is noticeable that it is possible for China to maintain a low CO_2 emissions growth rate compared with its rate of economic growth. The basic assumption for the analysis is that the annual economic growth rates are 9% from 1990 to 2000, 7.5% from 2000 to 2010, and 6% from 2010 to 2030, so the elasticity values for CO_2 emissions are 0.58, 0.53 and 0.37 for the market case. This represents a relatively low elasticity relative to the experience of developed countries.

A properly functioning market could contribute to the diffusion of technologies

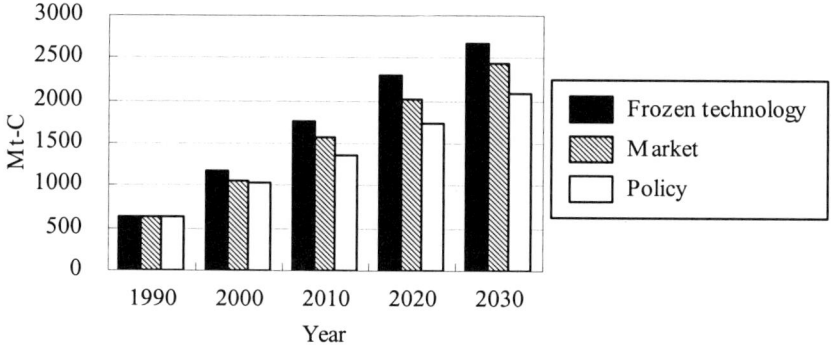

Fig. 1. CO_2 emissions in China

to reduce CO_2 emissions. Compared with the frozen technology case, there would be a 10.37% CO_2 emissions reduction for the market case in 2010 and 9.32% in 2030. This shows that establishing a better market mechanism is an efficient way to achieve advanced technology diffusion in China.

Industry is the major sector that emits CO_2 (Fig. 2). In 1990, the results show that CO_2 emissions from industry account for 75% of the total in the market case, and will generally decline to 70% in 2030, giving it the dominant role in future.

There is no significant reduction in CO_2 emissions in the transport sector (Fig. 3). Vehicle prices are very high compared to other countries, and the prices of gasoline and diesel for vehicle use are comparatively low. So the benefit of lower energy consumption from the use of efficient vehicles cannot compensate for the expenditure for the vehicles. Thus, efficient vehicles cannot be expected to be widely adopted. CO_2 emissions will increase in the rural residential sector for the market case compared to the no technological progress case. The major reason is that a substantial amount of biomass energy will be replaced by commercial energy based on the market along with the rise in income. However, there is consid-

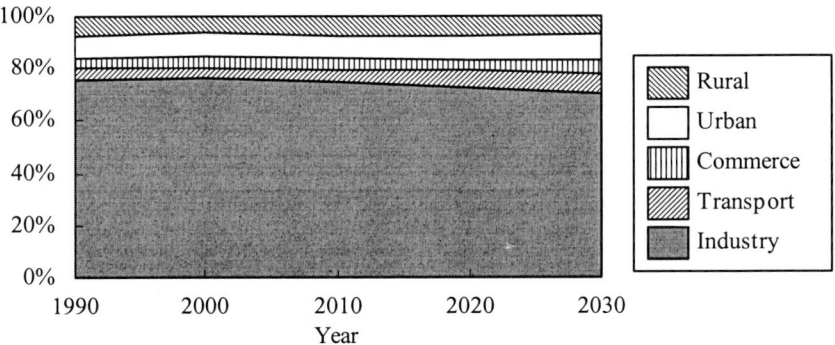

Fig. 2. Share of CO_2 emissions in China by sector for the market case

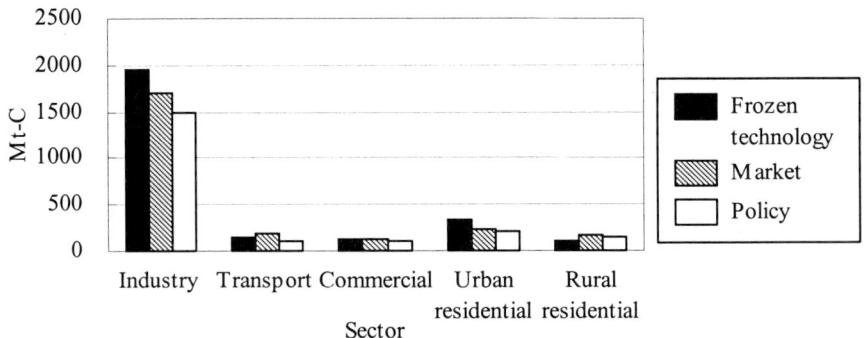

Fig. 3. CO_2 emissions in 2030 by sector

erable potential for CO_2 emissions reductions in the residential sector, including both urban and rural areas, if policy options are adopted.

CO_2 reduction policies are effective in reducing the rate of growth of CO_2 emissions. From Fig. 3, it can be demonstrated that adoption of intervention policies for CO_2 reductions using taxes and subsidies can provide greater reductions in CO_2 emissions compared with the market case. However, the reduction results differ according to the policy options. A combination of policies for taxes and subsidies are comparatively better than a taxes only policy.

The introduction of energy conservation technologies in the industrial, residential and services sectors will benefit from a carbon tax, but it is not such a substantial benefit for the transport and agriculture sectors. Due to the low energy prices and high vehicle prices, carbon taxes and subsidies have a very minimal effect in the transport sector. The reason for the lower impact of a tax in the agriculture sector is that some advanced technologies are already being selected in the market case.

Figure 4 gives an example of CO_2 emissions in the steel making sector in China. Significant emissions reduction was observed in this sector. Technological progress really plays a very important role in reducing unit energy use and, therefore, in CO_2 emissions reduction.

In the study, SO_2 emissions were simulated for these three scenarios. Since desulphurization technologies are not considered here, a reduction in SO_2 emissions is calculated from technological change among the scenarios. In order to provide better information on future CO_2 emissions and SO_2 emissions, a detailed inventory based on more than 2,300 counties in China was established. Figure 5 and Fig. 6 present the CO_2 emission intensity in 2010 and SO_2 emission intensity in 2010, respectively. The emission intensity is given in a rather simple way based on the future whole emissions in China and the extrapolation of the present intensity in each county.

Figure 7 presents the costs of emissions reductions by sector. In this figure, the vertical axis is the average cost for CO_2 emission reduction. Subsidies in four cost cases are described here. The horizontal axis is the rate of cumulative CO_2 reductions due to the subsidies beginning in the year 2000 to 2010 compared with the

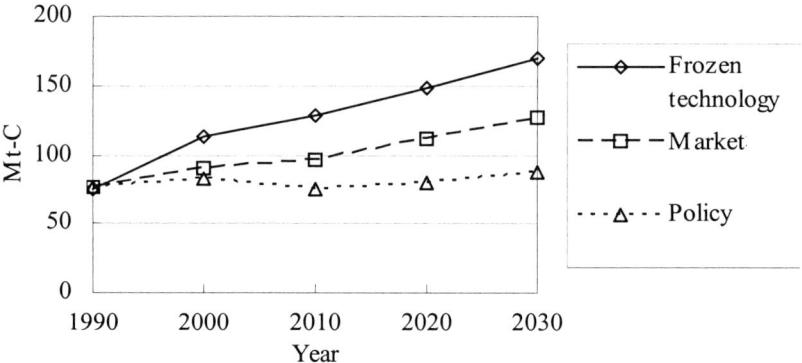

Fig. 4. CO_2 emissions in the steel making sector in China

market case.

From Fig. 7 it can be seen that the average CO_2 emissions reduction costs for the transport sector are very high. This is for the same reason that the amount of CO_2 emissions reduction according to the policy option is small. A rational way for the transport sector to reduce its CO_2 emissions is to encourage vehicle producers to introduce new technologies and decrease new vehicle prices to a suitable

Fig. 5. CO_2 emission intensity in China for 2010 (see color plates)

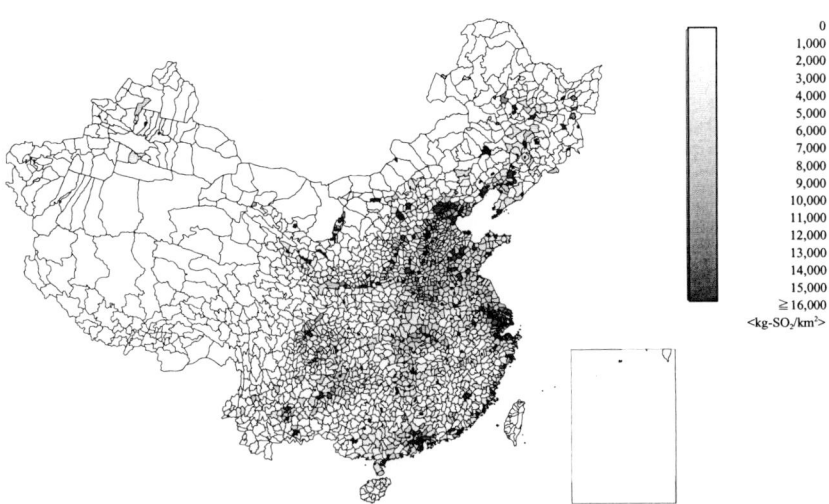

Fig. 6. SO_2 emission intensity in China for 2010 (see color plates)

level. In contrast, in the residential sector, CO_2 reduction costs are relatively low since the major energy utilization technologies such as refrigerators, washing machines, TVs and lighting devices, etc., are already in a competitive market. The prices for these are rational and they can provide a better service and conserve more energy, so it is easier to introduce energy saving technologies in the residential sector. Another factor is an increase in the standard of living that requires clean energy in the residential sector. The consumption of electricity, natural gas and other types of gas will increase while coal consumption declines. The costs for the industry sector stand at the mid point. Since there are many technologies in this sector, it is reasonable for the combined average costs for CO_2 reductions to maintain this level.

This analysis of average costs for CO_2 reductions presents a brief ranking in the CO_2 emissions reduction costs for each sector. The residential sector has the lowest costs, followed by the commercial sector, industry sector, etc.

Table 7 gives an example of technology selection for typical cases. These technologies are listed in Table 4. From this result, the technological process and the contribution of technological progress to CO_2 emissions reduction can be readily understood. It shows the effects of different cases (policies) on technology selection or progress. For these technologies, the effects of a policy of taxes and subsidies is better than that of a tax only policy, but this may not be true for technologies in other sectors or processes.

Table 8 shows the cost-effective technologies identified through this study.

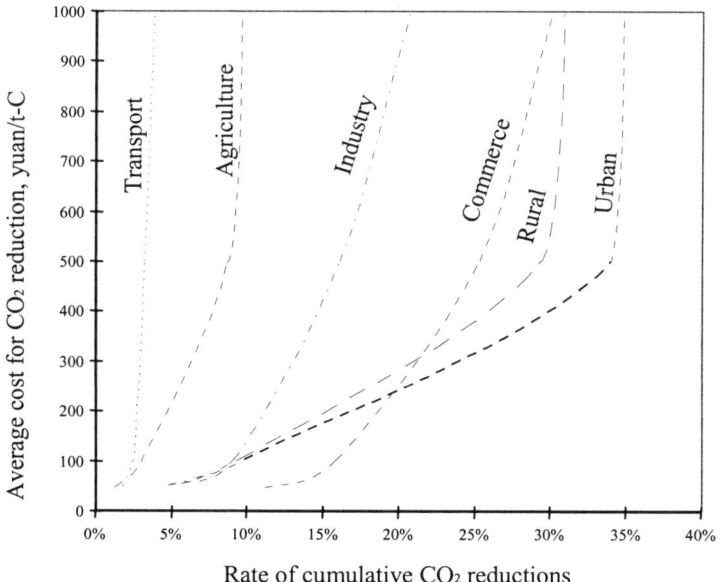

Fig. 7. Average cost for CO_2 reductions by sector

5.4 Findings and Conclusion

Based on this analysis, the conclusions were as followings:

1. China will continue a path of rapid economic development, and energy consumption and CO_2 emissions also will increase quickly to satisfy the demand for economic development. CO_2 emissions will rise 1.7, 2.5 and 3.8 times 1990 levels in 2000, 2010 and 2030 for the market case if no intervention policies are

Table 7. Technology selection for several cases in the coke making process in the steel industry

Technology	1990	2000		2010		2020		2030	
		Market	Policy	Market	Policy	Market	Policy	Market	Policy
Small size coke oven	59.9	18.0	-	1.1	-	0.0	-	0.0	-
Large size coking furnace	40.1	70.6	62.9	85.5	-	87.5	-	100.0	-
Large size coking furnace from Japan	0.0	11.5	37.1	13.4	35.3	12.5	30.0	0.0	-
New coking oven + coke wetting	0.0	0.0	-	0.0	64.7	0.0	70.0	0.0	100.0

Unit: %

Table 8. Cost-effective technologies

Sector	Technologies
Steel	Large-scale equipment (Coke oven, blast furnace, basic oxygen furnace etc.); Equipment for coke dry quenching; Continuous casting machine; TRT; Continuous rolling machine; Equipment for coke oven gas, OH gas and BOF gas recovery; DC electric arc furnace
Chemicals	Large-size equipment for chemical production; Waste heat recovery system; Ion membrane technology; Existing technology improvements
Papermaking	Co-generation system; Facilities for residual heat utilization; Black liquor recovery system; Continuous distillation system
Textiles	Co-generation system; Shuttleless loom; High-speed printing and dyeing
Nonferrous metals	Reverberator furnace; Waste heat recovery system; QSL for lead and zinc production
Building materials	Dry process rotary kiln with pre-calciner; Electric power generator using residual heat; Colburn process; Hoffman kiln; Tunnel kiln
Machinery	High-speed cutting; Electric-hydraulic hammer; Heat preservation furnace
Residential	Cooking using gas; Centralized space heating system; Energy-saving electric appliances; High-efficiency lighting
Services	Centralized space heating system; Centralized cooling and heating system; Co-generation system; Energy-saving electric appliances; High-efficiency lighting
Transportation	Diesel trucks; Low energy consumption cars; Electric cars; Natural gas cars; Electric railway locomotives
Technology used in common	High-efficiency boiler; FCB technology; High-efficiency electric motor; Variable-speed motors; Centrifugal electric fan; Energy-saving lighting

adopted.

2. Massive CO_2 emissions could be reduced by market mechanisms. The 182 million tons of carbon and 249 million tons of carbon will be reduced in 2010 and 2030 by comparing the market case and the frozen technology case. In order to reduce China's CO_2 emissions, market mechanisms should be improved to promote the distribution of efficient technology in order to have a better effect on energy conservation and CO_2 emissions reduction.

3. China can reduce CO_2 emissions further if some policy options are adopted. From the analysis it was noticed that there is a large potential for China to reduce CO_2 emissions by adopting different intervention policies. Levying a carbon tax and returning it as a subsidy is the most efficient way for CO_2 emissions reductions to be achieved among the policies assessed in this research.

4. The residential sector and services sector have comparatively low costs for CO_2 emissions reduction. China will be benefit from making efforts to reduce CO_2 emissions in the residential and services sectors, since it is not only a low cost strategy for CO_2 emissions reduction, but can also reduce SO_2 emissions and TSP (Total Suspended Particles) emissions, which will improve local air quality. The industry sector has a medium level of costs from the analysis, but this is an aggregated result. Some products in the industrial sector benefit from energy conservation, which should be considered for CO_2 emissions reduction at low cost.

5. There has been a rapid growth in energy consumption in the transport sector, and effective energy consumption measures are not significant. Thus this sector will be the most important for reducing CO_2 emissions. More countermeasures should be introduced, such as fuel taxes and the encouragement of competition to reduce vehicle prices and raise the efficiency of vehicles.

6. International technological collaboration is essential for energy conservation and CO_2 emissions reduction in China. Some advanced technologies from developed countries have very high efficiencies and can save substantial amounts of energy, and they are widely used in the developed countries. But there are still few of such technologies in China due to the cost and other factors. It is important for China to further collaborate in the introduction of energy saving technologies with the developed counties.

7. Further research activities on climate change should be promoted in China to achieve a better understanding of greenhouse gas (GHG) emissions and other aspects of climate change. In order to support China in developing its policy response to climate change, these types of research activities can help to establish a common knowledge base among researchers in China and other countries. This is an essential premise for future collaboration between developed countries and developing countries.

Base on these findings, the conclusions from this study are summarized as follows:

1. Technological progress plays a very important role in GHG emissions reduction and energy use in China, which will contribute significantly to global and local environmental conservation

2. The market mechanism is an efficient way to achieve the diffusion of advanced technologies in China
3. International collaboration on knowledge transfer is a key factor for the enhancement of technological progress in China
4. Sophisticated policies should be designed at the sectoral level, especially in transport, commerce and the chemical industry.
5. New environmental policies are required that are designed for the early stages of development in China in order to integrate strategies for both the local environment and global environment
6. International collaboration on technology should be promoted at the earliest opportunity to help developing countries to enhance their ability to ensure a clean environment.

From the analysis, it was also found that the analysis itself could be improved in future. For example, in order to ensure that the policy assessment is more accessible to policy makers, some macroeconomic analysis is necessary to feedback to the bottom-up end use model. By linking this with a top down model, some indirect costs could be analyzed when adopting policy options, with consideration for interrelationships among the sectors. The effect of insufficient investment should be analyzed to reflect the most likely energy consumption and CO_2 emissions patterns in China. In addition, more policy options should be assessed, especially for the non-intervention case, for example, renewable energy orientated policies, green lighting practices, shifting the energy supply pattern, etc. International collaboration also should be considered as an important approach to reducing CO_2 emissions in China.

References

AIM Project Team (1996) A guide to the AIM/Enduse model. AIM Interim Paper IP-95-05, Tsukuba
Bruce J, Lee H, Haites E eds. (1996) Climate change 1995 - Economic and social dimensions of climate change. WGIII of IPCC, Cambridge University Press, Cambridge
China AIM Project Team (1996) Application of AIM/Emission model in P.R. China and preliminary analysis on simulated results. AIM Interim Paper IP-96-02, Tsukuba
China statistical yearbook 1991, 1995, 1996. China Statistical Publishing House, Beijing
Cline WR (1992) The economics of global warming. Institute for international economics, Washington D.C.
Energy Statistical Yearbook of China 1986, 1989, 1991. China Statistical Publishing House, Beijing
Guo F (1994) Economy in China - Reforming and development. China Statistical Publishing House (in Chinese)
Hibino G, Kainuma M, Matsuoka Y, Morita T (1996) Two-level mathematical programming for analyzing subsidy options to reduce greenhouse gas emissions. Working Paper, IIASA, Laxenburg
Houghton JT, Filho LG, Bruce J, Lee H, Callander BA, Haites E, Harris N, Maskell K eds. (1995) Climate change 1994 - Radiative forcing of climate change and an evaluation of the IPCC IS92 Emission Scenarios. IPCC, Cambridge University Press, Cambridge

Jiang K, Hu X, Matsuoka Y, Morita T (1998) Energy technology changes and CO_2 emission scenarios in China. Environment Economics and Policy Studies 1, 141-160

Klaassen G (1996) Acid rain and environmental degradation. Edward Elgar Publishing Company, Brookfield

Matsuoka Y, Morita T (1996) Recent global GHG emission scenarios and their climatic implications. Global Warming, Carbon Limitation and Economic Development, 117-136

Morita T, Matsuoka Y, Penna I, Kainuma M (1994) Global carbon dioxide emission scenarios and their basic assumptions, 1994 Survey. CGER-I011-'94, Tsukuba

Nakicenovic N, Nordhaus WD, Richels R, Toth FL (1994) Integrative assessment of mitigation, impacts, and adaptation to climate change. Proceedings of a workshop held on 13-15 October 1993, IIASA, Laxenburg

Nakicenovic et al. (2001) IPCC special report on emissions scenarios (SRES). Cambridge University Press, Cambridge

Nordhaus WD (1994) Managing the global commons: The economics of climate change. The MIT press, Cambridge

Oates WE (1994) The economics of the environment. Edward Elgar Publishing Company, Brookfield

Pearce DW, Turner RK (1989) Economics of natural resources and the environment. The Johns Hopkins University Press, Baltimore

Risbey J, Kandlikar M, Patwardhan A (1996) Assessing integrated assessments. Climatic Change 34, 369-395

Rugider D, Poterba JM (1991) Global warming: Economic policy responses. The MIT Press, Cambridge

Zheng PY, Wei LQ, Miao FC (1995) Accelerating the change in economic growth patterns. Planning Press of China, Beijing, p20 (in Chinese)

Zhou FQ, Jiang KJ, Hu XL (1997) Make progress for modeling in China. In: Proceeding of the IPCC Asia-Pacific workshop on integrated assessment models, Tokyo

6. Application of AIM/Local Model to India using Area and Large Point Sources

Manmohan Kapshe[1], Amit Garg[2], and Priyadarshi R. Shukla[1]

Summary. India's emissions inventory estimates indicate that Large Point Sources (LPS) contribute above 60% of CO_2 and SO_2 emissions. Uneven distribution of energy resources, unbalanced regional development and the present high economic growth has led to emission patterns with dispersed hotspots. The policy making to address the environmental concerns thus rests on the assessment of future emissions and the options to mitigate them. The paper shows, using AIM/Local model with GIS interface, that Indian CO_2 emissions shall continue to rise steadily till 2030, whereas the SO_2 emissions shall decline after 2020, creating a natural decoupling of Greenhouse Gas (GHG) and local emissions. The carbon mitigation analysis, under three global policy regimes, indicates substitution of coal by gas, besides pushing energy efficient and low carbon technologies. Under all the scenarios, LPS contribute a major share of emissions, with industrial centers and large cities growing into major hotspots of emissions. Paper suggests that these spots would be the major focus of future emissions mitigation policy analysis for applications of formal tools like the AIM/Local model.

6.1 Introduction

Industrial development has contributed significantly to economic growth in India over last few decades; however, industrialization has not been uniform. Large and modern urban centers coexist with traditional rural and agrarian economy. The varying sectoral growth rates, consumption patterns and resource endowments have led to widely different regional and sectoral emission distributions. Some of the regions have experienced fast industrialization, and increasing air pollution in such areas is becoming an important environmental issue. Coal based thermal power plants, steel and cement plants have been major contributors to CO_2 and SO_2 emissions emanating from fossil fuel consumption. Transport sector is a major contributor to urban air pollution. Emissions from large industries are growing at a rate faster than the national average. High concentration of pollution in India is not due to a lack of sound environmental policy regime, but due to a lack of implementation at the local level. The LPS emissions are growing much faster than the national average due to growing population, increasing urbanization and higher consumption levels. Therefore, there is a need to estimate future emissions

---

[1] Indian Institute of Management, Ahmedabad 380015, India
[2] Winrock International India, New Delhi 110057, India

both from LPS and from area sources to prepare an implementation plan for emission mitigation.

This paper presents the analysis of Indian emissions using AIM/Local model. Indian emissions inventory for carbon dioxide (CO_2), methane (CH_4), nitrous oxide (N_2O), sulfur dioxide (SO_2) and nitrogen oxides (NO_X) at sectoral levels for 1995 and 2000 shows that LPS contributed to Indian CO_2 and SO_2 emissions to a large extent (above 60%) while for the other gases LPS contribution was below 10% (Garg *et al.* 2002). Therefore, we have projected CO_2 and SO_2 emissions temporally and regionally using AIM/Local model at area and LPS levels. 382 LPS and 466 area sources (Indian districts) are covered for the base year 2000. New LPS are added to reach 457 (2010), 523 (2020) and 587 (2030) based on present investment proposals for new plant installations, retrofitting and expansion options for the existing plants, and policy dynamics. The LPS share in Indian CO_2 emissions is projected to increase marginally from 64% to 65% in 2030 while that of SO_2 increases from 62% to 70% up to 2020 and decreases marginally to 66% in 2030. The total CO_2 emissions in India are projected to increase almost 3.15 times to 2945 Mt-CO_2 in 2030. However SO_2 emissions grow only about 1.15 times during the same period to 5.8 Mt-SO_2 in 2030 indicating a disjoint between GHG and local pollutant emissions in future. This is mainly due to adoption of flue gas desulfurisation and clean coal technologies in the power sector, and considerable reduction in sulfur contents in petroleum products. Some of these policy dynamics are already visible in India. The paper also demonstrates this Indian emission dynamics and would be useful for policy makers and researchers.

6.2 Methodology

The emission sources may be broadly classified as LPS, area sources and line sources (mainly for transportation sector). In the present analysis, we have combined the area and line sources since activity data for transport sector at line source level is not available. The combined representation is called area sources. The aggregate national emissions are the sum of LPS and area source emissions. For future emission projections at LPS and area source levels, AIM/Local model with Geographical Information System (GIS) support has been used.

The AIM/Local model follows the approach of linear programming to find an optimal solution by selecting a combination of technologies with the least cost while satisfying the given constraints of fulfilling the demand and meeting the environmental targets and/or energy supply constraints in the specific region (AIM Project Team 2002). This model estimates the emissions from the LPS and area source, which can be used to calculate the total emissions from a region. Emission calculation methodology used in AIM/Local model is in line with the recommended methodology of the Intergovernmental Panel on Climate Change (IPCC 1996) and follows a similar approach as used by (Li *et al.* 1999) and (Garg *et al.* 2001a, 2001b). A detailed description of the emission estimation methodology used in AIM/Local model is given in the Appendix.

Indian AIM/Local model is developed for five major sectors namely power generation, industrial, transport, residential and agriculture sectors. The choice of these sectors is based on their importance in the national energy consumption. The industrial sector covers fifteen major industry types. Each sector is modeled with considerable technological details about consumption of different energy forms, emission of various gases, cost components, and technological shares. Power generation and industrial sectors provide all the LPS for the analysis whereas the other sectors have been modeled as area sources.

The AIM/Local model, suitable for estimating future emissions from LPS and area sources, is demand driven. The enduse sectoral demands in turn depend upon national macro-economic growth projections. Based on the last 30-year time series GDP data, government projections and expert opinion, the Indian GDP is assumed to grow in real terms by 6% per annum on an average during 2000-2015, by 5% during 2015-2025, by 4% during 2025-2035 and by 3% in the later half of the 21st century under the reference scenario. The Indian GDP grows 4.8 times at an annual rate of 5.2% and population rises from the present 1 billion to 1.35 billion between the years 2000 to 2030 (UN 1998) indicating a four-fold increase in per capita income levels.

The industrial enduse sector demands saturate in the long run following a logistic model. This is divided into LPS and area demand. Excess demand over and above the capacity of the LPS has been taken as area source demand in case of industries where the total estimated national demand exceeded the total capacity of all the LPS considered. The autonomous energy efficiency improvements (AEEI) in enduse technologies supplying these demands capture improvements due to better management practices, learning curve, improved infrastructure, retrofitting for the existing demand technologies and incremental technological interventions.

The AIM/Local model also captures the present policy dynamics to reduce anthropogenic air pollution by various measures including fuel quality improvements, adopting cleaner technologies and stricter enforcement of emission regulations. Fuel quality improvement includes coal beneficiation, increased use of imported coal (lower ash and higher calorific value than the average Indian coal), and reduction in sulfur contents of petroleum products.

Production quantity for an LPS has been estimated on the basis of the sectoral demand elasticity and the past production trends of the plant, wherever available. Information regarding the new LPS till year 2010 has been taken from various data sources like CMIE (2002b) and policy documents of the government. New LPS locations beyond the year 2010 have been estimated based on past development trends, retrofitting expansion options for the existing plants, studies related to suitability of industrial locations in India, and present policy dynamics. These are however indicative and not conclusive since the actual LPS locations may vary as future unfolds.

6.3 Data Sources and Coverage

Many diverse data sources were utilized since there is no comprehensive database covering all the types of emitters for India. These included published documents of the Government of India, state governments, government organizations and institutions, industry federations and autonomous organizations covering various sectors and fuels (Garg *et al.* 2001a, 2001b, 2002). Future LPS data was mainly taken from published reports and databases like CMIE (2002a, 2002b). These provide status information of the various planned investment projects in India (power, refineries, cement, steel and fertilizer plants, etc.) till almost 2010. Coupled with the retrofitting and capacity augmentation options for the existing plants, present policy directives of the government and expert opinion, LPS information for the next 30 years was assimilated. We have tried to cross verify each existing LPS data using more than one data source providing a profound richness and robustness to the base data. We have projected CO_2 and SO_2 emissions in the present paper since LPS have a dominant share only for these two emission types. Table 1 provides the LPS coverage over the years.

6.4 Results

6.4.1 Reference scenario

The reference scenario assumes the dynamics-as-usual, i.e. continuation of macroeconomic (including structural changes in the economy), demographic and energy sector trends (such as autonomous energy efficiency improvements and penetra-

Table 1. Large point source coverage over the years for India

Sector	Sub-sectors	LPS covered			
		2000	2010	2020	2030
Energy	Power (coal & oil)	82	111	131	150
	Power (natural gas)	12	17	20	23
	Steel	10	16	22	28
	Cement	85	98	110	123
	Fertilizer	31	41	52	62
	Paper	33	38	43	48
	Sugar	28	28	29	30
	Caustic soda	19	21	23	26
Industrial processes	H_2SO_4 manufacturing	63	64	66	68
	Aluminium	3	4	5	5
	Copper ore smelting	8	9	10	11
	Lead ore smelting	5	6	7	8
	Zinc ore smelting	3	4	5	5
Total		382	457	523	587

tion of clean and renewable fuels and technologies), as well as government policy trends. There are no direct climate change policy interventions in the reference scenario.

The reference scenario analysis shows that during 2000-2030, energy use will grow three times and carbon emissions from energy will grow 2.7 times (Fig. 1). Under the dynamics-as-usual, the energy system shall continue to depend on fossil fuels, primarily the domestic coal (Fig. 2). Between years 2000 and 2030, carbon intensity of GDP declines at 1.9% per year (Fig. 3). The improvement in carbon intensity is contributed mainly by the decline in energy intensity at 1.5% annual rate and the rest is contributed by the substitution of the share of coal in energy by gas and marginally by renewable energy. Future emissions of CO_2 and SO_2 for the reference scenario are shown in Table 2. Although CO_2 emissions grow annually at 3.4%, the SO_2 emissions rise at much lower rates. This is due to policies that are already under implementation as well as increasing public pressure on national policy makers for local pollutants, which is not the case for the GHG emissions. The CO_2 and SO_2 emission trajectories move in closer bands till 2010 due to continuation of existing vintages. Thereafter, while CO_2 emissions continue to rise, the SO_2 emissions begin to decline following the Kuznets curve phenomenon (Kuznets 1958). The GHG and local pollutant emissions are thus decoupled in future. SO_2 emission reduction happens from mandatory use of Flue Gas Desulfurisation (FGD) in large coal power plants, introduction of low sulfur diesel, washing of coal and stricter enforcement of local air quality regulations. These policies however do not affect the GHG emissions.

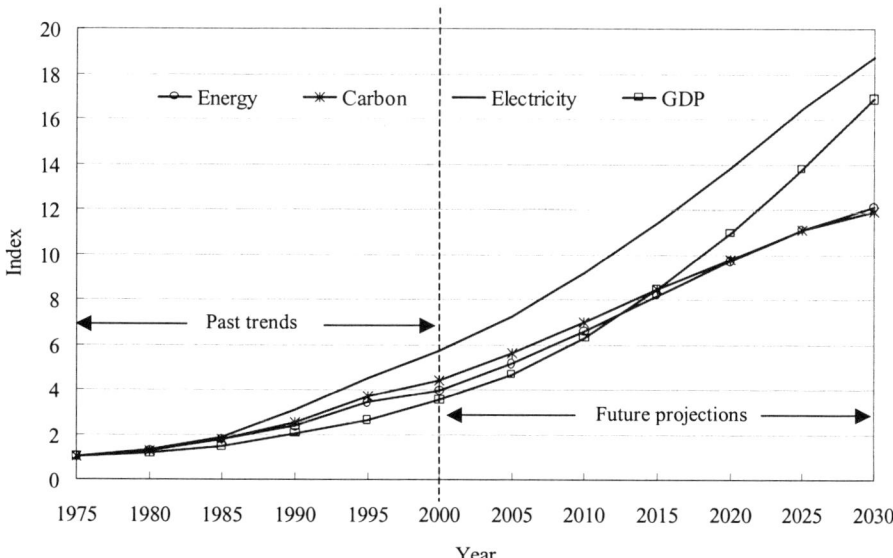

Fig. 1. Energy, carbon, electricity and GDP (reference scenario projections)

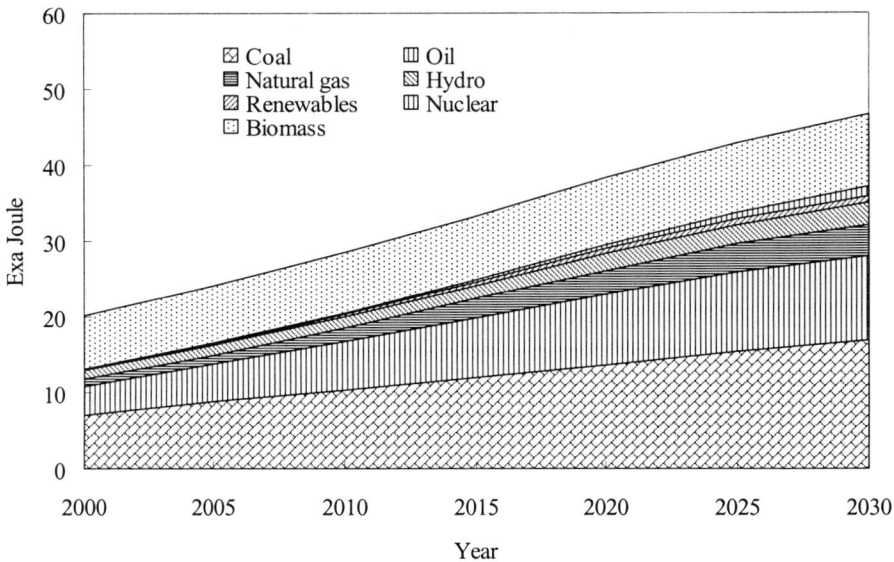

Fig. 2. Energy demand from 2000-2030 (reference scenario projections)

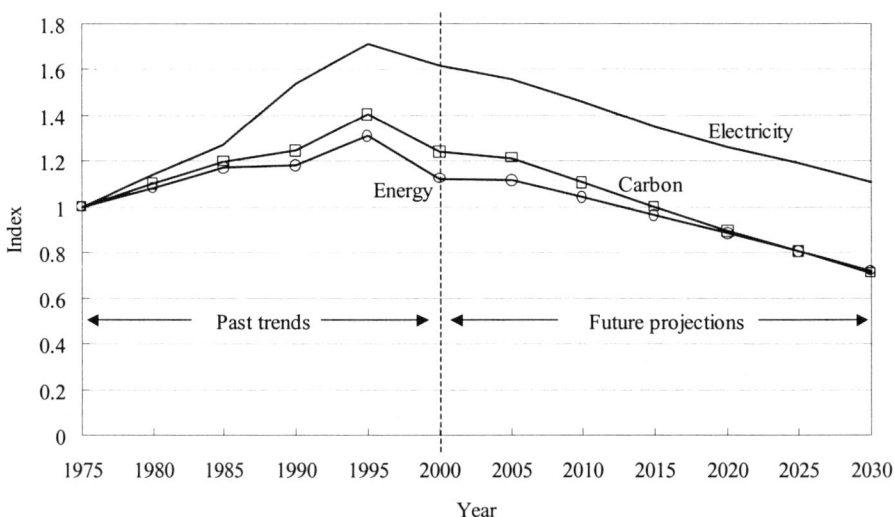

Fig. 3. GDP intensities of energy, electricity and carbon (reference scenario)

There is an interesting emission dynamics evolving in India and transport sector is at the center of this development. The sulfur content in the diesel oil supplied to the metropolitan cities (Delhi, Mumbai, Chennai and Kolkata) has been decreased during the year 2000 from 1% sulfur by weight to 0.25% by the Indian refineries as per the Indian government directives. The sulfur content has been further reduced to 0.05% by weight in Delhi by late 2001. This has resulted in an appreciable decrease in SO_2 emissions from the transport sector in these cities. These four cities account for almost 8% of all India diesel consumption in 2000. Diesel, in turn is almost 40% of the national petroleum product consumption and transport sector accounts for almost three fourths of this consumption. Although the effect of this emission dynamics may not be felt in the overall national SO2 emissions, which have continued to rise, the long-term implications are appreciable.

Besides these there has been recent Supreme Court judgment that has ordered Euro II standards to be followed for all new cars in India, which will be upgraded to Euro IV by 2005 (Mashelkar *et al.* 2002). While this is not very significant for SO_2 control, it shows the mind-set of the judiciary and policymakers to control local pollution. Moreover it is necessary to have low sulfur diesel for meeting the emission norms beyond Euro II. The reduction in SO_2 emissions would be further strengthened by another recent policy decision on mandatory washing of coal that is used 700 Km away from the mine mouth. This measure is aimed at reducing fly ash and also simultaneously reduces some sulfur. Since, over a third of coal is used beyond 700 Km, this measure is expected to start reducing SO_2 from coal use in near future. This policy dynamics manifests in a reduction in SO_2 emissions in future even though the absolute energy consumption, and therefore CO_2 emissions, continue to rise.

Coal remains the mainstay of the Indian energy system but its use becomes cleaner due to higher penetration of clean coal technologies (WB 1997). Coal consumption increases about 2.5 times during 2000-2030 from 310 Mt in 2000. About 90% of the Indian coal product consumption is by LPS for power generation and industry. Residential coal consumption (area sources) for cooking purpose is mainly limited to lower middle class households in semi-urban areas. The urban households normally use LPG and kerosene, while fuel wood, dung cakes and electricity are also consumed in small proportions. Commercial establishments like hotels and restaurants consume some coal for cooking but their consumption is miniscule in comparison to the LPS coal consumption. Biomass supplies the rural energy demand to a large extent with kerosene supplementing it partially.

The sectoral fuel consumption indicates continued dominance of power sector in coal use and transport in petroleum products with each having 70 percent share in 2030. Transport sector coal consumption is negligible for future years due to phasing out of coal based steam traction from Indian Railways. LPS dominate power sector consumption while transport sector has area source dominance. Power sector share in natural gas consumption increases to more than half from the present one third, caused by increasing competitiveness of Combined Cycle Gas Turbine technologies (CCGT) for electricity generation (Shukla *et al.* 1999). Gas consumption is also LPS dominated and rises rapidly in industries like fertilizers and petro-chemicals. While the share of gas in primary energy still remains low, the trends suggest a rising penetration of gas, most of which would have to be imported.

The distribution range of LPS emission in the total national emissions is indicated in Table 2. The largest 50 LPS contribute almost 50% of all India emissions in 2000, which decreases to 41% in 2030 as the emissions from smaller LPS increase (Table 3). The CO_2 and SO_2 emissions from LPS and area sources over the years for reference case are illustrated in Fig. 4 and Fig. 5 respectively.

Table 2. Share of LPS emission in all India emissions

	Emission details	2000	2010	2020	2030
CO_2	Number of LPS	303	374	435	495
	LPS emissions (Mt-CO_2)	630	989	1418	1912
	All India emissions (Mt-CO_2)	983	1556	2189	2945
	LPS/total (%)	64	64	65	65
SO_2	Number of LPS	368	437	499	559
	LPS emissions (Mt-SO_2)	3.12	3.96	4.35	3.83
	All India emissions (Mt-SO_2)	5.02	5.87	6.25	5.77
	LPS/total (%)	62	67	70	66

Table 3. Distribution range of LPS emissions (% of all India emissions)

Largest LPS	CO_2 emissions				SO_2 emissions			
	2000	2010	2020	2030	2000	2010	2020	2030
1 to 25	35.2	32.5	31.0	31.5	37.1	38.0	37.3	32.5
26 to 50	11.5	11.0	10.9	11.0	12.0	12.8	13.0	12.4
51 to 75	5.7	5.4	5.7	5.6	5.2	5.6	6.0	5.9
76 to 100	3.7	3.9	4.1	4.2	3.1	3.5	4.0	3.9
101 to 200	6.7	7.8	8.6	8.7	4.2	5.8	7.2	8.2
All others	1.3	2.7	3.0	3.9	0.6	1.8	2.2	3.4

Table 4. LPS contribution to CO_2 energy sector emission (% of all India emissions)

Sector	2000	2010	2020	2030
Power	41.3	41.5	43.0	48.0
Steel	12.4	11.3	10.2	8.0
Cement	9.0	9.6	8.8	7.5
Fertilizer	0.7	0.3	0.2	0.2
Paper	0.4	0.5	0.5	0.5
Aluminum	0.1	0.2	0.2	0.3

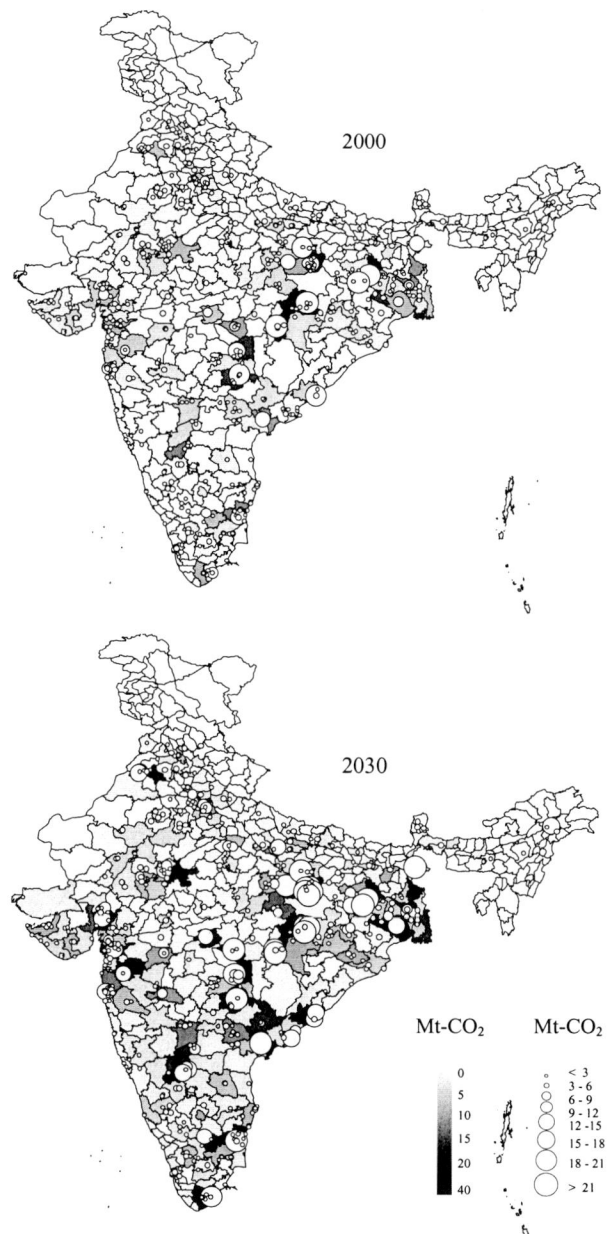

Fig. 4. Regional distribution of CO_2 emissions in India for 2000 and 2030 in reference scenario (see color plates)
Note: Circles show emissions from large point sources.

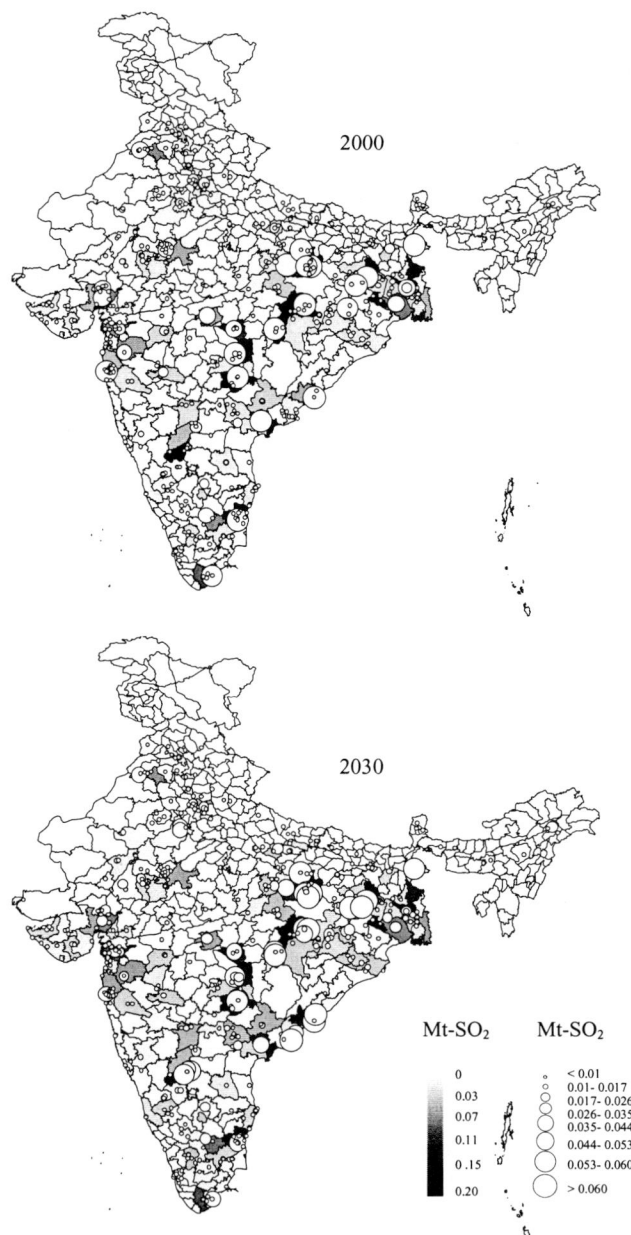

Fig. 5. Regional distribution of SO$_2$ emissions in India for 2000 and 2030 in reference scenario (see color plates)
Note: Circles show emissions from large point sources.

Table 5. LPS contribution to SO_2 energy sector emission (% of all India emissions)

Sector	2000	2010	2020	2030
Power	44.5	47.1	46.5	40.0
Steel	11.6	13.1	14.7	16.0
Cement	4.7	6.1	7.1	8.4
Fertilizer	0.5	0.2	0.1	0.1
Paper	0.5	0.7	0.9	1.2
Aluminum	0.3	0.3	0.4	0.5

The sectoral shares (Tables 4 and 5) indicate that LPS emissions would continue to be major contributors to national CO_2 and SO_2 emissions in future. The share of LPS in SO_2 emissions decreases marginally in the later years as compared to CO_2 emissions. This is because the SO_2 emissions from LPS as well as some area sources such as transport sector reduce considerably but the emissions from the small-scale industries like brick making, which depend on coal, continue to grow. These small establishments though essentially point sources are spread all over and have been classified as area source in the model. This reduction of LPS share in SO_2 emissions further indicates that the disjoint between the GHG and local pollutant emissions, discussed earlier, is unfolding gradually in India.

6.4.2 Climate change mitigation scenarios

India does not have any GHG emission mitigation commitments presently. However rising emission trends necessitate an understanding and analysis of carbon mitigation policy options for India. These policies prompt technology and fuel substitution by changing the relative costs of competing fuels in favor of those with lower carbon contents. Carbon mitigation would therefore require technological transformation of Indian energy system. Efficiency improvement measures, cleaner technology and cleaner fuels will penetrate faster and infrastructure to facilitate these has to be built. This will need investments in infrastructure to support the new technologies and reforms to remove the barriers for their penetration. Societies with strong reduction commitments would therefore have to start thinking differently.

The reference case provides a platform for analyzing implications of alternate carbon mitigation scenarios for India. The present analysis considers a reference scenario that assumes a world that is akin to the B2 scenario of Special Report on Emissions Scenarios (SRES) (IPCC 2000). Climate intervention scenarios are then considered that presume alternate Kyoto-plus regimes aimed at stabilizing long-run concentrations at 550, 650 and 750 parts per million volume (ppmv) of CO_2 concentrations. The global emission budgets are as per SRES and have been allocated between world regions, with around 7.5% share of global carbon emissions as the cumulative emission budgets for India during the 21st century. The present analysis is for the period 2000-2030.

Some important results are indicated in Table 6 and Fig. 6. Inertia of existing technological stock in the energy system prevents significant fuel substitution till the year 2010 for meeting carbon mitigation targets and coal continues to dominate the energy system. However coal consumption declines drastically in the long-term for 550 ppmv emission targets. Decline in gas consumption in medium and long term in this scenario is due to increasing penetration of carbon free technologies like renewable and nuclear for power generation. However coal is mainly substituted by natural gas in power sector. While natural gas exhibits early penetration, renewable penetration is late due to their high initial investment costs. Due to the rising gas prices with higher use of gas in later years, the renewable energy penetration increases with rising mitigation targets.

Table 6. Results of climate change policy scenarios

Parameter	750 ppmv			650 ppmv			550 ppmv		
	2010	2020	2030	2010	2020	2030	2010	2020	2030
Coal #	-2	-5	-18	-4	-17	-32	-9	-29	-51
Gas #	+1	+3	+13	+1	+10	+13	+1	+18	-9
CO_2 (Mt-CO_2)	1408	1907	2431	1327	1775	2229	1254	1621	1998
SO_2 (Mt-SO_2)	5.82	6.13	5.43	5.67	5.39	4.87	5.42	4.79	3.81

\# Percentage changes over the reference scenario consumption.

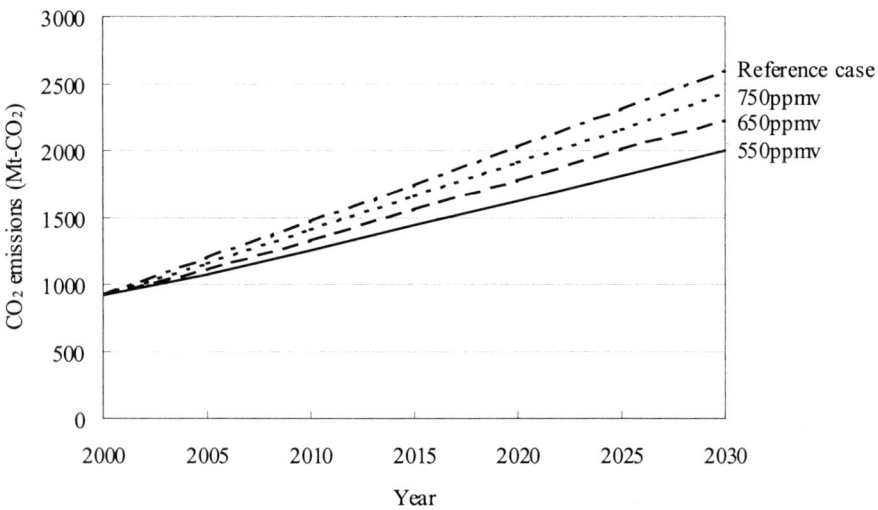

Fig. 6. Carbon dioxide stabilization scenarios (2000-2030)

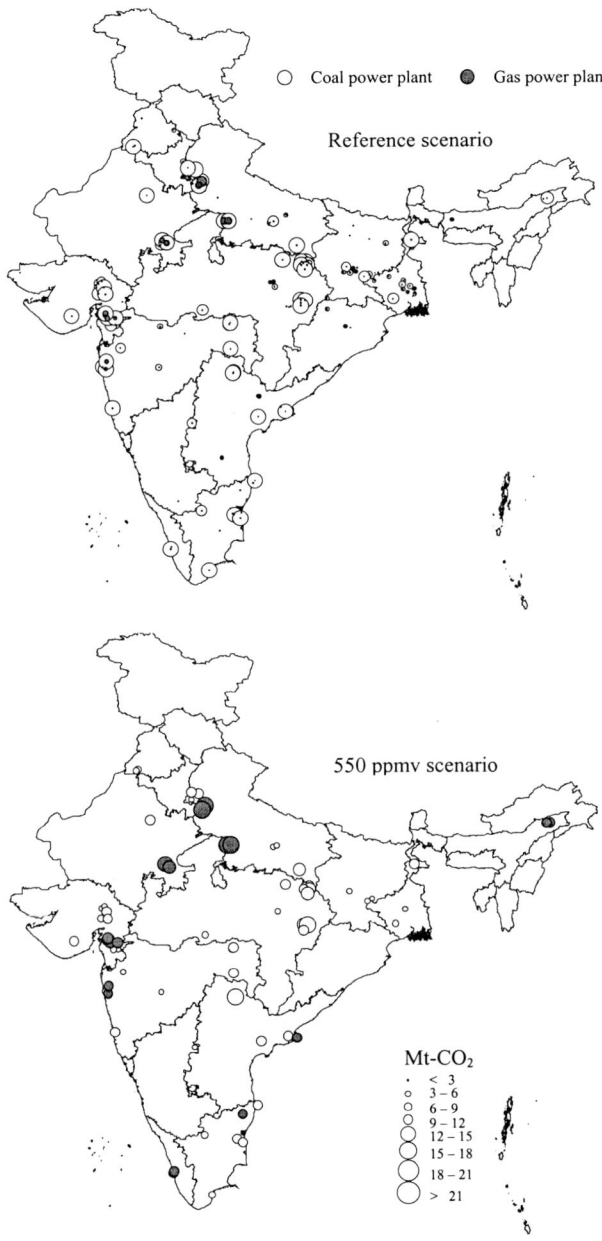

Fig. 7. Comparison of CO_2 emissions from large power plants in reference and 550 ppmv stabilization scenarios in 2030 (see color plates)

Table 7. Supply and demand side contributions to emission reductions (%)

Mitigation scenario		2010	2020	2030
750 ppmv	Supply side	32	53	71
	Demand side	68	47	29
650 ppmv	Supply side	36	56	72
	Demand side	64	44	28
550 ppmv	Supply side	40	63	75
	Demand side	60	37	25

Analysis of carbon mitigation scenarios provides some useful insights regarding energy supply and demand side contributions to emissions reductions (Table 7). In the early periods (up to 2010), the demand sectors show more flexibility than the supply side by contributing more to carbon emission reduction since enduse technological stock turnover is faster due to their relatively shorter lifetimes. But in later periods, the supply side contribution increases, finally reaching almost a three-fourth share in the total mitigation in 2030. These, however, include emission reduction due to lower power generation requirements as demand side energy efficiencies improve.

The inertia of supply side technology turnover is due to their much longer lifetimes, high investment requirements and longer gestation period for infrastructure development. The supply side mitigation options are primarily associated with fuel switching from coal to natural gas. Since India is not endowed with much of natural gas resources, most of this gas would be imported and therefore has national energy security issues attached. Long-term options include penetration of carbon-free technologies like nuclear and renewable technologies. Judged from a market perspective, there are large inefficiencies on the demand side due to capital shortages, risk, high transaction costs and weak financial markets. The supply side is better organized and has lower inefficiencies. This dynamics is demonstrated by the spread of power sector LPS emissions in future (Fig. 7). The fuel switching from coal to gas, especially in the coastal areas and in the demand centers of northern India is apparent. There are more gas-based power plants in the 550 ppmv case as compared to the reference case. Gas based power plants contribute 15% of power sector CO_2 emissions in 2030 in the reference case while it almost doubles in 550 ppmv case despite the fact that the emissions from an equivalent gas based power plant are 40% lower than a coal plant (Garg and Shukla 2002). SO_2 reduction is even stronger in the 550 ppmv case since gas emits no sulfur.

Carbon mitigation also acts as a push policy for cleaner technologies especially in the industrial sectors. Autonomous energy efficiency improvements provide the initial options for carbon mitigation in India since the average energy efficiencies in most sectors are well below international averages. Improved oxygen furnace with gas recovery replaces existing open hearth furnace for steel production, improved dry process with pre-heaters and pre-calcinators replaces/augments the existing wet/dry process in cement industry, vertical shaft and high draught kilns replace clamp type and Bull's trench kilns in brick industry, CNG and electric cars partially replace diesel and petrol cars etc.

6.5 Conclusions

Analysis of LPS emissions highlights sectors and plants where mitigation efforts should be targeted for cost-effectiveness. The main contributors to Indian emissions presently are about 70 LPS (50 power, 5 steel and 15 cement plants), thus offering a good opportunity for focusing mitigation efforts. Power sector is the predominant emission source for CO_2 and SO_2. Operational improvements (like heat rate reduction, excess air control etc.), better maintenance, reducing transmission and distribution losses in the power sector would go a long way in emissions mitigation. The other policy options are switching from coal to lower carbon content fuels like natural gas, sequestering the emitted carbon from LPS, increasing renewable and nuclear technology penetration etc. Carbon emission mitigation analysis under 750, 650 and 550 ppmv scenarios demonstrate this fuel shift.

Energy efficiency improvement measures in other sectors like steel, cement, caustic soda, sugar, brick making and fertilizer would improve productivity while reducing overall emissions. Demand side sectors contribute more to carbon emission mitigation in the short run while supply side (mainly power generation) contributes more in the long run. The longer life, higher investment requirements and longer gestation periods of supply side technologies as compared to the demand side explain this.

Apart from LPS, there are many small, moving and concentrated point sources like vehicles in urban transport. These are covered under area sources in the present analysis. Mitigation options have to be carefully planned for these sources since some of the above mentioned policy options, though useful for small but distributed sources as well, would not be cost-effective for them since implementation efforts would be substantial. For example, although transport sector sources contribute around one tenth to India's CO_2 equivalent GHG emissions and almost a third to NO_X emissions, their characteristics (large numbers and low emissions per source) necessitate huge mitigation efforts as compared to very focused mitigation efforts for LPS. However measures like improving diesel and gasoline quality and stricter vehicle emission norms will reduce local pollution levels and to a certain extent GHG emissions as well.

Transport sector emissions from Delhi are an interesting case for such analysis where introduction of low sulfur diesel in recent years reduced SO_2 emissions significantly. Subsequent shift to mandatory use of Compressed Natural Gas (CNG) in public vehicles further reduced sulfur and particulate emissions, besides reducing carbon emissions. These policies, introduced by the Indian Supreme Court orders, are not necessarily cost-effective. A superior mix of policies would minimize long-term cost for achieving the desired air quality standards at city level. The AIM/Local model is well suited to analyze future emission from all sectors and to derive such policies.

The results from our study have demonstrated that the LPS would continue to be responsible for considerable part of the Indian carbon emissions. Power generation and industry would contribute almost 75% of CO_2 emissions. Since these sectors have a natural dominance of LPS, mitigating emissions from LPS would become even more important in future. However, Indian GHG and local pollutant emission trajectories have a disjoint in future as SO_2 emissions decline due to deeper penetration of FGD in power sector, improved fuel quality resulting in lower sulfur and ash contents, stricter enforcement of emission regulations and efficiency improvements. Although GHG and local pollutant emission mitigation targets for a country are often useful as overall policy targets, the marginal mitigation cost for achieving each target varies across regions and sectors. The LPS analysis contributes to effectiveness of emissions mitigation by indicating the locations and sectors where controls can lead to maximum benefits. The present work is a step in this direction for India.

References

AIM Project Team (2002) AIM-Local: a user's guide. AIM interim paper, IP-02-01. National Institute for Environmental Studies, Tsukuba
Biswas D (ed) (1999) Parivesh news letter: June 1999. Central Pollution Control Board, Ministry of Environment and Forests, Government of India, New Delhi
CEA (Central Electricity Authority of India) (1996) Performance review of thermal power stations 1995-96. New Delhi
CMIE (1999) India's energy sector. Center for Monitoring Indian Economy, Mumbai
CMIE (2002a) India's Energy Sector. Center for Monitoring Indian Economy, Mumbai
CMIE (2002b) Prowess: the Indian corporate database. Center for Monitoring Indian Economy, Mumbai
Garg A (2000) Technologies, policies and measures for energy and environment future. Doctoral thesis, Indian Institute of Management, Ahmedabad
Garg A, Shukla PR, Bhattacharya S, Dadhwal VK (2001a) Sub-region (district) and sector level SO_2 and NO_X emissions for India: assessment of inventories and mitigation flexibility. Atmospheric Environment 35: 703-713
Garg A, Bhattacharya S, Shukla PR, Dadhwal VK (2001b) Regional and sectoral assessment of greenhouse gas emissions in India. Atmospheric Environment 35: 2679-2695
Garg A, Ghosh D, Shukla PR (2001c) Integrated energy and environment modelling and analysis for India. OPSEARCH of India 38
Garg A, Kapshe M, Shukla PR, Ghosh D (2002) Large point source (LPS) emissions from India: regional and sectoral analysis. Atmospheric Environment 36: 213-224
Garg A, Shukla PR (2002) Emissions inventory of India. Tata McGraw-Hill Publishing Company Ltd., New Delhi
Hu X, Yang H (2000) Disaggregate SO_2 emissions from national total to county level distributions for China. 5th AIM international workshop, National Institute of Environmental Studies, Tsukuba
IPCC (1996) Revised IPCC guidelines for national greenhouse gas inventories: reference manual, vol 3. Inter Governmental Panel on Climate Change, Bracknell
IPCC (2000) Special report on emissions scenarios. Cambridge University Press, Cambridge
Kainuma M, Matsuoka Y, Morita T, Hibino G (1999) Development of an end-use model for analyzing policy options to reduce greenhouse gas emissions. IEEE trans on systems, man, and cybernetics- part C: Applications and reviews 29-3: 317-324

Kainuma M, Matsuoka Y, Morita T (2001) CO_2 emission forecast in Japan by AIM/end-use model. OPSEARCH of India 38-1: 110-125

Kuznets SS (1958) Six lectures on economic growth. Free Press New York, New York

Li YF, Zhang YJ, Cao GL, Liu JH, Barrie LA (1999) Distribution of seasonal SO_2 emission from fuel combustion and industrial activities in Shanxi province, China, with $1/6°x1/4°$ longitude/latitude resolution. Atmospheric Environment 33: 257-265

Mashelkar RA, Biswas DK, Krishnan NR, Mathur OP, Natarajan R, Niyati KP, Shukla PR, Singh DV, Singhal S (2002) Report of the expert committee on auto fuel policy. Ministry of Petroleum and Natural Gas, Government of India, New Delhi

Shukla PR, Ghosh D, Chandler W, Logan J (1999) Developing countries and global climate change: electric power option in India. PEW Center on Global Climate Change, Arlington

UN (United Nations) (1998) World population projections to 2150. United Nations Department of Economic and Social Affairs Population Division, New York

WB (World Bank) (1997) A planner's guide for selecting clean coal technologies for power plants. World Bank Technical Paper No 387, The World Bank, Washington, D.C.

Appendix: AIM/Local Modeling Approach

The AIM/Local model, suitable for estimating future emissions from large point sources and area sources, is demand driven. The model uses linear programming approach to arrive at the optimal combination of technologies with least cost to satisfy the service demand while meeting the environmental targets and / or energy supply constraints in a specific region. It calculates emissions from LPS and area sources separately. The total emissions from a region are then obtained by summing up the LPS and area source emissions.

Emissions from large point sources (LPS)

The calculation of emissions from energy combustion may be done at three different levels referred to as tiers 1, 2 and 3 in the IPCC Guidelines (IPCC 1996). Tier-1 methodology, concentrates on estimating the emissions from the carbon content of energy kind supplied to the country as a whole or to the main energy combustion activities. This is a simple method and emissions from all sources of combustion are estimated on the basis of the quantities of energy kind consumed and average emission factors. Tier-2 estimation methodology is based on detailed energy/technology information covering stationary and mobile sources. It is more detailed than tier-1 methodology but uses the same concept of energy kind consumption based emission coefficients like CO_2 emissions per unit of coal combustion. Tier-3 is similar to tier-2 except that the emission coefficients are based on enduse demands like CO_2 emissions per unit of power generated. AIM/Local model uses a combination of tier-2 and 3 methodology of emission estimation. LPS emissions in the model are estimated by two different approaches. Model follows an approach similar to tier-2 for the estimates of emissions from the LPS for energy consumption. The data required includes information about the production quantity, production process, energy combustion by various technologies, emission coefficients for the energy kind and pollution removal technologies used. The emissions are estimated by multiplying the energy consumption by each technology with respective emission coefficients for that energy kind.

In this model, tier-3 approach is used for estimation of emissions from the industrial processes. The data requirement for this approach includes production quantity, production process and emissions factors per unit of production. Emissions are estimated by multiplying the production quantity by the corresponding emission coefficients.

Net emissions from an LPS in both the above approaches are calculated by accounting for the pollution removal factor due to the pollution removal technology.

Thus emissions from LPS for energy consumption and production processes are given by

$$Q_l^{LPS} = R_l^{LPS} \times \{\sum_k (E_{l,k}^{LPS} \times f_k) + \sum_v (V_{l,v}^{LPS} \times f_v)\} \tag{1}$$

Where,

Q_l^{LPS} :   Net emission from large point source l

R_l^{LPS} :   Release rate of pollutants after removal technology of large point source l

$E_{l,k}^{LPS}$:   Energy consumption of energy kind k for large point source l

f_k :   Emission coefficient of energy kind k

$V_{l,v}^{LPS}$:   Production quantity of production process v for large point source l

f_v :   Emission coefficient of production process v

k :   Energy kind

v :   Production process

Emissions from area sources (AS)

In the present study the start year of modeling has been taken as 1995. India had 25 States and 6 Union territories in 1995. These are further subdivided in 466 small administrative areas called districts. Due to reorganization of states and sub-division of districts, this number has changed over the years but for the modeling purpose administrative boundaries as of 1995 have been used. Information collection on the district level for all the required parameters is difficult due to limited data availability. AIM/Local model provides the flexibility of regional definition on two levels and facilitates estimation of aggregate emissions for a larger region and allocation of the same to the smaller areas of the region based on suitable allocation index for each sector.

In this model, an approach similar to the tier-2 method has been used for emission estimates from area sources. The emissions for a sector are estimated by multiplying the consumption for each energy kind in a technology with respective emission coefficient for the energy kind and the production quantity for each production process with respective emission coefficient.

This can be given by the following equation:

$$Q_j^{AS} = \sum_k (E_{j,k}^{AS} \times f_{j,k}^{AS}) + \sum_v (V_{j,v}^{AS} \times f_{j,v}^{AS}) \tag{2}$$

$$E_{j,k}^{AS} = E_{j,k} - \sum_{l\in\{\text{point sources belong to sector } j\}} E_{l,k}^{LPS} \tag{3}$$

$$V_{j,v}^{AS} = V_{j,v} - \sum_{l\in\{\text{point sources belong to sector } j\}} V_{l,v}^{LPS} \tag{4}$$

Where,

Q_j^{AS} :   Emissions from sector j from area sources

$f_{j,k}^{AS}$:   Emission coefficient of energy kind k from sector j, taking into account the effect of removal technologies

$E_{j,k}^{AS}$:   Energy consumption of energy kind k for sector j

$E_{j,k}$:   Total energy consumption of energy kind k for sector j

$f_{j,v}^{AS}$:   Emission coefficient of production process v from sector j, taking into account the effect of removal technologies

$V_{j,v}^{AS}$:   Production quantity of production process v for sector j

$V_{j,k}$:   Total production quantity of production process v for sector j

j :   Sector

The sectoral emissions estimated on the national level are then allocated to the districts based on a suitable allocation index like district population, area, road density etc., for which the information is available on district level. Suitable parameters, which were considered as the major drivers for emissions from a particular sector, were used to generate this index. The emissions from district i can thus be given by

$$q_i^{AS} = \sum_j (Q_j^{AS} \times \frac{I_{i,j}}{\sum_i I_{i,j}}) \tag{5}$$

Where,

q_i^{AS} :   Emission in district i from area sources

$I_{i,j}$:   Emission intensity index for sector j in district i

Total emissions

Equation (1) gives the emissions from the point sources and equation (2) gives the emissions from a sector in a region. The total emissions from sector j in a region can thus be given by

$$Q_j = Q_j^{AS} + \sum_{l \in \{\text{point sources belong to sector } j\}} Q_l^{LPS} \tag{6}$$

The total emissions at the district level can be calculated by summing up the allocated area source emissions and point source emissions from a district.

$$Q_i = q_i^{AS} + \sum_{l \in \{\text{point sources in district } i\}} Q_l^{LPS} \tag{7}$$

7. Methodology for Exploring Co-benefits of CO_2 and SO_2 Mitigation Policies in India using AIM/Enduse model

Rahul Pandey[1] and Priyadarshi R. Shukla[2]

Summary. This study illustrates a methodology to explore co-benefits of CO_2 and SO_2 mitigation objectives, along with initial results for India, using AIM/Enduse model. It is assumed for India that use of low-sulfur fuels in transport sector, rapid penetration of sulfur removal technologies in power sector and large industry boilers will enable early decoupling of the two emissions under the business-as-usual scenario. Two additional sets of scenarios – one for carbon taxes and the other for corresponding SO_2 constraints – were set up to analyze co-benefits. Initial results suggest that under the application of carbon tax there is a strong overlap among the economic options for reduction of CO_2 and SO_2 emissions over the business-as-usual level. However, under pure SO_2 mitigation targets over business-as-usual, the economic options for SO_2 mitigation and CO_2 mitigation are likely to get decoupled. AIM/Enduse, being rich in representation of technological processes, is an effective vehicle to analyze these effects.

7.1 Introduction

The Indian energy system, dominated by coal, accounted for over 250 million tons of carbon emissions from the country in 2000. Despite a decade old process of economic reforms and accompanied introduction of efficient technologies and practices in certain sectors like process industries, manufacturing, and transportation, carbon emissions continue to rise at an alarming rate. This is mainly because of three reasons: i) The Indian economy continues to grow at a high rate; ii) Cumulative shift away from coal remains insignificant, particularly in the electricity generation and industrial sectors, and iii) Inefficient technologies and practices still thrive in significant parts of the economy including several small and medium industries and services, agriculture, and the traditional sector. Although India does not have GHG emissions mitigation commitment under Kyoto Protocol to the United Nations Framework Convention on Climate Change (UNFCCC), it would be interesting to analyze the effect of future carbon mitigation commitment on Indian economy, energy, and emissions.

An important concern of policy makers in developing countries like India will be how to achieve a synergy between domestic environmental policy priorities and

---

[1] National Institute for Environmental Studies, Tsukuba 305-8506, Japan (on leave from Indian Institute of Management, Lucknow 226013, India)
[2] Indian Institute of Management, Ahmedabad 380015, India

GHG mitigation objectives to accrue the co-benefits. Concerns of controlling local pollution deservedly figure high on the list of domestic policy priorities. Since developing countries have scarce financial resources, it is essential for them to design policies that are aimed toward domestic priorities and simultaneously contribute to GHG mitigation objectives.

In this paper we present a methodology for analyzing above-mentioned synergy using AIM/Enduse model, with initial results for India. We analyze two sets of policies – (i) CO_2 emissions mitigation (ii) SO_2 emissions mitigation – and the extent to which the latter can contribute to the former and vice versa. Focus of our analysis is on technology and energy options available within different sectors of Indian economy to achieve these objectives. For this purpose we set up the AIM/Enduse model for India by treating entire country as a single area, in contrast to chapter 6 of this book which analyzes Indian area and large point source emission using AIM/Local model with spatial disaggregation.

7.2 Bottom-up Modeling Framework

Depending on the way a model captures interactions between energy and economy, it is classified as bottom-up or top-down. Zhang and Folmer (1998) have discussed different bottom-up and top-down economic modelling approaches used in the context of carbon dioxide emissions mitigation. Pandey (2002) has discussed this classification from the viewpoint of energy policy modeling and research concerns for developing countries. Bottom-up models contain detailed representation of the energy resources, technologies, and end-uses. They are better suited than top-down models to analyze sector and technology level policy options. Although top-down models have better characterization of impacts on economic growth, price feedback, and trade (Hourcade 1993), they are weak in representation of technology and energy details.

AIM/Enduse (Morita *et al.* 1996; Kainuma *et al.* 1999, 2000), MARKAL (Fishbone and Abilock 1981; Berger *et al.* 1987), and EFOM (Finon 1974) are examples of bottom-up models. AIM/Enduse scores over several other bottom-up models with respect to its representation of technological detail. On one hand, it permits modeling of technological processes as complex networks of devices through which energy and materials flow. On the other hand, it enables representation of SO_2 and NO_x pollution removal processes as attachments to regular industrial technologies. Refer to 'A Guide to AIM/Enduse model' in this book for further details.

Setting up country-level AIM/Enduse model for India comprised four steps: (i) Selection of sectors, services, technologies, reference year and discount rate, (ii) Estimation of data for services and technologies in the reference year, and (iii) Projection of service demands, technology shares, and technology improvements over 37 year time horizon, and (iv) Design of business-as-usual (BAU) and other scenarios for policy analysis. We summarize the salient features of AIM-India in Table 1.

Table 1. Salient features of AIM-India

Feature	Description
Time horizon	37 year horizon, from 1995 to 2032
Demand sectors[*]	18 sectors comprising agriculture, commercial, residential, road transport, rail transport, air transport, water transport, iron & steel, aluminium, cement, brick, nitrogenous fertilizer, pulp & paper, caustic soda, soda ash, cotton textiles, sugar, and other industries
Energy conversion and supply sectors	3 energy conversion and supply sectors comprising electricity, oil refining, and natural gas (refer to Sec. 4 of 'A Guide to AIM/Enduse model')
Services	Over 70 services including 33 final (or external) services
Energy	Over 20 energy kinds including 10 primary (or external) energy kinds
Technologies[*]	Over 190 devices in demand sectors, and over 25 devices in energy conversion sectors (refer to Appendices J and K of 'A Guide to AIM/Enduse model')
Sulfur removal processes	Pollution removal processes like coal washing, limestone injection, conventional and advanced flue gas desulfurization in electricity generation and process industries
Data estimation for reference year	Bottom-up methodology for estimation of data in reference year based partly on published sources and partly on standard assumptions (refer to Sec. 4 of 'A Guide to AIM/Enduse model')
Projection of service demands	Projection of service demands until 2032 based on a top-down methodology comprising projection of drivers using logistic regression (refer to Sec. 4 of 'A Guide to AIM/Enduse model')

[*] Technological processes in small and medium industries have not yet been modeled in the current version of AIM-India.

7.3 Design of Scenarios

7.3.1 BAU scenario

No GHG policy intervention was assumed in BAU over the 37 year horizon. Service demands were projected assuming GDP growth with compounded annual growth rate of 5% in 1995-2032, decreasing from 5.7% in 1995-2010 to 4.0% in 2020-2032. Annual discount rate was fixed at of 6%.

BAU scenario assumes SO_2 control measures that are already envisaged (for the near-term implementation) and others that can be anticipated (in the long-term) for a rapidly growing developing economy with low per capita present income. Mitigating SO_2 pollution in urban India is a domestic concern that has recently attracted attention of policy makers. Electricity generation, Iron & Steel industry, Road transport, and Biomass combustion comprised over 65% of India's SO_2 emissions in 1995 (Garg and Shukla 2002). Recent steps taken by policy makers include regulation targeted at reducing sulfur content of diesel in large cities accompanied with an elaborate vehicle inspection and certification system. Such measures have already led to a reduction of SO_2 concentration in traffic intersection areas in Delhi by over 50% from 1995 to 2000 (Sengupta 2001). Additionally, regulations for adoption of pollution removal technologies like coal washing and flue gas desulfurization in thermal power plants have been reasonably successful. In BAU, we assumed continuation of these trends.

7.3.2 Policy scenarios

Carbon tax scenarios

Carbon tax is one of the instruments for mitigating global CO_2 emissions that is widely discussed in international forums on climate change. Main economic advantage of emissions tax is that it limits the cost of reduction programme by allowing emission to rise if costs are unexpectedly high (IPCC 2001). Since we wanted to study the linkage between SO_2 mitigation and CO_2 mitigation objectives using AIM/Enduse, we first chose three levels of carbon tax and then set corresponding SO_2 mitigation targets for defining SO_2 constraint scenarios. AIM/Enduse permits application of sectorwise and energywise application of tax or constraints on each gaseous emission. In each scenario, a constant level of carbon tax was applied from 2010 onwards (no tax was applied before 2010). These levels are based on the likely ranges of tax indicated by results for 550 ppmv carbon mitigation target through 2100 from AIM/CGE global model (see Chapters 4 and 10 of this book). However, it must be noted that at this stage these tax levels are for illustrating the methodology. Following tax scenarios were assumed.

- C-Tax (US$50/t-C): Constant tax of US$ 50/t-C from 2010 onwards
- C-Tax (US$100/t-C): Constant tax of US$ 100/t-C from 2010 onwards
- C-Tax (US$200/t-C): Constant tax of US$ 200/t-C from 2010 onwards

SO_2 constraint scenarios

To analyze the co-benefits, we constructed three SO_2 constraint scenarios having SO_2 limitation equivalent to the SO_2 emissions trajectory for each of the three carbon tax scenarios. Table 2 shows the SO_2 emissions constraints for the three scenarios. The scenarios, named SO_2 Constraint1, SO_2 Constraint2, and SO_2 Constraint3, correspond to SO_2 emissions in C-Tax (US$50/t-C), C-Tax (US$100/t-C), and C-Tax (US$200/t-C) scenarios respectively.

The effect of a SO_2 constraint is different from the effect of a carbon tax even if the level of SO_2 emissions is the same in the two scenarios. While a carbon tax increases the price of an energy-kind in proportion to its carbon content, a SO_2 constraint imposes a hard limit on the total quantity of SO_2 emissions in a year. While the former is an example of a market-based intervention, the latter is that of a command-and-control regulation. Imposing an upper limit on SO_2 emission quantity will induce different players to chose technologies and fuels that are lesse sulfur intensive, independent of their carbon emission performance.

Table 2. SO_2 constraint scenarios

Scenario	1995	2005	2010	2020	2032
SO_2 Constraint1	4.93	5.44	5.87	6.06	4.74
SO_2 Constraint2	4.93	5.44	5.87	4.61	3.08
SO_2 Constraint3	4.93	5.44	5.79	3.98	2.30

Note: Units are in Million ton SO_2; These figures denote the upper bounds on quantity of SO_2 emission; Data between specified years are linearly interpolated.

7.4 Results and Analysis

Discussion of BAU scenario for India has been covered under 'reference scenario' in Chapter 6 of this book. We will confine our discussion to analysis of policy scenarios. Since the results are from the initial stage of our study, our discussion in this section is meant to illustrate the richness of analyses that is possible with AIM/Enduse model, rather than provide specific numbers for policy recommendation.

Figures 1 and 2 show the CO_2 and SO_2 emissions in BAU and carbon tax scenarios. In comparison to BAU, CO_2 emissions in 2032 under carbon tax scenarios of US\$ 50, US\$ 100, and US\$ 200 decline by 9%, 28%, and 35% respectively. Marginal reduction in CO_2 decreases with increasing carbon tax, indicating a corresponding increase in the marginal cost of carbon reduction.

Carbon taxes aimed at reducing CO_2 also induce significant SO_2 reduction. Explanation for this close association can be found in primary energy substitution as shown in Fig. 3. Under carbon tax scenario of US\$ 200, coal reduces by 286 Million tons oil equivalent (Mtoe), whereas natural gas increases by 91 Mtoe and renewables and nuclear energy (excluding biomass) increase by 32 Mtoe, in 2032, as compared to BAU. Most of the energy substitution occurs in electricity generation (higher penetration of natural gas and renewables), iron & steel production (higher penetration of electric arc furnace and direct reduction process), pulp & paper production (higher penetration of waste paper based process), sugar industry (greater use of cogeneration), and residential sector (greater use of fluorescent lamps). There is little change in supply of crude oil due to little switch in road transport sector.

Additionally, there is an increase in efficiency of technologies in various sectors leading to a decline in total primary energy supply during the period 1995-2032 from 29.0 Btoe in BAU to 27.6 Btoe in carbon tax US\$ 200 scenario. Substitution of coal primarily by natural gas and renewables, and increase in efficiency, result in a close association between reductions in CO_2 and SO_2. This phenomenon is also observed in case of other carbon tax scenarios.

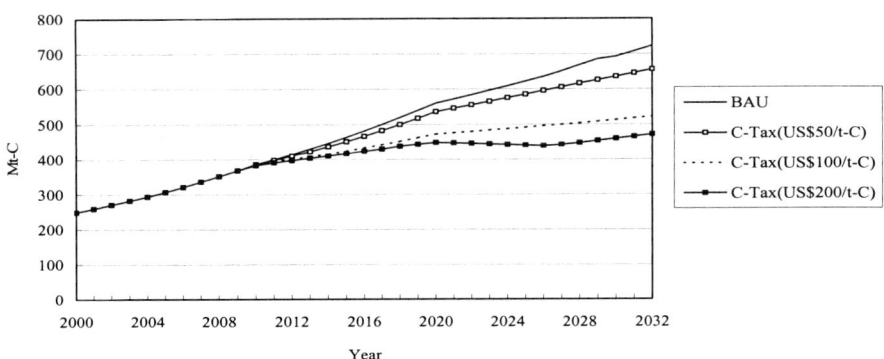

Fig. 1. CO$_2$ emissions in BAU and carbon tax scenarios

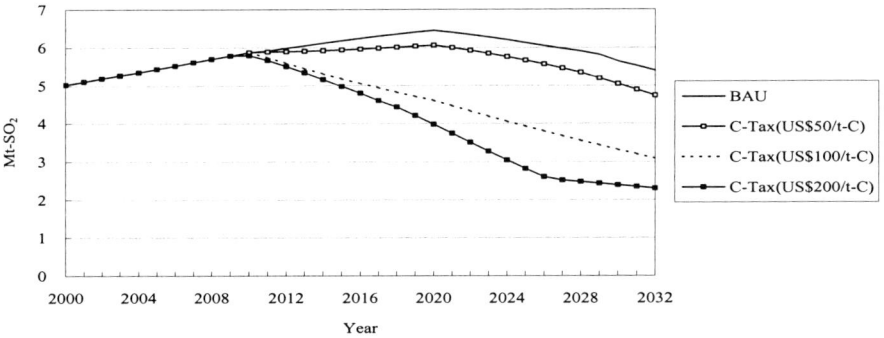

Fig. 2. SO$_2$ emissions in BAU and carbon tax scenarios

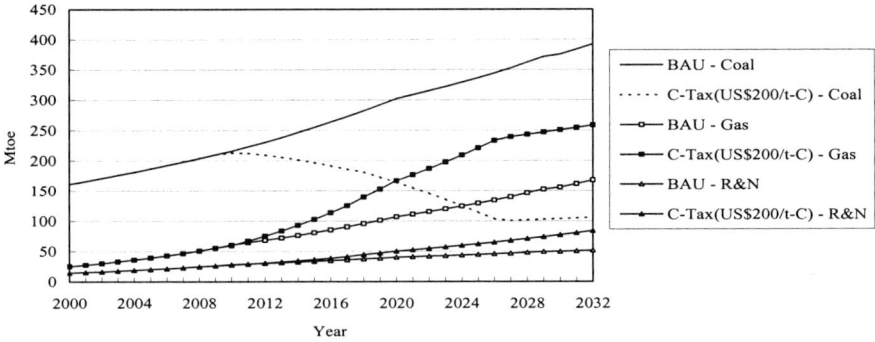

Fig. 3. Primary energy supply in BAU and carbon tax scenarios
Note: Coal includes coal and lignite; Gas includes natural gas; R&N includes renewables and nuclear energy excluding biomass.

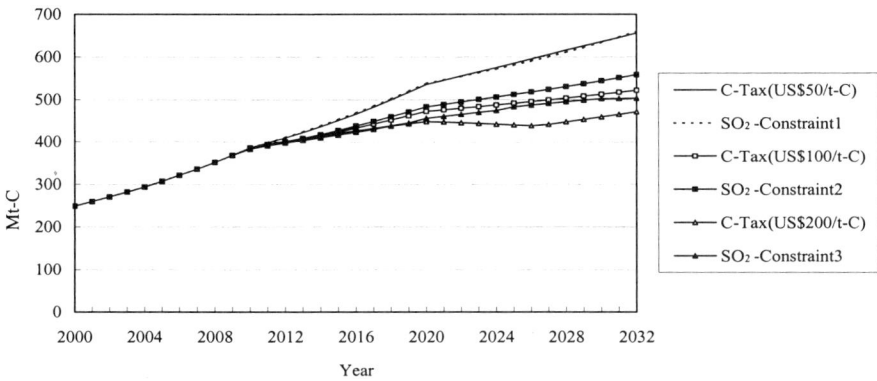

Fig. 4. Comparison of CO_2 emissions in carbon tax and SO_2 constraint scenarios

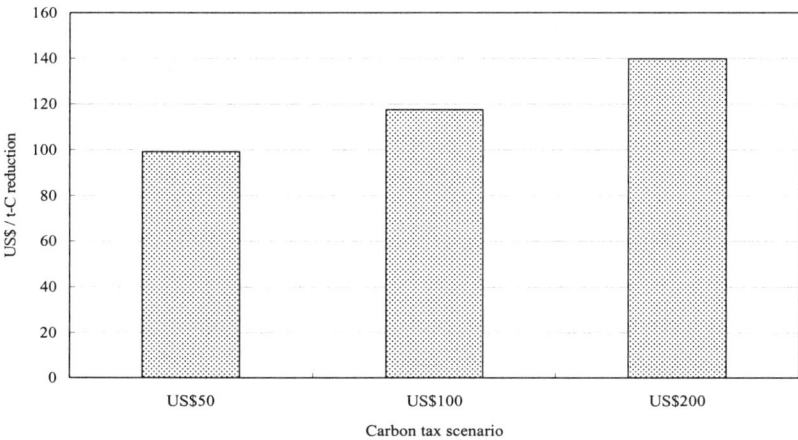

Fig. 5. Average undiscounted system cost over 2000-2032 period of CO_2 reduction under different levels of carbon tax
Note: The cost includes only initial investment cost in plant and machinery and energy costs; it does not include cost of land and building, wages and overhead costs, cost of carbon tax, and cost of implementing regulatory measures.

Figure 4 shows CO_2 emissions under carbon tax and corresponding SO_2 constraint scenarios. Figure 5 compares the average cost of CO_2 reduction over 2000-2032 under carbon tax scenarios. These figures are based on initial results, and in future they may change as we intend to consider more sulfur removal options especially in small and medium industries, and effect of on-going reforms in the power sector. Nevertheless, with the existing results, we can make the following observations.

- For the same trajectory of SO_2 reduction under both policies, carbon emissions are lower under carbon mitigation policy as compare to sulfur mitigation policy.
- Overall, a domestic regulation for reducing SO_2 emissions beyond the BAU is likely to result in some reduction in CO_2 emissions as well. This is because the majority of options exclusively for SO_2 removal available in power sector, transport sector, and large industries, have been selected in BAU itself. Marginal cost of SO_2 mitigation over BAU through such exclusive options is higher than the marginal cost of its mitigation through fuel switch options. This could be because of two reasons: (i) we have not considered SO_2 removal options in small and medium industries, and (ii) further advanced technologies for SO_2 removal in power sector and large industry boilers are expensive. We would expect more decoupling between CO_2 and SO_2 emissions if small and medium industrial processes are modeled and heavy subsidies are given to the advanced sulfur removal technologies.
- Average undiscounted cost of CO_2 reduction increases with the level of carbon tax (from US$ 99/t-C for US$ 50 tax to US$ 140/t-C for US$ 200 tax). This is due to increasing marginal cost of mitigation resulting from higher investment cost of renewable energy technologies and higher cost of supplying natural gas and other cleaner fuels.

7.5 Concluding Remarks

Using a simple methodology for analyzing co-benefits using AIM/Enduse model, our initial study demonstrates that under application of carbon tax, there is a strong overlap between most economic options for SO_2 mitigation and CO_2 mitigation. This is mainly because energy substitution from coal to natural gas and renewables offers an economic way of sharing costs to achieve both CO_2 and SO_2 mitigation. Even without any carbon tax some gas based technologies are proving to be economically viable (compared to coal based technologies) in electricity generation and a few other industries worldwide (Pandey 2002).

Although under a CO_2 mitigation policy regime, extents of SO_2 mitigation and CO_2 mitigation are strongly correlated, the two trajectories get decoupled under a SO_2 mitigation policy regime. It is difficult to comment on the extent of this decoupling because we have not considered several SO_2 removal options in small and medium industries in India.

These results have implication for sequencing of mitigation options over a long term planning period. Facilitating rapid penetration of conventional sulfur removal technologies for coal washing, limestone injection, and flue gas desulfurization, is an immediate domestic policy imperative (as is assumed under BAU), independent of GHG mitigation objective. However, for GHG mitigation policy, preparing institutions and infrastructure for facilitating medium-term penetration of natural gas is a robust GHG mitigation option. Although this strategy is economically desirable under GHG mitigation commitment, it will help in achieving medium-term reduction of both CO_2 and SO_2 emissions.

While the SO_2 limitations do generate co-benefits of carbon mitigation, the residual reduction of carbon vis-à-vis BAU would be relatively less significant compared to reduction of SO_2 under carbon control policies. The co-benefits are thus likely to be asymmetric, i.e. carbon limitation has much greater residual impact on SO_2 trajectory whereas SO_2 limitation has milder residual effect on carbon trajectory. The GHG mitigation policies for India therefore may have to be crafted for its own sake, in accordance with the global GHG mitigation dynamics.

Since our analysis in based on initial stage of the study, a few words of caution deserve mention here. Firstly, potential for exclusive SO_2 removal in India is far more than what we have considered in this study, especially in small and medium industries. We intend to enrich our technological database for these industries in future. This potential may lend economic credibility to strong decoupling of SO_2 and CO_2 mitigation objectives for a long period of time.

Secondly, we have not studied the effect of rapid changes going on in the policy regime, markets, and technological progress in the electricity industry. These are global trends and most countries including India have initiated power sector reforms, guided mainly by the reforms model of some of the more advanced countries like the UK. Several experts predict that changes in the structure and technologies in electricity industry worldwide will tilt the economic balance decisively in favor of smaller scale generation technologies like those based on natural gas (Pandey 2002; Patterson 1999). These trends, independent of GHG mitigation commitments, may lend economic credibility to substitution away from coal in the Indian power sector.

Since AIM/Enduse permits exhaustively detailed modeling of technological systems and their emission characteristics, it is an effective vehicle to analyze co-benefit policies. In the next stage, we intend to include small and medium industrial processes too in our study.

References

Berger C, Haurie A, Loulou R (1987) Modelling long range energy technology choices: the MARKAL approach. Technical paper, GERAD, Montreal

Edmonds J, Wise M, Pitcher H, Wigley T, MacCracken CN (1996) An integrated assessment of climate change and the accelerated introduction of advanced energy technologies. Pacific Northwest National Laboratory, Washington, D.C.

Finon D (1974) Optimization model for the French energy sector. Energy Policy 2(2): 136-151

Fishbone LG, Abilock H (1981) MARKAL, a linear programming model for energy systems analysis: technical description of the BNL version. International Journal of Energy Research 5: 353-375

Garg A, Shukla PR (2002) Emissions inventory of India. Tata McGraw-Hill Publishing Company Limited, New Delhi

Hourcade JC (1993) Modelling long-run scenarios: Methodology lessons from a prospective study on a low CO_2 intensive country. Energy Policy 21(3): 309-326

IPCC (2001) Climate change 2001: mitigation. Contribution of Working Group III to the third assessment report of the Intergovernmental Panel on Climate Change. Cambridge University Press, Cambridge

Kainuma M, Matsuoka Y, Morita T (2000) The AIM/End-use model and its application to forecast Japanese carbon emissions. European Journal of Operational Research 122: 416-425

Kainuma M, Matsuoka Y, Morita T, Hibino G (1999) Development of an end-use model for analysing policy options to reduce greenhouse gas emissions. IEEE Transactions on Systems, Man, and Cybernetics – Part C: Applications and Reviews 29(3): 317-324

Kainuma M, Matsuoka Y, Morita T (1998) Analysis of post-Kyoto scenarios: The AIM model. In: Economic modeling climate change: OECD workshop report. Organization for Economic Development and Cooperation, Paris

Morita T, Kainuma M, Harasawa H, Kai K (1996) A guide to the AIM/Enduse model - technology selection program with linear programming. AIM Interim Paper, National Institute for Environmental Studies, Tsukuba

Pandey R (2002) Energy policy modeling: agenda for developing countries. Energy Policy 30(2): 97-106

Patterson W (1999) Transforming electricity. Brookings Press, UK

Sengupta B (2001) Vehicular pollution control in India: technical and non-technical policy measures. Presented at Regional workshop on transport sector inspection and maintenance policy in Asia. ESCAP/UN, Bangkok, Dec 10-12

Zhang Z, Folmer H (1998) Economic modelling approaches to cost estimates for the control of carbon dioxide emissions. Energy Economics 20(1): 101-120

8. Application of AIM/Enduse Model to Korea

Tae Yong Jung[1], Dong Kun Lee[2], and Song Woo Jeon[3]

Summary. The AIM-Korea Model has been developed to analyze global environmental policies together with local ones in Korea. Various scenarios are considered for analyzing policy options to reduce CO_2 emissions. The major findings of this study can be summarized in terms of three aspects. Firstly, CO_2 emissions will continue to increase as long as energy demand increases. It could be difficult to introduce in the market by 2020 energy savings or devices with low CO_2 emissions in every sector without implementing any climate policy measures. Secondly, the policy implications of the mitigation of CO_2 emissions in Korea is that the adoption of energy-saving devices should be further encouraged, which is indicated by the simulation results showing that the scenario of providing subsidies leads to the most effective reduction in CO_2 emissions, as opposed to imposing only a carbon tax. Thirdly, the marginal cost of mitigating CO_2 emissions varies according to the sector. Hence, it is important to identify the potential sectors where CO_2 emissions reductions might be achievable at relatively low cost. According to the simulation results, the transportation, residential and commercial sectors could be potential candidates.

8.1 Introduction

In this study, the energy consumption patterns and the projection of these for the residential, commercial, transportation and industrial sectors were analyzed for Korea, and through this an attempt was made to find relevant CO_2 mitigation policies and measures for every end-use sector. For this purpose, AIM (Asia-Pacific Integrated Model), which was developed by the National Institute for Environmental Studies (NIES), Japan (Morita *et al.* 1994, 1995, 1996), was used. The present AIM-Korea Model is an application of the original AIM/End-use model. The application of a Mini-AIM in the residential sector is included. For the transport sector, further simulation was carried out to verify the possibility of introducing CNG buses into the market, which is one of the major climate policies of the Korean Ministry of the Environment. For the industrial sector, the business-as-usual (BaU) scenario was set up for other industries, such as food and beverages, pulp, and so forth. AIM-Korea is now run simultaneously for all sectors including power generation.

[1] Institute for Global Environmental Strategies, Hayama 240-0198, Japan
[2] Sangmyung University, San 98-20, An So-Dong, Chonan, Chung Nam, 330-180, Korea
[3] Office of Planning and Coordination, Korea Environment Institute, 613-2 Bulkwang-Dong Eunpyung-Gu, Seoul, 122-706, Korea

In Section 8.2, AIM-Korea Model and its data requirement are briefly described. In Section 8.3, the energy consumption patterns for every end use sector in Korea are analyzed according to various scenarios. The business-as-usual (BaU) CO_2 emissions projection and possible policy simulations are conducted to determine the potential mitigation of CO_2 emissions in Korea. A detailed analysis of the residential sector and the transportation sector is also presented in this section. Section 8.4 includes a discussion on policy implications.

8.2 AIM-Korea Model and Data

Table 1 summarizes the sectors and fields in the AIM-Korea Model. All end use sectors are included; residential, commercial, transport and industrial. Mini-AIM, which is a convenient tool for analyzing action programs to mitigate CO_2 emissions, is applied to the residential sector. The transport sector is classified into two fields; passenger and freight transport. In each field, the technical specifications for every specific transport mode are analyzed. The Korean Ministry of the Environment is particularly interested in introducing CNG vehicles into the market, which will contribute to mitigating CO_2 emissions: as well as SO_2 emissions. Further simulations are being made in this sector to assess the conditions that would allow the penetration of CNG vehicles in the transport market. In the industrial sector, energy intensive industries such as iron and steel, cement, and the petrochemical industry have been separated out. In this study, other industries, such as food and beverages, pulp, machinery, construction, and agriculture, so forth are also separated out for the BaU projections for CO_2 emissions. Since this model is an end use model, the CO_2 emissions from the energy conversion sectors, such as electric power generation and district heating, are distributed to each end use sector. This energy conversion sector is integrated into each end use sector, so that the AIM-Korea Model runs all sectors simultaneously. This model also reflects the unique circumstances of each end use sector. For example, in the residential sector, energy demand for primary heating and hot water are combined and account for almost 80% of all energy demand, since the winter in Korea is very cold and historically Koreans have had high levels of heating in winter. Hence, the combination of primary heating and hot water supply are separated out in the AIM-Korea Model.

The data for the AIM-Korea Model is quite detailed. Since this model assesses the selection of technologies, detailed data is required on energy sources, the calorific value of various types of fuel, the price of these fuels, and their CO_2 emission factors. Energy services represent the utility resulting from energy consumption and the units are defined according to the type of energy used.

Energy services technologies indicate the equipment and appliances that actually consume energy. In order to determine the dynamics of energy services, it is necessary to have detailed information. Two types of data are necessary. Firstly, basic data is required, such as the initial cost (purchase price) of energy services, number of units, number of households, and the amounts of fuel consumed or saved. Secondly, qualitative data is also required, such as the useful life of equipment (replacement period), market share, the introduction of specific types of equipment and appliances in the market, and the obsolescence of different technologies.

In the residential sector, one of the characteristics of the AIM-Korea Model is the separation of auxiliary heating from primary heating and the combination of primary heating with hot water supply. This change reflects the different situation in Korea compared with that of Japan, since the specific energy consumption patterns of any country are determined by the cultural and historical background, lifestyles and practices, as well as the weather conditions. The AIM-Korea Model therefore allows for more than one service to be included in the model from more than one service source simultaneously. For example, a kerosene pan heater provides auxiliary heating that requires both kerosene and electricity as energy sources. For example, Table 2 summarizes the factors that modify the requirement for energy services in the residential sector.

Table 1. Sectors and fields of the AIM-Korea model

Sectors	Field	
Residential	• Cooling • Primary heating and hot water • Auxiliary heating • Electricity/town gas	• Cooking • Lighting • Home appliances
Commercial	• Cooling • Primary heating and hot water supply • Auxiliary heating • Electricity/town gas	• Cooking • Lighting • Electrical appliances
Transportation	• Passenger transport	• Freight transport
Industrial	• Iron and steel industry • Cement	• Petrochemical industry • Other industries

Table 2. Factors that modify the requirement for energy services in the residential sector

Service	Factors
Cooling	Number of households × Floor area × Cooling intensity
Primary heating/Hot water supply	Number of households × Floor area × Heating/Hot water intensity
Auxiliary heating	Number of households × Floor area × Heating intensity
Cooking	Number of households × Floor area × Cooking intensity
Lighting	Number of households × Lux
Televisions	Number of households × Penetration × High technology
Refrigerators	Number of households × Penetration × High technology
Washing machines	Number of households × Penetration × High technology
Vacuum cleaners	Number of households × Penetration × High technology
Microwave ovens	Number of households × Penetration × High technology
Personal computers	Number of households × Penetration × High technology
Other end uses	Number of households

Table 3. Future energy services in the residential sector

Service	1995	2000	2010	2020
Number of households (thousand)	12501	13967	16561	18733
Number of households (1995=1.0)	1.0	1.117	1.325	1.490
Cooling	1.0	1.919	4.491	8.317
Primary heating/Hot water supply	1.0	1.150	1.490	2.002
Auxiliary heating	1.0	1.111	1.333	1.651
Cooking	1.0	1.156	1.390	1.651
Lighting	1.0	1.253	1.961	2.991
Televisions	1.0	1.297	1.843	2.318
Refrigerators	1.0	1.251	1.709	2.138
Washing machines	1.0	1.195	1.586	1.958
Vacuum cleaners	1.0	1.532	2.359	2.974
Microwave ovens	1.0	1.704	2.752	3.520
Personal computers	1.0	1.598	2.465	3.079
Other end uses	1.0	1.316	3.214	3.572

Table 3 shows the future energy services in the residential sector. Obviously, the most important factor determining the scale is the number of households, which will increase by more than 50% by 2020. The growth rate for this variable is much faster than that of the population, reflecting changing lifestyles. It is worthwhile noting that the intensity of cooling is becoming much greater due to the rapid diffusion of air conditioners, as family incomes rise.

In addition, primary heating and auxiliary heating will more than double, which is a greater increase than that of the number of households. Both cooling and heating services increase faster than the increase in the number of households, which implies that the cooling and heating intensity will increase as family incomes rise. As a home appliance, the personal computer will become more widely available, which is in accordance with one of the government's major policies so that Korea can become an 'Information Society' early in the new century.

Table 4 shows future energy services in the commercial sector, where the building floor area is the main factor determining the scale of energy consumption patterns. This variable will more than double by 2020, compared with the 1995 level.

Table 5 presents the transport sector. In this sector, various transport modes are considered. In particular, types of passenger vehicles are divided according to size. As per capita income rises, the demand for private vehicles will increase faster than other transport modes.

In Table 6, the data requirements are listed for the three energy intensive industries; iron and steel, cement and petrochemicals. In the iron and steel industry, the proportion of steel production using electric arc furnaces is important, besides the total production of crude steel, since the energy intensity of integrated steel making is much higher than that of the electric arc furnace. By the same token, in the cement industry, the proportion of ready mixed cement is important. In the petrochemical industry, ethylene production is the factor determining the scale of the various petrochemical products, for which the market is almost saturated.

Table 4. Future energy services in the commercial sector

Services	1995	2000	2010	2020
Building floor area (million m^2)	251.0	302.0	449.0	613.0
Building floor area (1995=1.0)	1.0	1.203	1.790	2.444
Cooling	1.0	1.203	1.790	2.444
Primary heating/Hot water supply	1.0	1.203	1.790	2.444
Auxiliary heating	1.0	1.203	1.790	2.444
Cooking	1.0	1.203	1.790	2.444
Lighting	1.0	1.203	1.790	2.444
Electrical appliances	1.0	1.203	1.790	2.444
Other end uses	1.0	1.203	1.790	2.444

Table 5. Future energy services in the transport sector (person km, ton km)

Service	Base Year	1995	2000	2005	2010	2020
Private passenger cars < 1500cc	1.06×10^{11}	1.000	1.274	1.925	2.453	3.075
Private passenger cars < 2000cc	4.57×10^{10}	1.000	1.278	1.928	2.451	3.085
Private passenger cars > 2000cc	5.52×10^{09}	1.000	1.277	1.920	2.464	3.080
Taxis < 1500cc (private)	5.09×10^{07}	1.000	1.202	1.424	1.542	1.554
Taxis < 1500cc (company)	2.25×10^{09}	1.000	1.204	1.427	1.542	1.556
Taxis > 1500cc (private)	3.15×10^{10}	1.000	1.203	1.425	1.543	1.556
Taxis > 1500cc (company)	4.63×10^{10}	1.000	1.203	1.425	1.542	1.555
Jeeps	8.92×10^{09}	1.000	0.980	1.233	1.413	1.379
Buses < 16 persons (private)	6.21×10^{10}	1.000	1.129	1.444	1.610	1.626
Buses > 16 persons (private)	2.85×10^{10}	1.000	1.130	1.446	1.614	1.625
Buses > 16 persons (company)	2.98×10^{11}	1.000	1.128	1.446	1.614	1.621
Inter urban railways (passenger)	2.03×10^{10}	1.000	0.709	0.685	0.719	0.818
Subways	2.47×10^{10}	1.000	1.522	1.721	2.142	2.696
Coastal water transport (passenger)	4.69×10^{08}	1.000	0.974	1.151	1.377	1.959
Air transport (domestic passenger)	4.56×10^{10}	1.000	1.180	1.496	1.853	2.982
Trucks < 1.0 ton	4.81×10^{09}	1.000	1.006	1.206	1.341	1.372
Trucks < 3.0 tons	3.56×10^{10}	1.000	1.006	1.205	1.340	1.371
Trucks < 5.0 tons	7.24×10^{09}	1.000	1.006	1.204	1.340	1.370
Trucks < 8.0 tons	6.49×10^{09}	1.000	1.005	1.203	1.339	1.370
Trucks < 12.0 tons	5.14×10^{09}	1.000	1.004	1.204	1.339	1.370
Trucks > 12.0 tons	1.17×10^{09}	1.000	1.009	1.205	1.342	1.376
Railway (freight)	1.38×10^{10}	1.000	1.174	1.254	1.312	1.377
Coastal water transport (freight)	4.38×10^{10}	1.000	1.132	1.288	1.429	1.751
Air transport (domestic freight)	8.18×10^{08}	1.000	1.169	1.443	1.724	2.274

Table 6. Future energy services in the industrial sector

	1995	2000	2010	2020
Iron and steel industry				
Crude steel production (1,000 tons)	36,772	39,200	42,700	42,700
Crude steel production (1995=1.0)	1.000	1.066	1.161	1.161
Proportion using the electric arc furnace (%)	37.8	38.8	35.6	35.6
Furnace (1,000 tons)	22871	24000	27500	27500
Electric arc (1,000 tons)	13901	15200	15200	15200

Table 6. Future energy services in the industrial sector (continued)

	1995	2000	2010	2020
Cement industry				
Cement production (1,000 tons)	51893	49310	63000	66200
Cement production (1995=1.0)	1.000	0.950	1.214	1.276
Proportion of ready mixed cement (%)	6.2	8.7	8.7	8.7
Petrochemical Industry				
Ethylene production (1,000 tons)	4,340	4,670	4,920	4,920
Ethylene production (1995=1)	1.000	1.076	1.134	1.134
Low density polyethylene	1,428	1,569	1,653	1,653
High density polyethylene	1,503	1,617	1,704	1,704
VCM	660	1,033	1,088	1,088
Polypropylene	2,602	3,176	3,347	3,347
PP	2,105	2,276	2,398	2,398
Octane	235	253	266	266
IPA	30	32	34	34
AA	160	172	181	181
BTX	5,114	5,518	5,813	5,813
Butadiene	601	687	932	932
Other petrochemicals	2,012	2,436	3,971	5,722

8.3 Simulation using AIM-Korea Model

8.3.1 Simulation setting and the BaU scenario

Based on the data for the year 1995, the simulations are set for up to the year 2020. The reason for setting the year 1995 as the base year is the availability of data. The Report on the Energy Census by the KEEI provides the most appropriate data for the present simulation and the most recent survey for this report is for the year 1995. (Ministry of Trade, Industry and Energy 1996). The year 2020 was selected as the end year for the simulation since the development of energy-saving technology could be predicted at least until this year.

The BaU scenario is one in which the current trends in energy consumption patterns will continue without any attempt to mitigate CO_2 emissions (KEEI 1998). However, expected new technologies, energy-saving, fuel switching, and existing plans are reflected in this scenario.

In Table 7 and Figure 1, CO_2 emissions by end use sector are shown. The total CO_2 emissions in 1995 amounted to 104.76 million tons of carbon (Mt-C), reaching 177.62 Mt-C in 2020 under the BaU scenario, which is a little higher than that of the previous study (Jung and Lee 1999), since the previous study reflected the financial crisis of Korea that occurred in December 1997. The GDP growth rate for 1998 was -5.9%. However, by 1999 the Korean economy had recovered with a more than 7% GDP growth rate. In this study, the current economic situation of Korea is considered, which results in a higher BaU projection for CO_2 emissions in this study. The proportion accounted for by the residential sector continues to increase from 17.4% in 1995 to 18.1% in 2020. The commercial sector shows the same trend. The proportion accounted for by this sector in 1995 was 8.3%, but it will reach 11.1% in 2020. It seems that the proportion of CO_2 emissions accounted for by the transport sector will be more or less stable after 2010, reflecting the saturation of the market for vehicle ownership. The transport sector will account for 22.1% of total CO_2 emissions in 2010.

In the industrial sector, CO_2 emissions are continuously increasing, but the proportion accounted for by this sector will decrease due to structural changes towards less carbon intensive industries. The proportion of CO_2 emissions accounted for by the industrial sector was 55.1% in 1995, while it will be 48.7% in 2020, which will be the lowest level. This trend is mainly due to the decrease in proportion of energy intensive industries. For example, the proportion accounted for by the iron and steel industry will drop from 17.8% in 1995 to 12.9%. The main reasons for the declining trend in this industry are production saturation and the change in the technology from the integrated process to the electric arc furnace. Obviously, energy efficiency improvements are also expected to occur. A similar trend is also observed in the cement and petrochemical industries. Hence, the share of CO_2 emissions accounted for by other industries will increase by more than 3% in 2020, compared with that of 1995. This type of industrial structural change in particular is rapidly developing in Korea. Among of the groups of booming industries in Korea are information technology (IT) related industries and the semiconductor industry, including computer parts, most of which are not energy intensive industries.

Table 7. Total CO_2 emissions by sector in Korea (BaU) (Mt-C, % in parentheses)

Sector	1995		2000		2010		2020	
Residential	18.21	(17.4)	20.39	(17.4)	25.73	(16.9)	32.16	(18.1)
Commercial	8.72	(8.3)	10.38	(8.9)	15.12	(9.9)	19.73	(11.1)
Transport	20.10	(19.2)	22.50	(19.2)	33.70	(22.1)	39.30	(22.1)
Industry	57.73	(55.1)	63.79	(54.5)	77.81	(51.1)	86.44	(48.7)
Iron and steel	18.60	(17.8)	20.28	(17.3)	22.62	(14.8)	22.95	(12.9)
Cement	11.59	(11.1)	11.17	(9.5)	14.26	(9.4)	14.99	(8.4)
Petrochemical	11.80	(11.3)	13.68	(11.7)	15.55	(10.2)	16.07	(9.0)
Other industries	15.75	(15.0)	18.67	(15.9)	25.38	(16.7)	32.43	(18.3)
Total	104.76	(100)	117.06	(100)	152.36	(100)	177.62	(100)

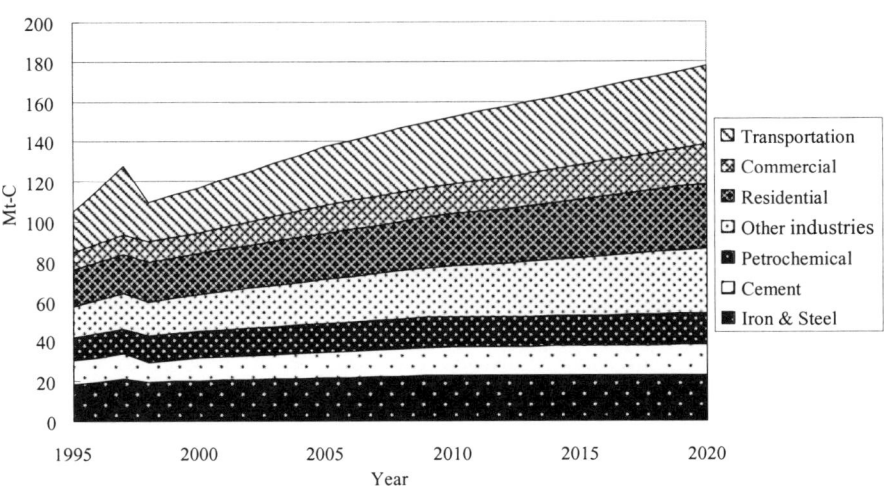

Fig. 1. Total CO_2 emissions by sector in Korea (BaU)

8.3.2 Policy scenarios and simulation results

Various scenarios on future energy consumption can be set up for the model simulation and the scenario sets for the present analysis are as follows.

The scenario for no technological changes (Scenario 1) is considered, which means that the current available technologies will continue to be used. This scenario seems to be unrealistic, but is necessary to calculate the upper bounds of CO_2 emissions. In Scenario 2 the introduction of a carbon tax (30,000Won/t-C) from 2000 is considered. In Scenario 3 a carbon tax (30,000Won/t-C) is collected, but is returned as subsidies to accelerate the introduction of solar energy for hot water or insulation, which is basically a form of tax recycling, starting from 2001. Since in AIM-Korea, the criteria for selecting technology is cost minimization, by extending the payback period, it is possible to reduce the annualized costs of energy-saving equipment, which usually requires high installation costs. These scenarios are applied to every end use sector. The scenarios mentioned above are summarized in Table 8. The results for these scenarios for all the end use sectors are presented next.

Table 8. Scenarios for all sectors

Scenario	Description	Short name
Scenario 0	Business as Usual	BaU
Scenario 1	No technological change	FIX
Scenario 2	Carbon tax: 30,000 Won/t-C from 2000 (1$ = 1,200 Won)	TAX
Scenario 3	Subsidy: 30,000 Won/t-C from 2001	SUBSIDY

Residential sector

Since the marginal costs of mitigating CO_2 emissions: in this sector are relatively low compared with the options in other sectors, it is concluded that mitigating options in this sector are feasible. Therefore, policy measures, such as energy labeling systems, rebates, and other incentives to encourage energy savings are highly recommended.

Figure 2 shows the projection of CO_2 emissions in this sector with the BaU scenario and others. In the BaU scenario, CO_2 emissions have an average annual growth rate of 2.3% (Table 7). By imposing a carbon tax of 30,000Won/t-C (about \$25/t-C), CO_2 emissions mitigation is not expected in this sector (TAX). If a carbon tax of 30,000Won/t-C is imposed and returned as subsidies (SUBSIDY), further CO_2 emissions reductions would be feasible, in which case, 2.3 Mt-C in 2010 and 7.4 Mt-C in 2020 would be possible. Hence, it is found that carbon tax recycling is more effective than a simple increase in the carbon tax rate in this sector.

Commercial sector

The main driving force for energy services in this sector is obviously the increase in building floor area, which is expected to more than double by 2020 compared to that in 1995. Especially in Korea, the structural shift from the manufacturing sector to the services sector is now occurring especially rapidly in Korea. At the same time, the contribution of this sector to the GDP is consistently increasing.

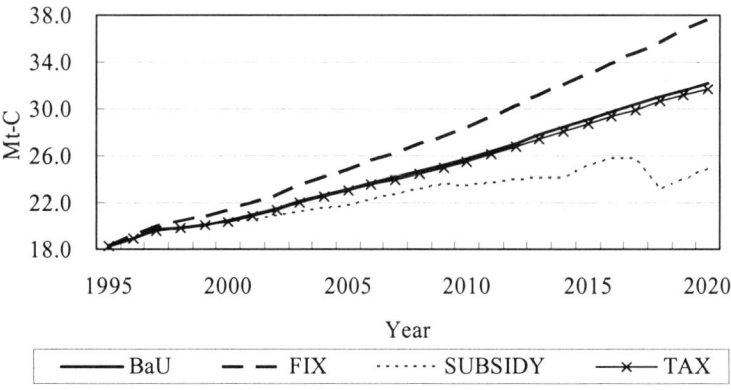

Fig. 2. CO_2 emissions projections in the residential sector

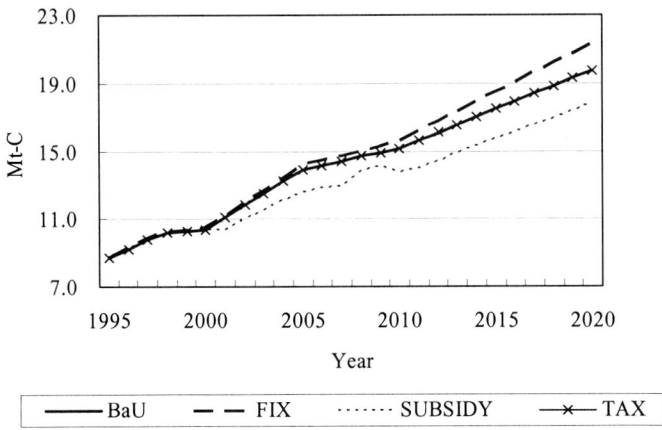

Fig. 3. CO$_2$ emissions projections in the commercial sector

Figure 3 shows CO$_2$ emissions projections for this sector with the BaU scenario and others. CO$_2$ emissions in 2020 with the BaU scenario would be 19.73 Mt-C, which is more than double that in 1995. In 2020, the proportion of CO$_2$ emissions accounted for by this sector will be 11.1% with the highest average annual growth rate of 3.3% between 1995 and 2020. By imposing a carbon tax of 30,000 Won/t-C (TAX), imposition of mitigation measures for CO$_2$ emissions in this sector is hardly effective, as with the residential sector. The introduction of a carbon tax is not sufficient to change technology selection in this sector. However, if the same amount of carbon tax is i mposed and then returned to this sector in the form of subsidies (SUBSIDY), further CO$_2$ emissions reductions would be expected, since energy efficient devices or LNG technologies, such as LNG boilers or co-generation, can be cost-effectively installed. It is expected that CO$_2$ emissions can be reduced by 1.9 Mt-C in 2020. Hence, the conclusion is that carbon tax recycling is more effective in this sector than increasing the tax rate. On the other hand, if current technologies are still in use in 2020, there will be an increase of 1.6 Mt-C in CO$_2$ emissions over the level under the BaU scenario.

Transportation sector

In the transportation sector, the demand for vehicles, especially passenger cars, has increased rapidly, as per capita income has increased. It is expected that this trend in increasing car ownership will continue, since it is projected that saturation of the passenger car market will not occur until around 2020. It is noted that due to the economic crisis, the energy demand and the demand for new passenger cars in 1998 temporarily declined, but in 1999 it rebounded with the recovery of the economy. Hence, in the long run, the trend towards increasing new car ownership

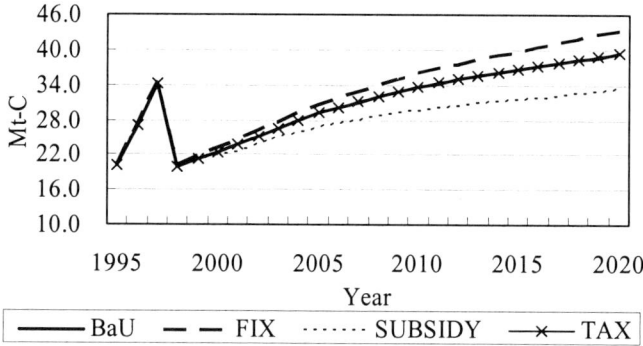

Fig. 4. CO$_2$ emissions projections in the transport sector

will continue until around 2020. Therefore, fuel substitution from carbon intensive to less carbon intensive fuels is an important option for mitigating CO$_2$ emissions in this sector. Also the secondary ben efits of reducing CO$_2$ emissions in this sector are important, since by reducing CO$_2$ emissions, it is also expected that SO$_2$ and other pollutants will be reduced, thus improving air quality, reducing congestion, and lowering related social costs.

Figure 4 shows CO$_2$ emissions projections in this sector with the BaU scenario and others. CO$_2$ emissions in 2020 for the BaU scenario would be 39.3 Mt-C, which is almost double than of 1995. The average annual growth rate of CO$_2$ emissions in this sector is 2.7% between 1995 and 2020. Again, the scenarios considered for this sector are to introduce a carbon tax and accelerate the introduction of new types of vehicles such as hybrid vehicles, as well as electric cars in the future. However, with a carbon tax of 30,000 Won/t-C (TAX), there can be no expectation of much reduction in CO$_2$ emissions in this sector, since the effect of this tax is not sufficient to change the cost structure, which is divided into fixed and variable costs. In other words, this tax scheme cannot ensure that new vehicles compete in the market, where the fixed costs of a new vehicle are still high. By imposing a carbon tax and providing subsidies (SUBSIDY), it is possible to mitigate CO$_2$ emissions by 5.9 Mt-C in 2020. It is found that the potential for CO$_2$ emissions reductions in this sector seems to be much greater than in other sectors. Therefore, policy measures in this sector should focus on how to slow down the trend towards car ownership, as well as to instigate fuel substitution and ensure that energy efficient vehicles are available in the market.

Industrial sector

In the industrial sector, the iron and steel, cement and petrochemical industries are examined. Since these three industries are energy-intensive and account for most of the CO$_2$ emissions in this sector, specific technologies in production processes of these industries are focused on. The share of the total CO$_2$ emissions accounted for by these three industries was about 40% for Korea in 1995. In this study, the BaU scenario was only included for other industries. The other industries are food and beverages, textiles, pulp, rubber, equipment assembly, machinery, electrical machinery, television production, automobiles, mining, construction and agriculture. The services included are boilers, district heating, electric power, and self-generated power.

Iron and steel industry. In the iron and steel industry, CO_2 emissions in 1995 were 18.6 Mt-C, which accounted for 17.8% of total CO_2 emissions in Korea. Figure 5 shows the CO_2 emissions projections for this industry for the BaU scenario and other scenarios. CO_2 emissions in 2020 with the BaU scenario would be 23.0 Mt-C, which is 1.2 times the level in 1995 with an average annual growth rate of 0.85%.

The results from the TAX scenario or the SUBSIDY scenario, assuming the introduction of a carbon tax from 2000, do not seem to be effective, since the facilities in this industry are too expensive for tax revenues to compensate for costs. Only 0.2 Mt-C of CO_2 emissions can be reduced in 2020. This finding implies that it is very difficult to reduce CO_2 emissions in this industry, as long as crude steel production keeps increasing and the share of production using the electric arc furnaces and integrated steel making processes is not reversed. According to the projections for crude steel production, the market will be saturated at around 47 Mt in 2010. Hence, after 2010, CO_2 emissions in this industry will be more or less stabilized.

Cement industry. In the cement industry, CO_2 emissions in 1995 were 11.59 Mt-C, which accounted for 11.1% of total CO_2 emissions in Korea. Compared with other sectors, various scenarios do not seem to be effective in mitigating CO_2 emissions in this industry, as long as cement production keeps increasing, which is similar to the situation for the iron and steel industry.

Figure 6 shows the CO_2 emissions projections in this industry for the BaU scenario and other scenarios. CO_2 emissions in 2020 with the BaU scenario would be 14.99 Mt-C, which would account for 8.4% of total CO_2 emissions in Korea. The average annual growth rate is 1.03% between 1995 and 2020. Like the iron and steel industry, the share of CO_2 emissions in this industry will be declining steadily, mainly due to the saturation of production. Figure 13 shows that with the SUBSIDY scenario, 1.18 Mt-C of CO_2 emissions can be reduced in 2020.

Petrochemical industry. In the petrochemical industry, the CO_2 emissions in 1995 were 11.8 Mt-C, which accounted for 11.3% of total CO_2 emissions in Korea. The CO_2 emissions in 2030 will be 16.07 Mt-C with an average annual growth rate of 1.24%. Scenarios such as a carbon tax of 30,000 Won/t-C (TAX) and carbon tax recycling (SUBSIDY) do not seem to be workable in reducing CO_2 emissions in this industry, compared with other sectors. For example, only 0.07 Mt-C of CO_2 can be reduced using subsidies in 2020. Figure 7 shows the CO_2 emissions projections in this industry for the BaU scenario and other scenarios. In the petrochemical industry as long as the production keeps increasing, there is little room for reducing CO_2 emissions, as in other energy-intensive industries.

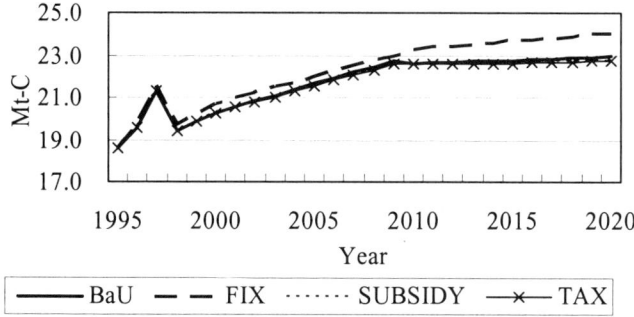

Fig. 5. CO$_2$ Emissions projections in the iron and steel industry

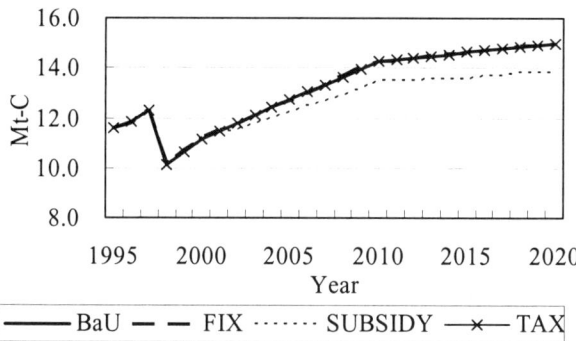

Fig. 6. CO$_2$ emissions projections in the cement industry

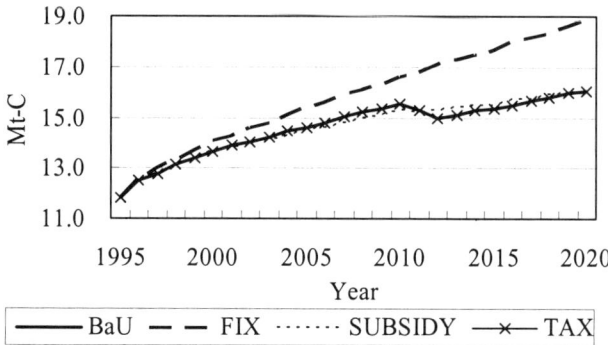

Fig. 7. CO$_2$ emissions projections in the petrochemical industry

8.3.3 Special scenarios for the sectors

Actions to reduce CO₂ emissions in the residential sector

Especially in the residential sector, there is a potential for CO_2 emissions reductions with changes in behavior and life style practices and the use of home appliances. This type of qualitative assessment can be applied to the Mini-AIM model*, which freezes technology selection in the BaU scenario and calculates the marginal change in CO_2 emissions due to possible action programs. For example, energy for cooking rice may be saved if pressure cookers are used to save cooking time. For dinner, if the whole family gets together, this also saves cooking time. The CO_2 emissions reduction potential of this type of action program is listed in Table 9. In Table 9, two cases are assumed, one in which the whole household undertakes the action program (100% action) and the other where 50% of households do (50%) so. This is compared with no action (0%, BaU scenario). For example, if the cooling temperature is increased by one degree or the filter is cleaned, a reduction of 34,000 t-C is possible with the full participation of households, which amounts to a reduction of 2.1% of CO_2 emissions for cooling services in the residential sector. It is worthwhile noting that for auxiliary heating, if insulation is installed, 20% of CO_2 emissions can be reduced, which is more than 60,000 tons of carbon. The marginal costs of these action programs are not so high, and most of them require minor changes to our behavior and practices.

Specific simulations in the transport sector

As pointed out, the transport sector is one of fastest growing sectors in terms of CO_2 emissions. Not only CO_2 emissions, but also other air pollutants cause local environmental problems. The Korean Ministry of the Environment has a special interest in mitigating emissions in this sector. Hence, in this study, more

Table 9. Actions in the Mini-AIM in the residential sector (Unit thousand t-C)

Action	50% Action	100% Action
Cooking	-69	-138
Cooling	-17	-34
Refrigeration	-22	-44
Auxiliary heating	-307	-614
TV	-27	-54
Lighting	-263	-525
Primary heating	-74	-149

---

* Mini-AIM is a part of AIM to assess the outcome of exogenously introducing certain technologies which are otherwise not chosen according to least-cost criteria. This model can simulate CO_2 emissions by changing the proportion accounted for by various technologies.

Table 10. Special scenarios for the transport sector

Scenario	Description	Abbreviation
Scenario 0	Business as Usual	BaU
Scenario 1	Extension of the Payback Period to 5 Years	Sc1
Scenario 2	Extension of the Payback Period to 8 Years	Sc2
Scenario 3	Carbon Tax - 300,000 Won/t-C from 2000	Sc3
Scenario 4	Subsidy - 15,000,000 Won per CNG Bus	Sc4
Scenario 5	Subsidy - 20,000,000 Won per CNG Bus	Sc5

simulations were set up in this sector to take into consideration the environmental policies of the Korean government. The scenarios considered for this simulation are different from the general scenarios, reflecting the particular features of the transport sector.

The BaU scenario is the same as the general one that assumes a payback period of three years. The next scenarios are extensions of the payback period to five years and eight years for all transport modes. These scenarios assume lower fixed costs for purchasing vehicles, which allows energy efficient vehicles to penetrate the market. Vehicles with advanced technologies are usually expensive, so the extension of the payback period can encourage consumers to select the new types of vehicles. The next scenario is to impose a carbon tax of 300,000 Won/t-C, which seems to be too high, but it is intended to be compared with a carbon tax of 30,000 Won/t-C. The last two scenarios are to apply a subsidy only for CNG buses of 15,000,000 Won per bus and then one of 20,000,000 Won. These scenarios are included to measure the effectiveness of the policy of the Koran Ministry of the Environment, which has a strong interest in introducing CNG buses as a form of mass transport with lower emissions. The scenarios mentioned above are summarized in Table 10.

Scenario 1 results in a reduction of CO_2 emissions by 5.02 Mt-C in 2020 which is a 12.7% reduction over the BaU scenario. This scenario allows all types of new vehicles into the market, as well as CNG buses. Scenario 2 makes an even further reduction possible. In 2020, the CO_2 emissions projection for this scenario is 3.30 Mt-C, which is a 16.0% reduction over the BaU scenario. The scenario involving imposition of a carbon tax of 300,000 Won/t-C (Sc 3) does not seem to be effective. As in the case of the 30,000 Won/t-C (TAX), the CO_2 emissions reductions do not occur.

Scenarios to provide subsidies for CNG buses (Sc4, Sc5) do not seem to be effective in reducing CO_2 emissions in this sector. These scenarios provide subsidies for CNG buses that carry more than 15 persons, which do not account for such a high proportion of transport modes. Hence, it is more critical to increase the actual proportion of CNG buses that meet passenger demand for this policy to be effective. Even with a subsidy of 20,000,000 Won, CO_2 emissions are only reduced by 0.64 Mt-C in 2020. However, in this scenario, the aspect of GHG emissions only was considered. If the benefits of reducing air pollutants are considered, this policy might be effective. In Figure 8, the CO_2 emissions projection of the further simulation is shown.

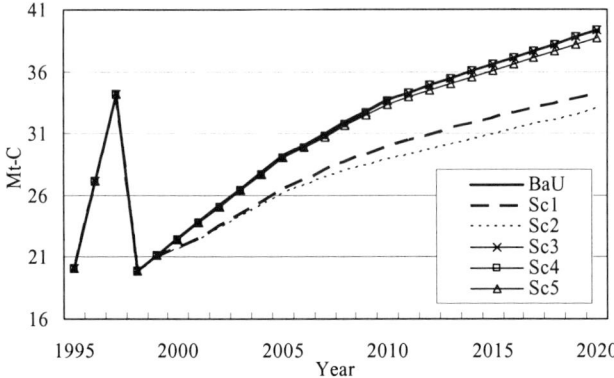

Fig. 8. CO$_2$ emissions projections in the transport sector (further simulation)
Note: For scenario names in the legend refer to Table 10.

8.4 Discussion

In this study, the CO$_2$ emissions projections for every end use sector in the Korean scenarios are conducted for the various scenarios, based on the AIM-Korea Model. Also, the conditions for the selection of new energy-saving technologies are assessed under the various scenarios in relation to the imposition of carbon taxes, the introduction of subsidies and all possible options. This process includes the estimation of the abated volume of CO$_2$ emissions due to the introduction of new energy-saving technologies into the market. This study is a further extension of research condu cted in fiscal 1998. The application of Mini-AIM in the residential sector is added. For the transport sector, further simulation was conducted to check the possibility of introducing CNG buses into the market, which is one of the major climate policies of the Korean Ministry of the Environment. For the industrial sector, a business-as-usual (BaU) scenario for other industries, such as food and beverages, pulp, and so forth has been set up. For every sector, power generation is included, which implies that AIM/Korea is now run simultaneously for all sectors.

The major findings of this study can be summarized from three aspects. Firstly, as shown in Section 8.3, in Korea, CO$_2$ emissions will continue to increase as long as the energy demand increases. Energy savings or low CO$_2$ emitting devices could be difficult to introduce into the market in every sector by 2020 without any climate policy measures. The marginally higher cost of new low CO$_2$ emitting devices is too large for them to penetrate the market. This finding holds only if consumers actually behave according to the assumption that they will follow the least cost principle.

Secondly, the policy implications of mitigating CO$_2$ emissions in Korea is to encourage the adoption of more energy-saving devices, which, as is shown from the simulation results, indicates that the scenario including subsidies ensures the most effective reduction of CO_2 emissions rather than imposition of a carbon tax alone. However, this involves a huge financial burden, and even moderate carbon tax rates cannot provide all the necessary funds. Furthermore, subsidies are considered to be incompatible with the polluter-pays-principle. Thus, it is recommended that subsidies for research and development on low emissions devices and other CO_2 abatement technologies, which would require a lower financial burden than direct payments to the consumers, are preferable to the sector. Hence, it is important to identify the sectors where CO_2 emissions reductions might be achievable at relatively low cost. Transport, residential and commercial sectors could be candidates, according to the simulation results, since not only is it possible to adopt new energy-saving technologies, but also the fuel switching may be possible in these sectors. The introduction of new devices alone is not effective in terms of climate policy as is shown in the case of the transport sector.

Finally, this study obviously has some limitations, which will require extensive CO_2 abatement devices and low emissions devices have not been assessed in this study. The analysis on these types of subsidies may provide more interesting insights into the issue of technological selection. Furthermore, this study does not cover the estimation of social costs incurred by implementing policy measures to mitigate CO_2 emissions. The AIM-Korea Model should be integrated into the top-down model in the future to identify the social costs of policy measures and provide more reliable information for policy formulation.

References

Jung TY, Lee DK (1999) Activities in Fiscal Year 1999 and Policy Design in Korea, National Institute for Environmental Studies, Japan

Korea Energy Economics Institute (1998) Action Plan for Mitigating GHG Emissions

Ministry of Trade, Industry and Energy (1987) Report on the Energy Census, Korea

Ministry of Trade, Industry and Energy (1990) Report on the Energy Census, Korea

Ministry of Trade, Industry and Energy (1993) Report on the Energy Census, Korea

Ministry of Trade, Industry and Energy (1996) Report on the Energy Census, Korea

Morita T, Matsuoka Y, Kainuma M, Lee DK, Kai K, Yamabe K, Yoshida M, Hibino G (1994) An Energy-technology Model for Forecasting Carbon Dioxide Emissions in Japan, F-64-'94/NIES, National Institute for Environmental Studies, Japan.

Morita T, Kainuma M, Harasawa H, Kai K, Lee DK, Matsuoka Y (1995) An energy technology model for forecasting carbon dioxide emissions in Japan. AIM Interim Paper, National Institute for Environmental Studies, Japan

Morita T, Kainuma M, Harasawa H, Kai K, Matsuoka Y (1996) A guide to the AIM/ENDUSE model. AIM Interim Paper, National Institute for Environmental Studies, Japan

9. Application of AIM/Enduse to Vietnam: A Study on Effects of CO$_2$ Emission Reduction Targets

Ram M. Shrestha[1] and Le Thanh Tung[1]

Summary. This study examines the effects of imposing CO$_2$ emission: reduction targets on the least cost energy resource requirements as well as emissions of CO$_2$, SO$_2$ and NO$_2$ from different economic sectors of Vietnam in 2020 using the AIM/Enduse model. The results show that in 2020 total primary energy requirement in the reference scenario would be 2,170 PJ as compared to 866 PJ in 2000. About 13.5 billion m^3 of natural gas would be required in the reference scenario in 2020 while hydropower capacity needed would be about 5,000 MW. The study also shows that the least cost energy mix for meeting emission reduction targets of 5% to 15% in 2020 would require the use of additional 3.6 to 3.9 billion m^3 of natural gas over and above the quantity of gas needed in the reference scenario. The incremental cost of reducing CO$_2$ emission by 5 to 15% of the reference scenario emission would be in the range of US$ 9.2 to US$ 58.3 per ton of CO$_2$.

9.1 Introduction

In recent years, Vietnam has recorded a relatively high economic growth rate (7%), particularly after it has moved towards a market based economy. According to Ministry of Planning and Investment, economic growth rate is projected to be around 6% during 2000-2020 (MPI 2001). Consequently, energy consumption and associated environmental emissions are expected to rise significantly in the coming decades.

Total energy consumption in Vietnam has been growing at an average annual rate of 8.5% recently. Fossil fuel consumption (coal, oil and natural gas) accounted for 75% of total primary energy requirement of the country in 2000 and was growing at an average annual growth rate (AAGR) of 9.2% during 1995-2000. As a result, CO$_2$ emission from energy use in the country had increased at the rate of 8.7% per annum during the period (Institute of Energy 1999).

Developing countries like Vietnam are facing increasing pressure for a meaningful participation in global efforts for greenhouse gases (GHGs) emission reduction. A reduction of GHG emission would, however, affect not only the energy supply cost but also the structure of energy system and technology mix in both supply and demand sides. In this paper, we analyze the implications of introducing CO$_2$ reduction targets for energy-mix, technology-mix, emission of local/regional pollutants (SO$_2$ and NO$_2$) and total energy supply costs in Vietnam in meeting

---

[1] Asian Institute of Technology, PO Box 4, Klong Luang, Pathumthani 12120, Thailand

projected demand for energy services in 2020. We also calculate the incremental cost of CO_2 abatement at selected CO_2 emission reduction targets.

9.2. Structure of AIM/Enduse Model for Vietnam

9.2.1 Basic structure

The AIM/Enduse model is a bottom-up linear programming optimization model. It accounts for the flow of energy resources from resource extraction to enduse through energy conversion/refining processes considering energy characteristics of the technologies involved in each stage. The general structure of the model is shown in Fig. 1. The model determines least cost combination of energy resources and technologies needed to meet the projected demand for energy services by various sectors in a study year.

There are several constraints used in the model. These are constraints on old stocks of energy devices, reformable devices, newly introducible devices, energy service demands and energy resource availability (see Part IV, Manual, this volume).

9.2.2 Energy supply sectors

Coal. In Vietnam, there are two kinds of technology in use for coal extraction i.e., open pit technology and underground technology. The options of using coal in electricity generation as well as in industry, residential and commercial sectors are considered in the model. The model also includes the options for import and export of coal.

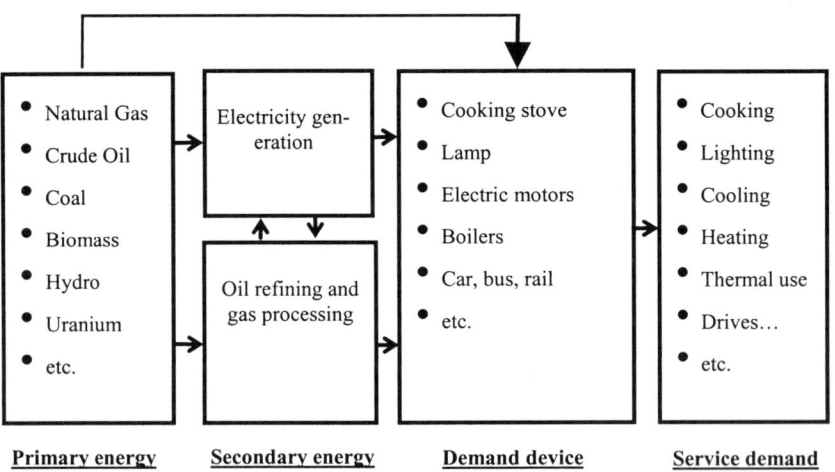

Fig. 1. Structure of the AIM/Enduse model

Crude oil and petroleum products. At present, all crude oil produced in Vietnam is exported, as there is no petroleum refinery in the country. For the energy system development in 2020, the model allows for not only domestic crude oil production but also crude oil export/import as well as petroleum refining and oil product import.

Natural Gas/LPG. The model allows for extraction and transportation of natural gas, transformation of the gas to liquefied petroleum gas (LPG) as well as the use of gas in power generation and industry sectors. The use of LPG is considered in commercial and residential sectors. It also has a provision for gas import.

Renewable energy. Renewable energy resources in Vietnam include hydropower, biomass, solar, geothermal and wind. Use of biomass is considered only for residential cooking while solar, geothermal and wind energy options are considered for power generation.

Electricity. For power generation, options considered in the model include nuclear, coal (both conventional and clean coal technologies), diesel, gas (gas-turbine and combined cycle technologies), hydropower and renewable energy (geothermal, wind, solar). Provisions for export and import of power are also made in the model.

Table 1. Enduse technology options considered in this study

Enduse service	Sector/Sub-sector	Technology option
Lighting	Residential and commercial	Incandescent, fluorescent and compact fluorescent lamp
Cooling	Residential and commercial	Conventional and energy efficient air conditioners
Cooking	Residential	Biomass, kerosene, electricity, and LPG cook stoves.
Hot water	Residential and commercial	Electrical water heaters
Thermal use	Commercial	Kerosene, coal and gas boilers
Electrical drives	Industry (all sub-sectors)	Standard and energy efficient motors
Process heating		
#1	Cement	Coal and natural gas kilns
#2	Iron and steel	Oil fired and electrical furnaces
Steam	Fertilizer and chemical, Iron and steel, pulp and paper and other industry	Coal, natural gas and oil burning boilers (conventional and energy efficient)
Other electrical appliances	Residential and commercial	Electrical appliances
Road	Passenger transport	Existing and new: two-wheelers, cars, vans, jeeps, electric cars, bus-diesel, bus-CNG and electric buses.
Rail	Passenger transport	Existing and new: diesel & electric locomotives.
Air	Passenger transport	Existing and new aircrafts.
Road	Freight transport	Existing and new: light and heavy trucks
Rail	Freight transport	Existing & new: diesel & electric locomotives.
Air	Freight transport	Existing and new cargo aircrafts
Water	Freight transport	Existing and new ships

9.2.3 Energy services

The model covers energy service demand of residential and major production sectors, i.e., commercial, industrial and transport sectors. Agriculture sector is not included here partly because of unavailability of all end-use data and partly due to the sector's small share (2%) in total final energy consumption of the country in 2000 (Institute of Energy 1999). Industry sector is classified into five sub-sectors, i.e., cement, fertilizer and chemical, iron and steel, pulp and paper and others. Transport sector comprises passenger and freight transport services. Passenger and freight transport services are both categorized further into three types (i.e., road, rail and air). Technology options considered in this study for each type of end-use/energy service demand in different sectors are presented in Table 1.

9.3 Scenarios

9.3.1 Reference scenario

The reference scenario in this study is considered as the "business as usual" scenario, i.e., without introduction of any emission mitigation policy. Data on energy service demands in this case are based on the energy demand forecasting study of Institute of Energy carried out using MAED and MEDEE-ENV models (Institute of Energy 1999). Technology data on existing power generation plants are based on Institute of Energy (2000) and that on candidate plants are based on Institute of Energy (2000), IEA (1998) and ADB (1998). The data on oil refineries and natural gas processing are obtained from Petrovietnam (2001a, 2001b, 2002). The data on end-use technology options are taken from COSMO (1999), ADB (1998), Shrestha *et al.* (1998), MOSTE (1997), Hanoi University of Technology (2000) and Minh (2000).

9.3.2 Emission reduction scenarios

Besides the reference scenario, we analyze three different emission reduction (ER) cases in which targets for reducing the CO_2 emission in the reference scenario in 2020 by 5%, 10% and 15% are considered. Hereafter, we call these emission reduction cases as ER5, ER10 and ER15 cases respectively. All others things in the ER cases remain the same as that in the reference scenario.

9.4 Data and Assumptions

An annual discount rate of 6.0% is used in the study. Emission factors for CO_2, SO_2 and NO_2 used in this study are based on Intergovernmental Panel on Climate Change (IPCC).

Table 2. Maximum allowable limits on renewable energy use in Vietnam in 2020

Resource	Availability (GWh)
Hydropower	62,000
Solar PV system	1,450
Geothermal	1,400
Wind power	1,500

9.4.1 Energy resources

Oil and petroleum products. In 2000, oil and natural gas production were about 12.5 million tons and 1.4 billion m^3 respectively. The total proven reserve of oil is estimated to be about 250 million tons (Petrovietnam 2001b). At present, all petroleum products are imported and their prices are determined by the Government Pricing Committee.

Natural gas. The estimated reserve of natural gas discovered so far in Vietnam is about 1,000 billion m^3. An optimistic scenario of natural gas supply potential is estimated to be about 15 billion m^3 per year (bcm/year) in 2020 (Petrovietnam 2002). This is also the maximum allowable limit considered in this study.

Coal. Total coal reserve is estimated to be over 6 billion tons. Most of it is anthracite. In recent years coal production has fluctuated between 4 to 8 million tons per year. According to Vietnam General Coal Company (1998), coal production could exceed 20 million tons in year 2020 if there is enough demand. We have used this figure as the maximum available quantity in this study.

Nuclear. In this study, maximum allowable capacity of nuclear power generation is set at 1,500 MW.

Renewable energy resources. In this study, renewable energy resources are considered to be available for use in 2020 up to the limits specified in Table 2.

9.4.2 Enduse data

Industry sector. The consumption of useful energy (defined as energy output delivered in the form of end-use energy services) of this sector was 127 PJ in 2000 and its average annual growth rate (AAGR) during 2000-2020 is estimated to be 7.8%. In 2000, the levels of useful energy consumption (UEC) of cement, fertilizer and chemical, pulp and paper, iron and steel and other sub-sectors were 18.0 PJ, 11.3 PJ, 13.2 PJ, 32.6 PJ and 52.0 PJ respectively. The AAGR of UEC of these sub-sectors during 2000-2020 are projected to be 6.6%, 8.7%, 8.5%, 8.5% and 7.4% respectively. Consequently, the share of iron and steel in 2020 would be 29.0% while that of pulp and paper, cement, fertilizer and chemical, and other sub-sectors would be 12.0%, 11.0%, 10.0% and 38.0% respectively (Institute of Energy 1999)

Commercial sector. Total energy consumption in the commercial sector was 23.1 PJ in 2000 and it is estimated to increase at an average annual growth rate of 6.6% during 2000-2020. The shares of different energy services in total useful energy consumption of the sector in 2000 were as follows: thermal use 27.7%, hot water 16.0%, air conditioning 11.7%, lighting 5.2% and other appliances 39.4%. In terms of useful energy demand, lighting is projected to have the highest AAGR of 8.6% followed by air conditioning, other appliances, hot water and thermal use with 7.8%, 7.1%, 6.1% and 4.8% respectively (Institute of Energy 1999).

Transport sector. The end-use demand data for the transport sector in 2000 were obtained from the General Statistical Office (2000). The average annual growth rate of service energy demand in the transport sector during 2000-2020 is assumed to be 7.2%. Table 3 presents the end-use demand for and corresponding share of various types of transport services.

Residential sector. Total useful energy consumption (TUEC) in this sector was 29.5 PJ in 2000 and it is estimated to increase at an AAGR of 7.9% during 2000-2020. The shares in the sector's TUEC of different energy services (i.e., end-uses) in 2000 are as follows: cooking 42%, air conditioning 10.5%, water heating 8.8%, lighting 5.8% and 32.9%. The AAGRs of useful energy demand for cooking, air conditioning, water heating, lighting and other end-uses are estimated to be 4.6%, 8.2%, 14.3%, 7.7% and 9.0% respectively (Institute of Energy 1999).

Table 3. Demand for transport services

Service Type		2000		2020	
		Value	Share	Value	Share
Passenger-transport	Road	18,857	70.5%	74,595	70.5%
$(10^6$ p-km)	Rail	3,462	12.9%	13,695	12.9%
	Air	4,428	16.6%	17,516	16.6%
Sub total, 10^6 p-km		26,747	100%	105,806	100%
Freight- transport	Road	4,799	12.0%	18,984	12.4%
$(10^6$ ton-km)	Rail	1,921	5.0%	7,599	5.0%
	Air	200	1.0%	791	0.5%
	Water	31,706	82.0%	125,422	82.1%
Sub total, 10^6 ton-km		38,626	100%	152,796	100%

Source: General Statistical Office (2000)

9.5. Simulation Results

9.5.1 Primary energy

The primary energy mix in 2020 under the selected emission reduction scenarios under the reference scenario is presented in Table 4 along with that in 2000. As can be seen from the table, total primary energy requirement (TPER) in Base and ER cases in 2020 would be over 2,140 PJ, i.e., about 2.5 times the figure in 2000. The use of natural gas, hydro and nuclear energy would increase with ER target while coal and oil consumption would decrease. In 2020, natural gas requirement would be 13.5 billion m^3 in the reference scenario. Natural gas use as a percentage of TPER in 2020 would increase from 21.9% in the reference scenario to 29.1% in ER15 case. In order to meet emission targets of 5%, 10% and 15%, additional quantities of natural gas of 3.6, 3.7 and 3.9 billion m^3 over and above that needed in the reference scenario would be required. The share of coal would fall from 36.5% in the reference scenario to 28.2% in ER15 case while that of oil would decrease from 28.7% to 26.7%. It should be noted here that the reduction in coal use is mainly because of considerable fuel switching from coal to natural gas in power generation. The share of hydro energy would increase from 5.7% in the reference scenario to 6.6% in ER15 case.

Table 4. Least cost primary energy requirements in different scenarios, %

Energy Resource	2000 Reference scenario	2020 Reference scenario	ER5	ER10	ER15
Biomass	7.1	3.5	3.5	3.6	3.6
Coal	37.8	36.5	31.3	29.4	28.2
Natural gas	8.2	21.9	27.7	28.4	29.1
Crude oil for refinery	-	29.3	25.9	23.2	21.3
Hydro energy	7.1	5.7	5.6	5.8	6.6
Nuclear	-	2.6	2.6	2.7	3.1
Geothermal	-	0.2	0.2	0.2	0.2
Solar	-	0.2	0.2	0.2	0.2
Wind energy	-	0.1	0.1	0.1	0.1
Import of oil product	39.6	-	2.8	5.8	5.4
LPG import	0.1	-	-	-	0.4
Electricity import	-	-	-	0.7	1.9
Total, PJ	866	2,170	2,199	2,163	2,145

9.5.2 Electricity

Electricity generation

The structure of power generation in the emission reduction scenarios is presented in Table 5. As can be seen from the table, total electricity generation in reference scenario in 2020 would be 3.62 times that in 2000. Furthermore, total electricity generation in 2020 would increase with CO_2 emission reduction target i.e., from 89.4 TWh in reference scenario to 103.3 TWh in ER15 case.

Table 5 also shows that the share of natural gas based power generation in 2020 would increase from 21.3% in reference scenario to 47% in ER15 case while that of coal fired power generation would decrease from 13.5% in reference scenario to 2.5% in ER10 case and 0% in ER15 case. Furthermore, there would be no oil based power generation in 2020. The share of hydropower would be 35.0% in the reference scenario and would remain almost at that level in the ER cases. The combined share of other renewable power generation sources (i.e., wind, geothermal and solar) would be about 3% in Base as well as the ER cases considered. Electricity import is found cost effective under ER10 and ER15 cases and accounts for 4.2% and 10.0% of total electricity supply respectively in 2020. The share of nuclear power generation would decrease with CO_2 emission reduction target although the level of generation in absolute term would remain unchanged in ER5 and ER10 cases and increase slightly in ER15 case.

Power generation capacity

As can be seen from Table 6, total power generation capacity in 2020 in the reference scenario would be 2.7 times that in 2000. Total electricity generation capacity would increase with CO_2 emission reduction target: The capacity would increase from 12,856 MW in reference scenario to 13,517 MW in ER5 case and to 14,552 MW in ER15 case. The share of natural gas based generation capacity

Table 5. Electricity generation by type of power plant in different cases, GWh

Type of generation	2000 Reference scenario	2020 Reference scenario	ER5	ER10	ER15
Hydro	13,056	31,389	31,667	31,944	36,111
Coal	3,333	13,333	6,111	2,222	0
Natural gas	5,278	37,222	45,278	45,833	48,611
Oil	3,000	0	0	0	0
Nuclear	-	5,000	5,000	5,000	5,556
Geothermal	-	894	906	914	1,022
Solar	-	894	894	894	894
Wind	-	717	725	731	767
Import	-	-	-	3,806	10,333
Total	24,722	89,444	90,556	91,389	103,333

would increase from 43.1% in reference scenario to 48.8% in ER15 case while that of coal fired generation capacity would decrease from 15.0% to 9.5%. The share of hydropower in total generation capacity is about 33.5% in the reference scenario and would remain almost at that level at the ER cases considered. In ER15 case, 4,873 MW of hydropower capacity would be installed which amounts to 65.0% of hydro energy potential. The combined share of other renewable sources (i.e., wind, geothermal and solar) would be about 2.3% in Base and ER cases. The share of nuclear power generation capacity would be about 5.3% in all cases considered.

Table 6. Total power generation capacity in different cases, MW

Power Plant	2000 Reference scenario	2020 Reference scenario	ER5	ER10	ER15
Hydro	2,866	4,350	4,258	4,312	4,873
Coal	700	1,932	1,687	1,576	1,382
Natural gas	1,150	5,530	6,517	6,580	7,127
Oil	670	39	39	39	39
Nuclear	-	669	677	684	772
Geothermal	-	109	111	112	125
Solar	-	109	109	109	109
Wind	-	88	89	89	94
Total	5,386	12,856	13,517	13,531	14,552

Table 7. Energy input used for power generation, PJ

Energy Type	2000 Reference scenario	2020 Reference scenario	ER5	ER10	ER15
Hydro	62.2	123.0	124.0	125.0	142.0
Coal	49.7	169.0	77.0	29.0	0
Natural gas	65.6	322.0	410.0	413.0	432.0
Oil	38.0	-	-	-	-
Nuclear	-	57.0	58.0	58.0	66.0
Renewable	-	9.8	9.9	10.0	10.5
Import	-	-	-	15.0	41.0
Total	215.5	680.8	677.9	650.0	691.5

Energy mix in power generation

Energy inputs used in electricity generation in 2020 under different cases are shown in Table 7 along with the corresponding numbers in 2000.

In reference scenario, total amount of natural gas used for power generation in 2020 would be 4.9 times of that in 2000. The amount of gas used for power generation in 2020 is found to increase with ER target: It would be 8.5 billion m^3 per year (bcm/year) in reference scenario and 10.8, 10.9 and 11.4 bcm/year in ER5, ER10 and ER15 cases respectively. The power sector share in total natural gas consumption in the country in reference scenario would be 67.7%.

9.5.3 Final energy consumption in sectors

Total consumption of final energy (defined as the quantity of energy delivered or used by final consumers) in reference scenario in 2020 would be 3 times that in 2000. Total final energy consumption (TFEC) in 2020 was found to decrease slightly with ER targets i.e., from 1,648 PJ in reference scenario to 1,633 PJ in ER15 case (Table 8). Industry sector would have the largest share (47.0%) in TFEC in reference scenario in 2020 followed by transport (29.0%), residential (16.3%) and commercial (7.6%) sectors. The shares would remain almost the same under the ER cases considered.

9.5.4 Emissions

CO_2 emissions

The sectoral contributions to total CO_2 emission in 2020 under different cases are shown in Table 9 together with the figures in 2000. CO_2 emission in reference scenario in 2020 would be about 34 million tons, which is 3 times the corresponding figure in 2000.

In 2000, the transport sector was the largest contributor to CO_2 emission (38.8%) followed by the industry, power, commercial and residential sectors. In 2020, the industry sector would be the largest contributor accounting for about 46.0% of the total CO_2 emission in reference scenario followed by the transport, power generation, residential and commercial sectors. With the emission reduction targets, the industry sector would still maintain the highest share in total CO_2 emission. It should also be noted that the shares of the industry and transport sectors in total CO_2 emission would increase in the ER cases considered while that of the power sector would fall. There would be only slight changes in the shares of the commercial and residential sectors.

SO_2 emissions

CO_2 emission reductions are also found to result in mitigation of SO_2 emission due to changes in energy mix. Table 10 gives the levels of SO_2 emission by sector.

Table 8. Final energy consumption by sector, PJ

Sector	2000 Reference scenario	2020 Reference scenario	ER5	ER10	ER15
Commercial	38	125	125	125	125
Industry	189	775	775	774	762
Residential	98	270	268	268	268
Transport	229	478	478	478	478
Total	554	1,648	1,646	1,645	1,633

Table 9. Sectoral shares in total CO_2 emission, %

Sector	2000 Reference scenario	2020 Reference scenario	ER5	ER10	ER15
Commercial	2.6	1.2	1.2	1.3	1.3
Industry	30.5	46.1	47.8	48.1	48.4
Residential	2.1	1.5	1.3	1.3	1.4
Transport	38.8	26.0	27.0	27.6	29.5
Power generation	26.0	25.2	22.7	20.7	19.4
Total, 10^3 tons	10,993	34,099	32,394	30,689	28,984

Table 10. Sectoral shares in total SO_2 emission, %

Sector	2000 Reference scenario	2020 Reference scenario	ER5	ER10	ER15
Commercial	3.0	1.0	1.1	1.2	1.3
Industry	43.5	62.3	70.5	75.6	78.2
Residential	2.6	0.5	0.3	0.4	0.4
Transport	24.0	14.2	16.0	17.2	18.8
Power generation	26.9	22.1	12.0	5.6	1.3
Total, 10^3 tons	198.4	648.6	572.7	533.1	489.0

Table 11. Sectoral shares in total NO_2 emission, %

Sector	2000 Reference scenario	2020 Reference scenario	ER5	ER10	ER15
Commercial	1.9	0.6	0.7	0.7	0.7
Industry	31.7	48.1	50.1	51.1	51.3
Residential	1.9	0.4	0.2	0.2	0.2
Transport	37.8	24.7	25.7	27.2	28.2
Power generation	26.7	26.2	23.3	20.8	19.6
Total, 10^3 tons	120.5	377.8	362.2	347.8	330.9

SO_2 emission in reference scenario would be about 648 thousand tons in 2020, which is 3.3 times the corresponding figure in 2000. In 2000, the industry sector was the largest contributor to SO_2 emissions (43.5%) followed by the power, transport, commercial and residential sectors. Industry sector would also be the largest contributor to SO_2 emission in 2020 accounting for about 62.3% of total SO_2 emission in reference scenario followed by the power, transport, residential and commercial sectors. The industry sector share in total SO_2 emission would not only continue being the highest but would also increase with ER target. Similarly, the share of the transport sector would increase with the ER targets while the share of power generation would fall. SO_2 emission in ER5, ER10 and ER15 cases would decrease by 13.2%, 21.7% and 32.6% cases respectively as compared with that in reference scenario.

NO_2 emissions

NO_2 emission in reference scenario in 2020 would be 378 thousand tons which is about 3 times the corresponding figure in 2000 (Table 11).
Transport sector was the largest emitter of NO_2 followed by the industry and power sectors in 2000. In 2020, the industry sector would be the dominant contributor accounting about 48.1% of total NO_2 emission. NO_2 emission in ER5, ER10 and ER15 cases would be reduced by 4.3%, 9.2% and 12.5% respectively as compared to the reference scenario mission. Furthermore, NO_2 emission shares of the industry and transport sectors would increase in ER cases while that of the power sector would decline.

9.5.5 Cost for CO_2 reduction

Table 12 presents the total cost of energy supply and incremental cost of CO_2 reduction in Base and ER cases. The incremental cost increases from $9.2 per ton of CO_2 in ER5 case to $58.3 per ton of CO_2 in ER15 case. Figure 7 shows the CO_2 emission abatement cost curve in year 2020. At the incremental abatement cost (IAC) of 10$/ton of CO_2, about 1.8 million tons of CO_2 could be mitigated while 29 and 46 million tons of CO_2 emission could be reduced at the IAC of $20 and $50 per ton of CO_2 respectively.

Table 12. Total cost of energy supply and incremental cost of CO_2 emission reduction in 2020 at constant prices of 2000

Scenario	Total cost (Million $)	Incremental cost of CO_2 reduction ($/ton-$CO_2$)
Reference scenario	16,835	-
ER5	16,885	9.2
ER10	17,116	25.7
ER15	17,792	58.3

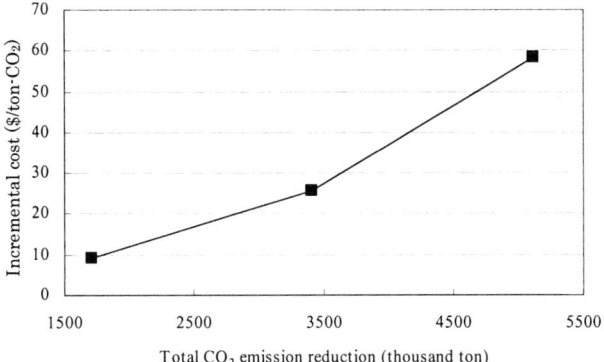

Fig. 2. Incremental CO_2 emission reduction cost curve

9.6 Conclusions

This study has examined the implications of introducing CO_2 emissions targets for the structure of energy use and the cost, as well as SO_2 and NO_2 emissions in Vietnam in 2020 by using the AIM/Enduse model.

Total primary energy requirement in 2020 in Base and ER cases would be in the range of 2,145 PJ to 2,170 PJ as compared to 866 PJ in 2000. The requirements of natural gas, hydro and nuclear energy were found to increase with emission reduction target while that of coal and oil consumption would decrease. In 2020, 92.5% of annual natural gas supply availability and 65% of total hydropower potential would be used up in ER15 case. Additional quantities of 3.6, 3.7 and 3.9 billion m^3 of natural gas would be required in order to meet emission targets of 5%, 10% and 15% respectively in 2020 over that needed in reference scenario.

Total electricity generation capacity requirement for the whole system would increase with CO_2 emission reduction target. Power generation capacity would increase from 12,856 MW in reference scenario to 14,552 MW in ER15 case. Natural gas based generation capacity would account for 43.1% in reference scenario and 48.8% in ER15 case. The share of coal based generation capacity would decrease from 15% in reference scenario to 9.5% in ER15 case.

Industry sector would be the largest contributor to total CO_2 emission in Vietnam in 2020 followed by the transport and power sectors. The power sector is found to be the main contributor to the reduction of CO_2 as well as SO_2 and NO_2 emissions.

Incremental cost of CO_2 abatement would be 9.2 US\$/ton CO_2 at the emission reduction target of 5% while it would increase to 58.3 US\$/ton CO_2 at the higher reduction target of 15%.

It should be noted that AIM/end-use is a static optimization model in that energy system optimization is carried out for a chosen year. Furthermore, energy production capacities (e.g., power generation capacity) are treated as continuously divisible decision variables. However, in fact, some energy investments (e.g., hydropower, nuclear) are lumpy in nature. Thus, the results of this study could differ somewhat from that of a study based on a dynamic energy system model, that considers explicitly the lumpiness of energy sector investments.

References

Asian Development Bank (ADB) (1998) ALGAS: Asian Least-cost Greenhouse Gas Abatement Strategy: Vietnam, ADB, Manila, Philippines

COSMO Oil Co., Ltd. and COSMO Engineering Co., Ltd (1999) The study on Energy Efficiency Improvement for Boiler and Co-generation System in Vietnam, Hanoi, Vietnam

General Statistical Office (2000) Statistical Year Book, Statistical Publishing House, Hanoi, Vietnam

International Energy Agency (IEA) (1998) Regional Trend in Energy Efficient: Coal-Fired Power Generation Technologies, http://www.iea.org.pubs/studies/files/regional/default.htm

Hanoi University of Technology (HUT) (2000), Database in energy produce situation from 1985 to 1999, HUT, Hanoi, Vietnam

Institute of Energy (IOE) (2000), the Fifth Master Plan on Electricity Power Development in Vietnam, Hanoi, Vietnam

Institute of Energy (IOE) (1999) Energy demand forecasting period 2000-2020, Hanoi, Vietnam

Ministry of Planning and Investment (MPI) (2001) Social-economics growth rate forecast for the period 2000-2020, Hanoi, Vietnam

Ministry of Science, Technology and Environment (MOSTE) (1997) Energy saving and efficiency use in Vietnam, MOSTE, Hanoi

Minh PSB (2000) Identifying clean development mechanism projects and assessments of their greenhouse gases mitigation potential from long-term energy planning perspective in Vietnam, Unpublished M.Eng. Thesis, Asian Institute of Technology, Bangkok, Thailand

Petrovietnam (2002) The Potential for Natural Gas Development in Vietnam, Petrovietnam Monthly Magazine No.1. pp26-36

Petrovietnam (2001a) Petrovietnam's downstream activities, Petrovietnam Review, Vol. 3, pp42-44

Petrovietnam (2001b) Oil and gas industry in the world, Petrovietnam Monthly Magazine No. 3, pp26-35

Shrestha RM, Biswas WK, Shrestha R (1998) The implication of efficient electrical appliances for CO_2 mitigation and power generation: the case of Nepal. International Journal of Environment and Pollution, 9(2/3):237-252

Vietnam General Coal Company (1998) General Scheme and Strategy for Coal Development up to 2010, Hanoi, Vietnam

10. Application of AIM/Enduse Model to Japan

Mikiko Kainuma[1], Yuzuru Matsuoka[2], Go Hibino[3], Koji Shimada[2], Hisaya Ishii[3], Shigekazu Matsui[3] and Tsuneyuki Morita[1]

Summary. The AIM/Enduse model has been frequently applied to Japanese policy making processes. This chapter illustrates three of these applications. First, the effects of carbon tax and subsidies to reduce CO_2 emissions in Japan are introduced. CO_2 emissions through 2010 are estimated for four cases: reference case, market case, carbon tax case, and carbon tax+subsidies case. It is shown that the subsidy scheme is a useful option to reduce CO_2 emissions. Second, ancillary benefits of CO_2 reduction for regional environmental quality are presented. A module to analyze ancillary benefits of CO_2 reduction is added to the original AIM/Enduse model. Emissions of CO_2, NO_x and PM in Aichi prefecture, Japan, are estimated through 2010 and it is shown that countermeasures to reduce CO_2 emissions can also reduce air pollutants. Third, CO_2 emission scenarios in Japan are quantified based on the four narrative scenarios for Japanese society and economy by referring to IPCC SRES scenarios.

10.1 Introduction

The AIM/Enduse model has been applied extensively in Japan (Kainuma *et al.* 1999, 2000, 2001). As shown in Table 1, various stakeholders who belong to various organizations such as the national government, local governments, a public corporation, private companies, and an NGO, have used AIM/Enduse model to evaluate countermeasures to reduce CO_2 emissions. The most recent application for the national environmental policy is introduced in Section 2 as an example of analyzing effects of introducing carbon tax and subsidies. Next, the ancillary benefits of countermeasures to reduce CO_2 emissions are presented in Section 3 as an example for regional environmental quality control programs. A module to estimate emissions of air pollutants such as NO_x and PM is added to AIM/Enduse model for analyzing ancillary benefits. This model was applied to Aichi prefecture for estimating reduction of air pollutants as a consequence of countermeasures taken for reducing CO_2 emissions. Finally Section 4 presents future CO_2 emission scenarios in Japan. Four narrative scenarios consistent with the Special Report on Emissions Scenarios (SRES) of IPCC (IPCC 2000) are developed. Based on these narrative scenarios, national energy consumption and CO_2 emissions through 2020 are estimated by using AIM/Enduse model and the causal relationships between national socio-economic development patterns and CO_2 emissions are studied. For

---

[1] National Institute for Environmental Studies, Tsukuba 305-8506, Japan
[2] Kyoto University, Kyoto 606-8501, Japan
[3] Fuji Research Institute Corporation, Tokyo 101-8443, Japan

Table 1. Application of AIM/Enduse model to Japan

Year	Application	Reference
1994	Analysis of effects of carbon tax	JEA (1994)
1996	Analysis of effects of carbon tax and subsidies	JEA (1996a)
1996	Simulation of long-term CO_2 emissions	JEA (1996b)
1997	Development of scenarios for reducing CO_2 emissions	
	Analysis of policy options concerning carbon tax	JEA (1997)
	Analysis of effects of key technologies	Tsuchiya (1997)
1998	Analysis of effects of countermeasures for Fukushima and Chiba prefectures	
	Estimation of long-term energy demand	
1999	Analysis of effects of countermeasures for Aichi prefecture	
	Analysis of ancillary benefits of regional climate action plan for Nagoya city, Aichi Prefecture and the whole country	Kouken Association (1999) Shimada *et al.* (2000)
2000	Analysis of effects of carbon tax and subsidies	JEA (2000)
2001	Analysis of countermeasures in Japan	MOE (2001)
	Development of scenarios for reducing greenhouse gas emissions	Tsuchiya (2001)
	Development of CO_2 emission scenarios based on IPCC SRES	Hibino (2001)

each narrative scenario, one reference emission scenario and one countermeasure emission scenario were developed. In case of the reference scenarios, CO_2 emissions increase by more than 10% in 2010 as compared to 1990. CO_2 emissions in three of the four countermeasure scenarios decrease by 2-4% in 2010 as compared to 1990. Difference in CO_2 emissions between 'global market-based' scenario and other countermeasure scenarios increases over time because of the difference of the basic assumptions.

10.2 Application to Analysis of the Effects of Carbon Tax and Subsidies in Japan

10.2.1 Background

During these 10 years, the AIM/Enduse [Japan] model has been applied to the national environmental policy making. Year by year, the focus of the study moved, however, the main part of the argument is how to reduce the national CO_2 emission, which fulfills the Kyoto protocol. A feasible and concrete program which suits the political situation is required. Economic instruments such as carbon tax and subsidies on energy efficient devices were considered as one of the possible options, and were often analyzed from the point of their effectiveness and economic feasibility.

10.2.2 Estimation of CO_2 emissions using AIM/Enduse [Japan]

Simulation assumptions

The effects of carbon taxes and government subsidies have been analyzed. Subsidies were introduced by using the carbon tax revenues to invest in energy-conserving technologies. The assumptions of the simulation, such as GDP, population, commodity production, and energy prices, were based on the projections of the Japanese government. Future emissions were estimated for four cases: (1) reference case, (2) market case, (3) carbon tax case, and (4) carbon tax + subsidies case.

It was assumed that current technologies will be used in the reference case. There are several reasons that current technologies will be used even though there are economic benefits in changing the technologies. For example, a lack of economic understanding and social reasons may lead to preference for old technologies. Market case assumed competition among technologies based on economic criteira. No countermeasures such as carbon/energy taxes or subsidies were assumed in the market case. Technology selection in this case is based solely on a cost. A carbon tax of ¥30,000 per ton of carbon was introduced in the carbon tax case. In the carbon tax + subsidies case, a carbon tax of ¥3,000 per ton of carbon was introduced and the tax revenues were used to subsidize energy-saving technologies to lower total CO_2 emissions.

Simulation results

In the reference case, total CO_2 emissions in 2010 increase by 18% compared to 1990, as shown in Table 2. The rates of increase in the residential, commercial, and transportation sectors exceed 30%.

Table 3 shows comparisons of CO_2 emissions in different scenarios in 2010. Technology selection in market case reduces total CO_2 emissions by 6.3% in 2010. In the carbon tax case, CO_2 emissions decrease by 3% between 1990 and 2010. An examination of sector-wise CO_2 emissions shows that compared to the reference case, emissions decrease by 14.8% in the industrial sector, 34.1% in the residential sector, 22.9% in the commercial sector, 11.5% in the transportation sector, and 18.6% in the energy conversion sector. The carbon tax therefore contributes the most to the reduction of CO_2 emissions in the residential sector.

Table 2. CO_2 emissions in reference case

Sector	1990	2000	2010	(1990=100)
Total	287.3	318.4	340.4	118
Industrial	141.0	144.8	147.3	104
Residential	37.7	44.7	49.5	131
Commercial	33.4	39.9	43.4	130
Transportation	58.2	69.6	80.4	138
Energy conversion	17.0	19.5	19.8	116

Unit: Mt-C

Table 3. Comparison of CO_2 emissions to reference case in 2010

	Reference case	Market case		Carbon tax case		Carbon tax + subsidies case	
Total	360.7	318.4	(-6.3)	278.9	(-18.6)	282.0	(-18.6)
Industrial	147.3	136.7	(-7.2)	125.5	(-14.8)	127.6	(-13.4)
Residential	49.5	44.1	(-11.0)	32.6	(-34.1)	33.3	(-32.8)
Commercial	43.4	39.5	(-9.0)	33.5	(-22.9)	33.9	(-22.0)
Transportation	80.4	79.6	(-1.0)	71.1	(-11.5)	71.1	(-11.5)
Energy conversion	19.8	18.6	(-6.3)	16.1	(-18.6)	16.1	(-18.6)

Unit: Mt-C (Figures in parenthesis indicate percentage change relative to reference case)

Table 4. Subsidized energy saving equipment

Sector	Energy saving equipment
Industrial	Waste plastic recycling for blast furnace, high-efficiency continuous annealing line, high-performance naphtha cracking device, high-performance industrial furnace, regene boiler, repowering for electric generation, combined cycle power plant
Residential	High-efficiency air conditioner, energy saving house, inverter type fluorescent lighting device, high-efficiency refrigerator, high-efficiency television, high-efficiency VCR, high-efficiency stereo, standby power saving, latent heat recovery type water heater, water heater with CO_2 refrigerant
Commercial	High-efficiency air conditioner, HF inverter lighting device, light system with sensor, high-efficiency mainframe, high-efficiency duplicator, high-efficiency personal computer, pumping power with VAV control, standby power saving, latent heat recovery type water heater
Transportation	High-efficiency small private passenger car, high-efficiency regular private passenger car, high-efficiency commercial passenger car

In the carbon tax case + subsidies case, CO_2 emissions decrease by 2% between 1990 and 2010. The effect is almost the same as that of the carbon tax of ¥30,000 per ton of carbon, in spite of the lower tax rate. The tax revenues are used to subsidize the energy-saving equipment shown in Table 4.

10.2.3 Discussion

With a carbon tax-only policy, it is necessary to have a tax of ¥30,000 per ton of carbon to decrease CO_2 emissions to 2% below the 1990 level by 2010. However, if all carbon tax revenues are used to subsidize energy-saving equipment and technologies, the 2% reduction can be achieved at a tax rate of ¥3,000 per ton of carbon.

It is shown that the subsidy scheme is a useful option to reduce CO_2 emissions. However, the subsidy scheme may differ due to certain characteristics of reality. One of these is the difficulty of distributing funds efficiently. The AIM model has all the necessary information concerning energy equipment and can find the optimal distribution of the tax revenues to lower CO_2 emissions. In view of this, it should be noted that a carbon tax rate of ¥3,000 per ton is roughly the lower limit for attempting to cut CO_2 emissions through national measures. Another point is that the subsidy may not conform to the polluter-pays principle. Since the subsidies proposed in this simulation are entirely funded by carbon tax revenues. Hence the objections and disputes arising from non-conformance to the polluter-pays principle are not considered.

10.3 Application of Analysis of Ancillary Benefits in Prefectures

10.3.1 Background

Many of the countermeasures to prevent global warming have beneficial side effects that improve air quality by reducing such pollutants as nitrogen oxides and other particulates. The costs of mitigating global warming are generally considered to be high. Uncertainty in benefits of measures for mitigating global warming adds to the reluctance against such investments. Since the improvements in local air environment are easily measured and seen as high priority by local communities, quantifying the ancillary benefits of global warming related investments is important.

Additionally, in order for developing countries to retain incentives for the implementation of global warming countermeasures, the countermeasures should be compatible with their economic growth while still helping to resolve the domestic problems–such as air and water pollution–that these developing countries face. In other words, there is an opportunity to integrate the environmental and resource-related policies of developing countries into an international cooperative system for global warming countermeasures (Yang *et al.* 2001).

While bearing such domestic and international trends in mind, models to quantitatively estimate the effects of measures to prevent global warming on regional air environments are being developed. These developments are likely to assist in comprehensive policy making process by integrating various countermeasures to prevent global warming and preserve the air environment.

10.3.2 Development of estimation model for ancillary benefits

Estimation model

We created an estimation model for the ancillary benefits of countermeasures to prevent global warming. Using the model, we estimated emissions of CO_2, NO_x, and PM, and analyzed the effects of global warming countermeasures in 2010. The emissions were estimated in accordance with the flowchart shown in Fig. 1. The possession rate by vintage of device indicates the rate of possession of energy service technology by device type and operating years. The usage duration ex-

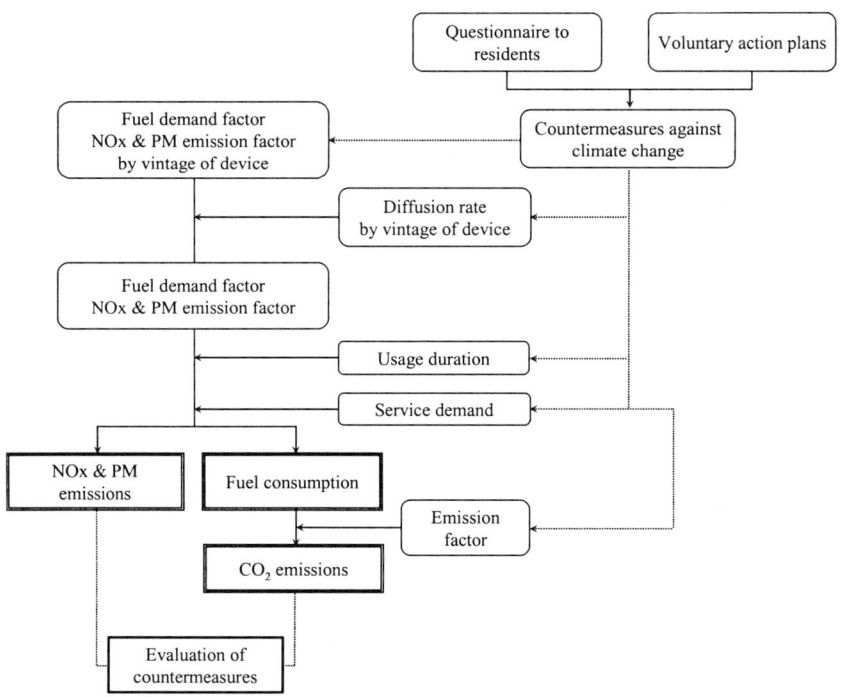

Fig. 1. Flowchart of estimation model for ancillary benefits

presses the number of usage hours per day and usage days per year. The service demand, such as the amount of production and the volume of transportation, is the driving force for energy consumption. The first point regarding the features of this model is that an average emission factor has been estimated by multiplying the diffusion rate and the factor by device/year type. The second point is that a questionnaire to residents and individual voluntary action plans by companies has been reflected in the countermeasures to prevent global warming.

The emissions were estimated in four sectors: the industrial (including energy conversion sector), transportation, commercial, and residential sectors. Furthermore, we classified the industrial sector into 28 types of business, the transportation sector into 28 types, the residential sector into 45 types, and the commercial sector into 23 types of energy service technologies. As a general rule, the years used for estimation in each sector were 1990 as the base year and 2010 as the target year.

Countermeasures against global warming

Major countermeasures that have been adopted are shown in Table 5. In terms of the industrial sector, a reduction plan concerning energy consumption or CO_2 emissions for 2010 was input based on responses from companies. For companies on which research was not carried out, the Keidanren Voluntary Action Plans on the Environment were adopted. For other companies to which these did not apply, an estimate was made assuming a 1% annual energy consumption reduction in accordance with the Energy Conservation Law.

With regard to the transportation sector, the standard values of the Energy Conservation Law were used for improved fuel efficiency. Furthermore, the introduction of advanced-technology vehicles was set up by referring to the target values of Aichi prefecture. For the residential and commercial sectors, improvements in equipment efficiency were set up in accordance with the Energy Conservation Law, and improvements in efficient equipment selection rates were estimated in accordance with the prefectural plan. Furthermore, with reference to the transportation sector, infrastructural improvements such as easing of traffic congestion and promotion of modal shift for the countermeasure case were considered based on the assumption of the estimated results for automotive usage in the reference case. For the residential sector energy saving dwellings were considered, while measures such as shift to energy saving buildings were also considered for the commercial sector. Targeted numerical values for electric utilities in this area as of 2010 were used for the CO_2 emissions factor for electricity, which is common to all sectors.

Some of the countermeasures against global warming through lifestyle changes of the general public in the prefecture estimated for the countermeasure case are shown in Table 6. This was created based on responses from residents of the prefecture to questionnaires issued in October 1999. The number of valid responses was 1,364 out of a total of 2,000 questionnaires distributed. The existing imple-

Table 5. Major countermeasures against global warming

Sector	Countermeasures
Industrial	Voluntary plan of chemical industry
	Voluntary plan of transport machinery industry
	Voluntary plan of steel industry
Residential	Improvement in efficiency of air conditioners
	Improvement in efficiency of fluorescent lighting
	Installation of solar hot water systems
	Insulation with pair glass in existing houses
	Strengthening the structure of insulation in newly built houses
	Installation of synthetic energy-saving houses
Commercial	Improvement in efficiency of air conditioners
	Improvement in efficiency of personal computers
	Improvement in efficiency of lighting systems
	Installation of HF inverter lighting systems
	Installation of high-efficiency emergency lights
	Installation of synthetic energy-saving houses
Transportation	Improvement in efficiency of privately owned gasoline cars
	Improvement in efficiency of ordinary diesel freight vehicles
	Introduction of privately owned gasoline hybrid cars
	Modal shift from truck to rail

mentation rate indicates the ratio that have already taken the actions concerned, while the expected implementation rate indicates the ratio that intend to carry out such actions among those who responded and includes those previously listed under the existing implementation rate.

10.3.3 Estimation results

Emissions by sector

Figure 2 shows CO_2, NO_x, and PM emissions in the reference and countermeasure cases in the base, latest, and target years.

CO_2 emissions in the target year would be reduced by 7% from the base year by establishing the countermeasures drawn up in the prefectural plan, while NO_x emissions would be reduced by 24% and PM emissions by 32%. As for NO_x and PM emissions in the transportation sector, it was decided that the planned values under the national regulations would be specified in both the reference and countermeasure cases. Furthermore, emissions produced in the generation of electricity were allocated in the demand sector in accordance with the amount of electricity consumption.

The emission ratios for the countermeasure cases using 100% as the reference case are shown in Fig. 3. When the total emissions of the sectors are compared between the reference and countermeasure cases in the target years, the amount of CO_2 emissions is reduced by 23%, but at the same time, it can be seen that NO_x emissions fall by 11%, as do PM emissions. When the results are analyzed by sector, NO_x drops by 11% and PM by 8% in line with a 19% reduction in CO_2 from the industrial sector. As for the transportation sector, CO_2 falls by 21%, although NO_x is only reduced by 8% and PM by 5%. In terms of the residential sector, CO_2 is reduced by 33%, NO_x by 30%, and PM by 38%. In the commercial sector, NO_x is reduced by 26% and PM by 31%, broadly in line with a 34% drop in CO_2.

Table 6. Selected countermeasures against global warming through lifestyle changes

Sector	Countermeasure	R1	R2
Trans-portation	Reduction of sudden acceleration from rest 10 times a day	37	87
	Removal of any unnecessary luggage (10 kg) prior to driving	27	78
	Reduced idling of cars by five minutes a day	33	86
	Driving with appropriate tire pressure	40	60
Residen-tial	1°C reduction in temperature setting of air conditioners	24	88
	1-hour reduction in heater use	0	60
	Reduction in use of lights	40	98
	Reduction in shower running time (3 minutes a day)	26	94
Com-mercial	Changing heater temperature setting to more appropriate one	42	92
	Changing cooler temperature setting to more appropriate one	42	91
	Turning off lights during lunch breaks	40	92

R1: Existing implementation rate (%), R2: Expected implementation rate (%)

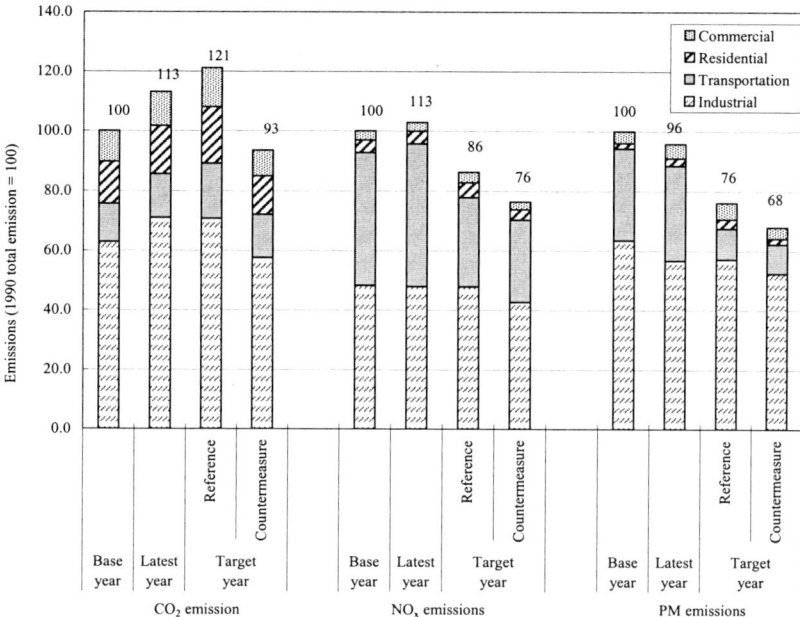

Fig. 2. CO$_2$, NO$_x$, and PM emissions in Aichi prefecture

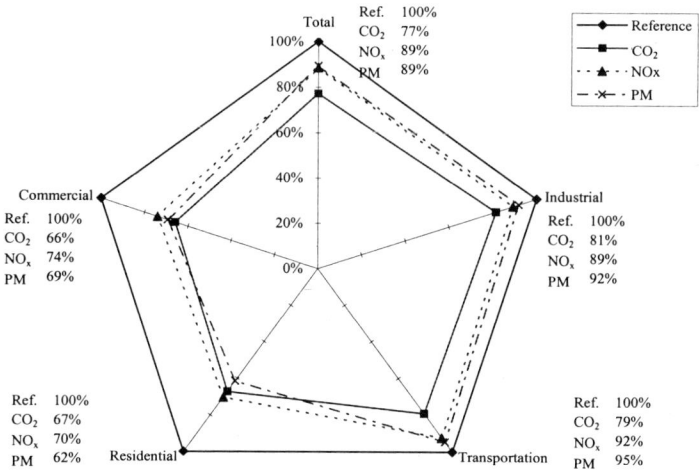

Fig. 3. Comparison between reference and countermeasure cases

Analysis of CO₂, NOₓ, and PM reduction effects by countermeasure type

Table 7 shows the CO_2, NO_x, and PM reduction contribution rates of each countermeasure type. The reduction contribution rate mentioned in this section indicates the ratio of the amount of reduction for each action with respect to emission amounts in the reference case for the target year. The reduction effects on each gas by global warming-related countermeasures against CO_2, NO_x, and PM are analyzed as follows.

Major reduction contribution rates for CO_2 result from the voluntary plan of the steel industry (6.3%), improved fuel efficiency for privately owned gasoline cars (6.6%), and improved efficiency of domestic air conditioners (heaters) (3.2%). Major reduction contribution rates for NO_x result from the voluntary plan of the steel sector (5.1%), modal shift promotion (2.4%), and installation of solar hot water systems (6.4%). Major reduction contribution rates for PM result from the voluntary plan of the chemical industry (3.4%), modal shift (2.6%), and improved efficiency of domestic air conditioners (heaters) (5.4%). Furthermore, high reduction contributions are shown for a reduction in the emission factor of electricity in the residential and commercial sectors; that is, 5.1 to 5.9% for CO_2, 3.2 to 3.7% for NO_x, and 3.8 to 8.7% for PM.

Table 7. Contribution of each countermeasure to total reduction

Sector	Countermeasure	CO₂	NOₓ	PM
Industrial	Voluntary plan of chemistry industry	1.2%	2.4%	3.4%
	Voluntary plan of steel industry	6.3%	5.1%	3.1%
	Voluntary plan of transport machinery industry	1.5%	0.4%	-0.6%
	Reduction in emission factor of electricity	3.3%	0.8%	1.1%
Trans-portation	Improvement in efficiency of privately owned gasoline cars	6.6%	0.0%	0.0%
	Introduction of privately owned gasoline hybrid cars	1.7%	0.9%	0.0%
	Reduction of sudden acceleration from rest 10 times a day	2.0%	1.1%	0.8%
	Driving with appropriate tire pressure	1.4%	0.8%	0.5%
	Modal shift from truck to rail	1.2%	2.4%	2.6%
Residen-tial	Improvement in efficiency of air conditioners (heaters)	3.2%	2.0%	5.4%
	Installation of solar hot water systems	3.0%	6.4%	0.0%
	1-hour reduction in heater use	2.4%	1.8%	1.6%
	Reduction in use of lights	2.2%	1.4%	3.7%
	Reduction in shower running time (3 minutes a day)	3.0%	5.5%	0.9%
	Reduction in emission factor of electricity	5.1%	3.2%	8.7%
Com-mercial	Improvement in efficiency of air conditioners	2.1%	1.3%	1.3%
	Improvement in efficiency of lighting systems	5.2%	3.3%	3.4%
	Improvement in efficiency of personal computers	3.3%	2.1%	2.1%
	Changing heater temperature setting to more appropriate one	2.5%	4.2%	5.8%
	Installation of synthetic energy-saving houses	2.0%	3.0%	3.2%
	Reduction in emission factor of electricity	5.9%	3.7%	3.8%

10.3.4 Discussion

In order to quantitatively understand the effects of countermeasures against global warming on the regional air environment, we developed a model and conducted estimates/analyses by applying this model to Aichi prefecture as a case study. The results obtained are summarized as follows.

- When the reference case was compared to the countermeasure case in the target year after carrying out the measures specified in the prefecture's regional promotion plan for global warming-related countermeasures, NO_x and PM emissions were both reduced by 11% against a 23% reduction in CO_2 emission.

- When the reference case was compared to the countermeasure case for the target year after implementing measures to reduce CO_2 emissions, equivalent or higher percentages of NO_x and PM reductions could be achieved in the residential and commercial sectors. In the industrial and transportation sectors, reductions in NO_x and PM of between a quarter and a half of those for CO_2 could be achieved.

- The contribution of each action to the amount of CO_2 reduction in the countermeasure case as compared to the reference case has also been analyzed. In terms of CO_2, the voluntary plan of the steel industry, improvement in the efficiency of gasoline engines in privately owned cars, and improvement in the efficiency of lighting systems demonstrate comparatively large reduction contribution rates. As regards NO_x, the voluntary plan of the steel industry, reduced running of showers, and changing heater temperature settings in offices to more appropriate ones produce comparatively high reduction contribution rates. For PM, the voluntary plan of the chemical industry, improvement in the efficiency of domestic air conditioners (heaters), and changing heater temperature settings in offices to more appropriate ones offer comparatively high reduction contribution rates. Furthermore, reductions in the emission factor of electricity in the residential and commercial sectors significantly contribute to reductions in CO_2, NO_x, and PM emissions.

An ancillary-effects estimation model was developed and applied to Aichi prefecture. Through this study of co-benefits, we also looked at the implication for future environmental policies. Our observations are as follows.

- It has become possible to quantitatively understand air pollutant emission reduction effects from countermeasures to prevent global warming by developing a model that can simultaneously estimate greenhouse gases and air pollutants. Its use allows comprehensive measures that functionally link global warming and air preservation-related measures to be studied. For example, the importance of the industrial and transportation sectors has been clarified in terms of CO_2, NO_x, and PM emissions in the case of Aichi prefecture. It has also been clarified that measures taken by the residential and commercial sectors are important from the perspective of ancillary effects.

- CO_2, NO_x, and PM reduction characteristics by measure are shown simultaneously, which could be useful in deciding the priority for measures in accordance with the regional characteristics of local public organizations. Through these estimates, the voluntary plan of the steel sector and reductions in the emission factor of electricity may be seen as measures or areas in which significant reductions in the amounts of NO_x and PM emissions, as well as CO_2, can be realized. Understanding the correlations of these ancillary reduction effects will be useful for determining future measures.
- Analysis of the effects of area separations required to study any countermeasures against air pollution onsite will be possible by linking this estimation tool to the geographical information system (GIS).
- As suggested by the Eco Policy Linkage that was proposed in Environment Congress for Asia and the Pacific, this estimation model can be utilized in developing countries as a tool to promote global warming countermeasures. Developing countries need an approach in which the effects of countermeasures against local pollutants such as sulfur oxide on greenhouse gas emission reductions are analyzed. This model can also be used for this purpose.

10.4 Development of CO_2 Emission Scenarios in Japan

10.4.1 Background

With the Kyoto Protocol due to come into effect soon, specific measures to ensure its compliance are a matter of urgency in Japan. Furthermore, as a preliminary step for the next and future negotiations, we are at a stage where we need to thoroughly discuss issues ranging from emission forecasts to the effectiveness of the measures for the second commitment period starting in 2013. In order to estimate future greenhouse gas emissions–a prerequisite for such discussions–we first require social and economic scenarios as a basis of these estimates for future greenhouse gas emissions. However, since many aspects concerning the future course of society and the economy are unclear, it is thought desirable to hold discussions using data on future emissions derived from simulations under various possible scenarios, rather than a single scenario. Four alternative scenario families of future emissions and their driving forces have been assumed by IPCC (IPCC 2000). Accordingly, the Working Group to Draw up Japanese Emissions Scenarios organized by the Ministry of the Environment has presented four narrative scenarios. The authors also estimated Japan's future CO_2 emissions for each narrative scenario using the AIM/Enduse [Japan] model. From these results, consideration was given to the causal relationship between Japan's growth pattern and the significance of measures to control greenhouse gases.

10.4.2 Japan's narrative scenarios and synopses

The Working Group to Draw up Japanese Emissions Scenarios presented four alternative scenarios for Japan, with due consideration for consistency with the emissions scenarios formulated by IPCC (Fuji Research Institute Corporation 2001). An outline of each is given below.

A1 scenario (global market-based scenario)

In the A1 scenario, the Japanese economy shifts toward a market-based economic system that attaches greater importance to the economic rationale for survival under a global market economy. Investments are targeted to increase productivity, with the aim of fueling economic growth. Also, employment opportunities will increase for the elderly, women, and foreign nationals—in other words, there will be more equal opportunities—indicating a shift toward merit-based employment regardless of age or gender in such a competitive business world.

In terms of lifestyles, active consumer spending based on high purchasing power will be brought about. Time saved by outsourcing housekeeping tasks resulting from a focus on economic efficiency will be spent on recreation and education.

Moreover, the population and capital will be centralized into a megalopolis. Transportation networks will be developed featuring railways and automobiles in the central area, and automobiles in the suburbs.

In the energy industry, price competition will intensify in line with reductions in the cost of electricity following deregulation. As a result, demand for electricity derived from fossil fuels (mainly coal and oil) will increase.

A2 scenario (scenario featuring regions and traditions)

In the A2 scenario, economic development follows the same trends as the existing social and economic systems, instead of altering these systems rapidly. Accordingly, the economy will remain relatively depressed, and society will be inwardly stable without social and economic structural reform.

In terms of employment, thanks to the retention of the orthodox Japanese management system, employment levels will be maintained despite the stagnant economy. However, there will be no significant increase in employment opportunities for the elderly, women, or foreign nationals. Furthermore, the working hours per person will be longer than those of any other scenario because there is no material enhancement in labor productivity.

In terms of lifestyle, consumption will remain at current levels. In terms of urban structure, the population and capital will be more diversified across a number of core cities. Thus, automobiles will be the main form of transportation for both cargo and passenger services across a road network connecting these cities, and also owing to active public investment in rural areas.

In terms of energy, electricity will become heavily reliant on nuclear power under an expansion of the former energy security policy.

B1 scenario (environmental-technology-led scenario)

This is a highly technology-oriented scenario that tends to favor both economic development and dematerialization through technological innovation. Investment in environmental preservation is preferred, and thus dematerialization will be advanced. Moreover, the economic growth rate will rise (though not as much as in the A1 scenario), led by the environmental industry and environment-related investments.

Employment opportunities for women and the elderly will increase. In this scenario, women's employment will rise, supported by outsourcing of child-care and nursing, in line with the growth of a welfare industry suited to an aging society with a declining birthrate, due to a focus not only on environmental preservation but also on welfare.

In terms of housing and transportation, an environmentally harmonious society will be built without any reduction in service demand, through resource optimization led by technological development. Items such as super heat pumps, fuel cells, etc., will become more popular, as well as long-life houses and advances in insulation through technological development. Also, inner-city transportation will be created using light rail transit (LRT) and urban monorails as cities will become more compact and diversified across the nation, rather than being congregated into a megalopolis. Fuel cells will become more popular not only in housing but also in automobiles.

In terms of energy, natural gas will be the main source following the adoption of natural gas thermal power stations and fuel cells.

B2 scenario (new independent areas scenario)

B2 is a scenario in which regions coexist symbiotically with independent and sustainable production areas for each. City structures will be very compact, in which communities will be the key decision-making elements in the socioeconomic system. Nongovernmental organizations (NGOs) and non-profit organizations (NPOs) will play important roles under such a socioeconomic system.

In the B2 scenario, a cyclical economy will be formed with arterial and venous industries cooperating in independent production areas. It is a scenario intended to harmonize the economy with the environment just as the B1 scenario does. However, such an environmentally harmonious society will be formed on the basis of changed lifestyles due to an altered sense of values in B2, not by technology as is the only solution in B1. For example, purchase-cycles for housing, furniture, appliances, automobiles, etc., will all be extended, so that expenditures on consumer goods will decrease, while on the other hand, consumer demand for longer lasting goods will increase despite their higher cost. Moreover, service industries such as the refurbishment and repair industry will be bolstered to support longer life cycles for these products. Manufacturers will also be transformed from companies that simply produce and sell goods into service-lease companies. As a result, they will strive for environmental harmony in all stages of their goods' life cycles, from production through usage to disposal.

In the realm of transportation, public transport and bicycles will take the place of automobiles thanks to the close proximity of living and business areas.

In employment terms, opportunities for women and the elderly will increase thanks to work-sharing schemes to promote social involvement of various types of people. Although the welfare system to support them will be enhanced, the role played by local communities under such welfare systems is important. As a result of work sharing, fewer hours will be allocated per person, and recreational periods will be spent doing various activities in the local community.

In terms of energy, service demand in the energy industry will be lower than in any of the other cases, due to a transition toward more environment-friendly lifestyles. Also, since people will be far more conscious of environmental matters and decision-making by local communities will be emphasized, environment-friendly energy systems will be adopted although they cost more. Accordingly, no new nuclear power plants will be built, while the building of new thermal power plants will be kept to a minimum—with an emphasis on natural gas power plants—because demand for large-scale power plants will decrease following the adoption of fuel cells based on local energy sources such as biomass energy, in addition to falling electrical demand due to changes in lifestyles.

10.4.3 Estimation of CO_2 emissions using AIM/Enduse [Japan]

Simulation assumptions

Using AIM/Enduse [Japan], we estimated the long-term profile of CO_2 emissions based on the four aforementioned scenarios. An estimation of the energy service demand per sector assumed on the basis of these four scenarios is shown in Table 8. Future CO_2 emissions in Japan were estimated using AIM/Enduse [Japan] for the following reference and countermeasure cases.

a. Reference case (fixed technology case)

   CO_2 emissions until 2020 were estimated on the assumption that the usage shares of energy technologies will not change.

b. Countermeasure case (carbon tax case)

   CO_2 emissions were estimated assuming the adoption of energy saving and renewable technologies with a marginal cost limited to ¥30,000, which will be realized by charging a ¥30,000 carbon tax per ton of carbon used for secondary energy.

Table 8. Energy service demand assumed in AIM/Enduse [Japan]

		2010 A1	2010 A2	2010 B1	2010 B2	2020 A1	2020 A2	2020 B1	2020 B2
General	Economic growth rate (%/year)	2.1	0.9	1.6	0.5	1.9	0.6	1.4	0.6
	Population (million)	127	126.3	128.4	127.6	123.6	121.4	126.4	124.1
Industrial	Crude steel (10,000 t)	8,887	9,860	9,120	9,119	7,876	9,584	8,265	8,268
	Cement (10,000 t)	8,398	8,194	8,264	7,727	8,667	8,245	8,422	7,161
	Ethylene (10,000 t)	668	701	640	645	616	687	573	598
	Paper and paperboards (10,000 t)	3,052	3,311	3,010	3,004	2,891	3,366	2,836	2,848
	Share of tertiary industry (%)	65.9	64.1	65.4	64.8	68	64.3	67.1	66.4
Residential	Households (million)	49.6	49	49.4	48.8	50	48.4	49.7	48.2
	Heating service per household (2000 = 100)	100	92	97	89	101	86	88	77
	Cooling service per household (2000 = 100)	145	137	145	133	145	145	145	143
	Information appliances per household (2000 = 100)	128	116	122	116	154	127	139	127
	Fuel cell cogeneration (million kW)	0	0	1	0	0.5	0.5	5	2.5
Commercial	Floor space (million m^2)	1,804	1,710	1,796	1,702	1,957	1,749	1,922	1,766
	Fuel cell cogeneration (million kW)	0	0	1	0	0.5	0.5	5	2.5
Transportation	Passenger transportation (mil. pass.-km)	1,387	1,343	1,367	1,323	1,402	1,325	1,377	1,291
	Fuel cell vehicles (%)	0	0	0	0	0	0	20	10
	Freight transportation (million t-km)	593	544	559	506	660	552	584	478
Power generation	Nuclear power plants (MW)	53,248	57,546	49,580	44,917	53,819	63,819	49,521	41,190

Results of simulations

CO_2 emissions of simulation cases are shown in Table 9 and Fig.4. CO_2 emissions will be largest, increasing by 33% compared to 1990, in the A1-reference case that emphasizes economic efficiency in a highly competitive market society. When we look at changes in GDP per capita as shown in Fig. 5, although the economic scale is larger than that of the other scenarios and energy demand is expected to increase, energy intensity drops dramatically in accordance with an industrial shift toward services. Furthermore, this scenario offers the worst situation in terms of carbon intensity, due to the maintenance of oil/coal-based energy systems with a worldwide stabilization in the energy status. In transportation terms, the development and adoption of fuel cell-powered vehicles does not advance due to relatively low gasoline prices. Coal and oil would also likely be chosen by electric utilities under free market conditions as low-cost energy sources. Moreover, in the A1-countermeasure case, CO_2 emissions increase by 2% in 2010 and by 5% in 2020, compared to 1990.

Energy intensity is largest in the A2 scenario. An industrial shift toward services cannot advance much because factors such as the comparative advantages of Japanese manufacturers remain the same under these circumstances in which a global economic market does not penetrate to a high degree, or stable demand for civil engineering and construction work led by old-style public investment maintains the current level of domestic steel and cement production, all of which result in a high energy intensity. Conversely, carbon intensity shows a certain drop. This

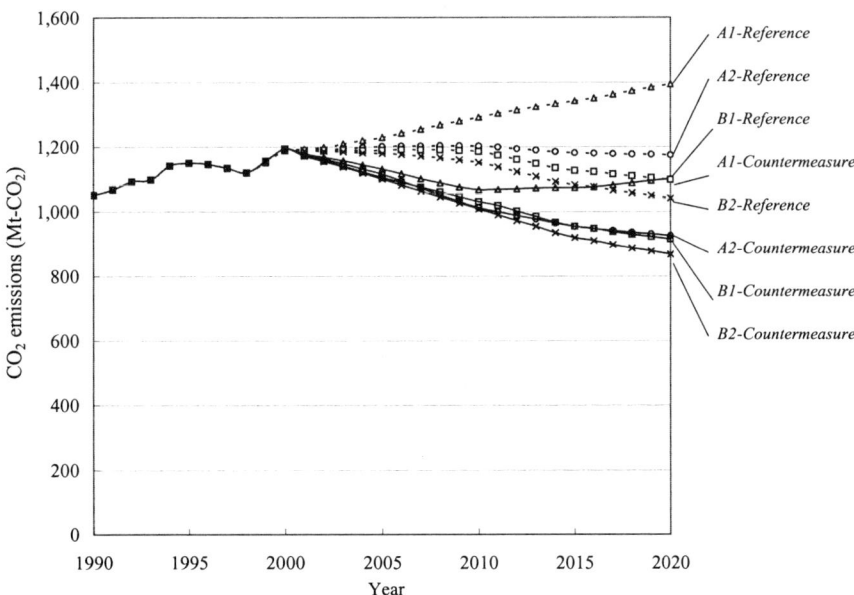

Fig.4. CO_2 emission profiles of simulation cases

is because the number of nuclear power plants continues to increase under circumstances that require energy security due to global instability in terms of energy supplies. In this scenario, CO_2 emissions would dramatically decrease from 1990 levels, showing drops of 4% by 2010 and 12% by 2020.

In the B1 scenario, relative compatibility between economic growth and dematerialization will be achieved by development and distribution of environmental technologies. Technologies to reduce the amounts of waste for disposal (i.e., eco-cement, biodegradable plastic) will have been developed and adopted for the industrial sector, as well as technologies for comfortable lifestyles (such as insulation systems) for the residential sector, countermeasure technologies against air pollution and noise (i.e., fuel cell vehicles and next-generation public transportation) for the transportation sector, and natural gas-based systems as countermeasures against air pollution (natural gas thermal power generation, fuel cell cogeneration) in the energy field. As a result, CO_2 emissions will be reduced. Both energy intensity and carbon intensity will be smaller, and CO_2 emissions will be significantly small even though the scale of economy is comparatively larger. If countermeasures to reduce CO_2 emissions are implemented, a significant reduction from the 1990 level will be possible as in the A2 scenario, with a reduction of 2% in 2010 compared to 1990, increasing to 13% by 2020.

The B2 scenario emphasizes local issues and equitability. From the viewpoint of revitalization and local equitability, job-sharing enables both a reduction in working hours and guaranteed employment at the same time. Direct decision-making by residents will be adopted for local issues, and decisions that might result in pollution or destruction of the local environment or even prove fatal will be replaced. In concrete terms, energy diversification rather than the new development of nuclear power plants, the purchase of recycled or long-lasting goods rather than building waste incinerators, and the use of fuel cell vehicles rather than gasoline-powered vehicles that cause air pollution will be chosen. The scale of economy will be relatively small and the speed of development and adoption of environmental technologies will be slower than in the B1 scenario, which stimulates the environmental industry under a market economy. However, CO_2

Table 9. CO_2 emissions of simulation cases

	1990	1998		2010		2020	
A1 - Reference	1,051	1,120	(107)	1,292	(123)	1,394	(133)
A2 - Reference	1,051	1,120	(107)	1,202	(114)	1,174	(112)
B1 - Reference	1,051	1,120	(107)	1,186	(113)	1,099	(105)
B2 - Reference	1,051	1,120	(107)	1,152	(110)	1,041	(99)
A1 - Countermeasure	1,051	1,120	(107)	1,068	(102)	1,102	(105)
A2 - Countermeasure	1,051	1,120	(107)	1,012	(96)	926	(88)
B1 - Countermeasure	1,051	1,120	(107)	1,030	(98)	914	(87)
B2 - Countermeasure	1,051	1,120	(107)	1,008	(96)	867	(83)

Unit: Mt-CO_2, (1990 = 100)

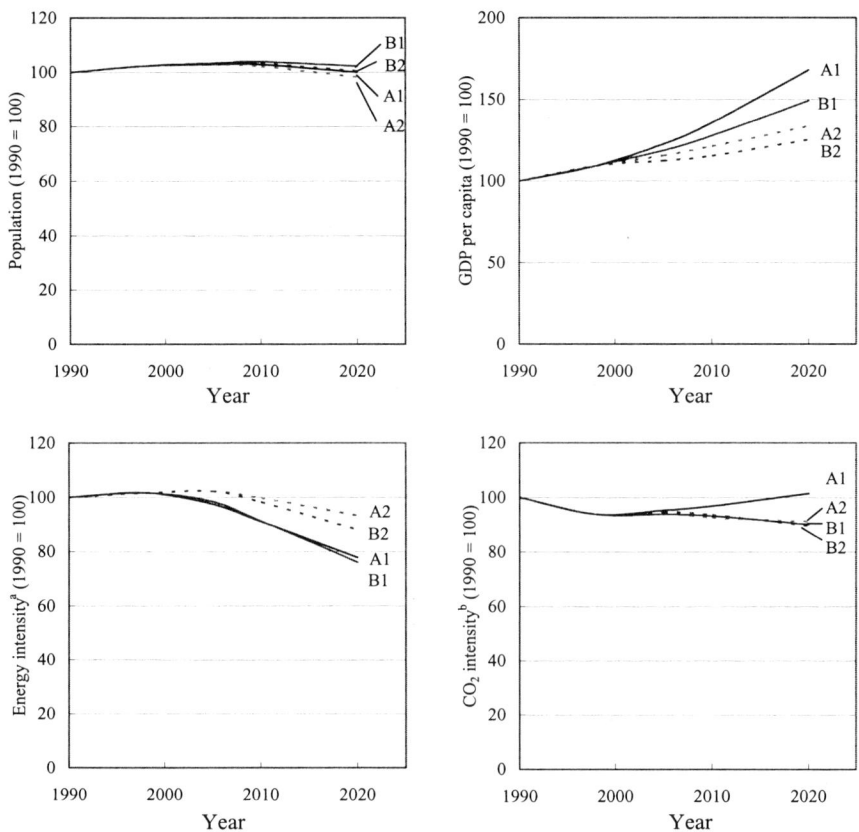

Note: a: Energy intensity = Energy consumption / GDP, b: CO_2 intensity = CO_2 emission / Energy consumption

Fig. 5. Population, GDP per capita, energy intensity, and CO_2 intensity in reference cases

emissions will still be at the same level as in B1, in the event that society and the economy are headed toward the end indicated in the B2 scenario. Over the long term, CO_2 emissions are the smallest among the four scenarios provided that greenhouse effect countermeasures are adopted, resulting in decreases of 4% in 2010 and 17% in 2020 compared to 1990. CO_2 emissions in Japan are smallest under the B2 scenario; however, the same is true under the B1 scenario in SRES by IPCC. This is because there is no significant difference in population increase between the Japanese scenarios, and the difference in economic growth between B1 and B2 directly affects the emissions, in contrast to B2 of SRES, in which the population increase causes larger emissions.

10.4.4 Discussion

As described above, energy consumption and CO_2 emissions in Japan up to 2020 were estimated using AIM/Enduse [Japan] on the assumption of four narrative scenarios of Japan's society and economy. When we compare the CO_2 emissions with the proviso that measures to reduce CO_2 emissions are adopted, possibilities for reductions to below the level of 1990 are demonstrated in three of the four cases, with the A1 scenario being the exception. Each case can be characterized as follows. The A2 scenario reduces carbon intensity by increasing nuclear power output. The development and adoption of environmental technology in the B1 scenario leads not only to economic growth but also to a reduction in greenhouse gases. In the B2 scenario, a social system is formed on a regional basis, with bottom-up decision-making that can be followed by reduction of greenhouse gas emissions. The means of achieving the reduction differ greatly from one scenario to the other. Furthermore, the emissions in 2010 when countermeasures are implemented will be almost the same for A2, B1, and B2, but will differ in 2020 (it should be noted that the scale of economy in B1 and A2 are significantly different, although their estimated emissions in 2020 are similar). In other words, when we examine long-term (i.e., 20 to 30 years) countermeasures against global warming, there must be sufficient discussion to understand which development pattern will be adopted and whether or not the countermeasures that are being planned or implemented are appropriate to that pattern, rather than merely discussing global warming countermeasures by themselves.

10.5 Concluding Remarks

Various decision-makers have used the AIM/Enduse model. Due to wide use of the modeling approach for dealing with the global warming problem, it is expected that the application of AIM/Enduse will spread to various involved parties in Japan. The following are some expected applications of AIM/Enduse in the near future.

- Governments will develop an integrated management system for the atmospheric environment with linkages between emission inventories and AIM/Enduse so as to address both air pollution and global warming simultaneously.
- Governments will analyze the effects of various policy combinations, such as voluntary plans, energy efficiency regulation, emission regulation, emission trading, energy tax, and emission tax.
- Private companies will analyze the impact of their own products on global warming using AIM/Enduse linked with their environmental accounting system.
- Private companies will develop strategies for emission trading using AIM/Enduse.

- Citizens will analyze their activities using a system linking environmental household account books and AIM/Enduse.

In order to meet these requirements, we need to develop additional modules for AIM/Enduse such as a user-friendly interface, expansion of target gases, linkages with material and top-down models, linkages with statistical tools and emission inventories, and so on.

References

Fuji Research Institute Corporation (2001) Report of the working group to draw up Japanese emissions scenarios. 150pp (in Japanese)

Hibino G (2001) Japanese narrative scenarios and CO_2 emissions. Prepared for Special Session of Society for Environmental Economic and Policy Studies. 9pp

JEA (1994) Second interim report of the study group on economic system to mitigate global warming. Japan Environment Agency, pp19-24 (in Japanese)

JEA (1996a) Third interim report of the study group on economic system to mitigate global warming. Japan Environment Agency, pp47-63 (in Japanese)

JEA (1996b) Report of the meeting on environment problems of global scale, special committee on global warming. Japan Environment Agency, pp47-63 (in Japanese)

JEA (1997) Final report of the study group on economic incentives by environmental tax and surcharge. Japan Environment Agency, pp31,41-42 (in Japanese)

JEA (2000) Report of the study group on economic instruments in environmental policies. Japan Environment Agency, pp70-74, 95-96 (in Japanese)

Kainuma M, Matsuoka Y, Morita T, Hibino G (1999) Development of an end-use model for analyzing policy options to reduce greenhouse gas emissions. IEEE Trans. Man and Cybern. Part C, 29(3):317-324

Kainuma M, Matsuoka Y, Morita T (2000) The AIM/end-use model and its application to forecast Japanese carbon dioxide emissions. European Journal of Operational Research, 122:416-425

Kainuma M, Matsuoka Y, Morita T (2001) CO_2 emission forecast in Japan by AIM/end-use model. OPSEARCH, 38(1):109-125

Kouken Association (1999) Ancillary-effects estimation model for local governments to improve their comprehensive environment. 207pp (in Japanese)

MOE (2001) Interim report of the subcommittee on scenarios to achieve targets, global environment section. Central Environment Council, Ministry of the Environment, pp188-191 (in Japanese)

Shimada K, Mizoguchi S, Hibino G, Matsuoka Y (2000) A study on the effects on local air quality of greenhouse gas mitigation measures. Environmental Systems Research 28: 77-84 (in Japanese)

Tsuchiya H (1997) Key technology policies to reduce CO_2 emissions in Japan. WWF Japan, pp7-32

Tsuchiya H (2001) WWF scenario for solving the global warming problem index for 2010 and 2020. WWF Japan, pp29-39

Yang H, Kainuma M, Matsuoka Y (2001) Modeling the clean development mechanism: direct benefits, co-benefits and priorities. IFAC workshop on modeling and control in environmental issues, Yokohama, 103-108

11. AIM/Material Model

Toshihiko Masui[1], Ashish Rana[1], and Yuzuru Matsuoka[2]

Summary. The AIM/Material model has been developed for assessing policy impacts not only on CO_2 emissions reduction but also other environmental issues, at present mainly focusing on solid waste management. Some activities for CO_2 emissions reduction are related to solid waste management. By using this model, both global climate policies and the domestic environmental policies can be assessed simultaneously. Other features of this model are as follows; not only an economic balance but also a materials balance is reproduced, and environmental industry and environmental investment to reduce environmental burdens are introduced. This model has been applied to Japan and India. In the case of Japan, the reduction of CO_2 emissions and final disposal of solid wastes in 2010 based on the Kyoto Protocol and government targets, respectively, will lead to a 0.5% loss in the GDP compared with the reference scenario. On the other hand, the introduction of environmental investment and other policies such as technology improvements and green consumption will mitigate more than half of the GDP loss. As for India, since toxic waste is a much more serious environmental problem than final disposal, this model is applied to the reduction of toxic waste.

11.1 Introduction

The AIM/Material model is a component of the AIM family and is used mainly for assessing the domestic macro economic impacts of policies on not only CO_2 emissions reduction, but also other environmental issues. AIM/Material has been developed in order to support solutions to the various environmental problems that each country may have. In the case of Japan, greenhouse gas emissions reduction is one of the most important global environmental issues, and moreover, the waste management problem is one of the most serious domestic environmental issues due to the scarcity of final disposal sites for solid wastes. In order to solve this solid waste problem, the "Basic Law for Establishing a Recycling-based Society" was enacted in May 2000, and the "Basic Plan for Establishing a Recycling-based Society" will be established in 2003 and will be reviewed every five years. In developing countries, not only climate change, but also domestic pollution problems such as water and air pollution problems are important for policy makers. Besides environmental policies, the environmental industry and environmental investment hold the key to realizing both economic development and environmental conservation. At present, this AIM/Material model can deal with both CO_2 emissions reduction and solid waste generation and treatment

---

[1] National Institute for Environmental Studies, Tsukuba 305-8506, Japan
[2] Kyoto University, Kyoto 606-8501, Japan

simultaneously. In this chapter, the structure of the model and its application to Japan and India are introduced.

11.2 Structure of AIM/Material Model

11.2.1 Overview

The AIM/Material model is a top-down type macroeconomic model based on the computable general equilibrium model for each country. This model has recursive dynamics year by year. One of the features of this model is that it represents a consistent approach not only to the economic balance, but also the materials balance. Due to this feature, solid waste generation and treatment can be dealt with. This section explains the structure of the AIM/Material model based on its application to Japan.

In this model, three economic agents are taken up: the production sector, the household sector, and the government. AIM/Material applied to Japan has 41 economic sectors and 49 commodities. Table 1 shows the economic sectors and commodities in the Japan model. Figure 1 represents the overall model structure. The production sectors produce economic goods through inputs of capital, labor, energy, other intermediate inputs, and pollutants. Pollutants as an input factor means the inputs necessary for the treatment of the generated pollutants, such as the cost of introducing environmental control and monitoring equipment and its operation, since it is assumed that the discharge of pollutants should not exceed the related environmental standards. That is to say, below the environmental standard, the generated pollution can be discharged into the environment, but any excess beyond the standard should be treated in appropriate ways. In the production sectors, the solid waste management sectors are different from normal production sectors. Solid wastes are mainly disaggregated into two types: industrial waste and municipal waste. Industrial waste is defined as waste from production processes, and municipal waste is from the household and business sectors. Each type of waste is disaggregated into more detailed waste categories. Table 2 shows the waste categories in this model. Each waste category of both industrial waste and municipal waste is disaggregated into 3 subsectors; direct reuse, direct final disposal, and intermediate management, such as incineration. Each sub-sector manages the solid wastes by inputting capital, labor, energy, and other intermediate inputs. The total quantity of the direct reused wastes and the reusable residual materials after intermediate management are supplied to the market as recycled materials. The total quantity of direct disposal wastes and the residue from intermediate management are dumped. In order to maintain the consistency of the materials balance, the share of recycled materials input in each sector is fixed in advance based on the capital, as shown in Table 2. The environmental industry is disaggregated from the other production sectors. In this model, the environmental industry is defined as the sector producing the environmental equipment to manage pollutants. The output of the environmental industry is supplied as an environmental investment in the other sectors.

The household sector retains labor and capital, and supplies it to the production

Table 1. Sectors and commodities in AIM/Material [Japan]

	Sector		Commodity
agr	Agriculture, forestry and fisheries		
min	Mining except energy		
m_c	Coal mining	mcc	Coking coal
		msc	Coal for general use, lignite, anthracite
m_o	Crude oil mining		
m_g	Natural gas mining		
fod	Manufacture of food		
tex	Manufacture of textile mill products		
plp	Manufacture of lumber, wood products, pulp, paper and paper products		
chm	Manufacture of chemical and allied products		
pls	Manufacture of plastic		
nmm	Manufacture of ceramic, stone, and clay products		
stl	Manufacture of iron, steel, ferrous metals and products		
nsm	Manufacture of non-ferrous metals and products		
fmt	Manufacture of fabricated metal products		
mch	Manufacture of general machinery		
elm	Manufacture of electrical machinery, equipment and supplies		
tre	Manufacture of transportation equipment		
pri	Manufacture of precision instruments and machinery		
oth	Miscellaneous manufacturing industries		
cns	Construction		
het	Steam and hot water supply		
wtr	Water supply		
sal	Wholesale and retail trade		
fin	Finance and insurance		
est	Real estate		
trs	Transportation and communications		
pub	Education, research, medical service, health & hygiene, and social welfare		
rnt	Goods renting and leasing		
rep	Car and machine repairing		
prs	Other service		
gov	Government service		
emc	Environmental industry		
sew	Sewage service		
mwm	Municipal solid waste treatment service		
iwm	Industrial solid waste treatment service		
col	Manufacture of coal products	cck	Coke
		ccg	Other coal products
		cbf	Paving materials
oil	Manufacture of petroleum	ogl	Gasoline
		ojf	Jet fuel oil
		okr	Kerosene
		olo	Light oil
		oho	Heavy oil
		onp	Naphtha
		olp	LPG
		oot	Other petroleum refinery products
gas	Manufacture of gas	gtg	Town gas
the	Thermal power generation		
hyd	Hydro power generation	ele	Electricity
nuc	Nuclear power generation		

Note: In the model, one sector can produce plural commodities using V matrix (make matrix).

sectors. As a result, households receive income as wages and rent from the production sectors. Under the constraints of income, households demand commodities based on the demand function. In the demand function, savings are also included as demand for investment goods. The demand for investment goods is determined by the fixed capital matrix.

The government sector imposes taxes on the production sectors and household sector. In this model, the tax is aggregated into 4 types: capital tax, labor tax,

Table 2. Classification of solid waste and assumption of recycling flow

		Commodity substituting to recycled waste									
		MIN	FOD	TEX	PLP	CHM	NMM	STL	NSM	CCK	OOT
ASH	Ash	○	○								
SLD	Sludge	○	○							○	
WOL	Slush, waste oil	○									○
WAC	Waste acid					○					
WAL	Waste alkali					○					
WPL*	Waste plastics					○					○
WPP*	Waste paper				○						
WWD*	Waste wood			○						○	
WTX*	Waste fiber and textile			○							
WAP*	Animal and plants wastes		○								
WRB*	Waste rubber					○					
SCM*	Metal trash, scrap metal							○	○		
WGC*	Waste glass						○				
SLG	Slag	○									
WCT	Construction and demolition waste	○									
DST	Dust, soot	○									
EXC	Animal excrement		○								
CRC	Animal carcass		○								
AGR			●								
TEX				●							
PLP					●						
CHM						●					
PLS						●					
NMM							●				●
STL		●							●		
NSM									●		
OTH						●					
CNS		●									
COL										●	
OIL											●

(Left labels: "Waste" applies to the upper rows; "Sector utilizing waste material" applies to the lower rows AGR–OIL.)

Note: Refer Table 1 for names of sector and commodity.
* in waste category shows the classification of both municipal and industrial waste. Others are for only industrial waste.
○ means the waste indicated by row can be used as the commodity indicated by column.
● shows that the reused material indicated by column is input to the sector indicated by row.

production tax, and import tax. In the case of a reduction in CO_2 emissions and the final disposal of solid waste, the related taxes, such as a carbon tax and solid waste tax, are imposed on the agents. Based on tax revenues, the government sector provides public services and public investment.

Although this model is a country model, international trade is very significant. Especially in Japan, the importation of fossil fuels is essential to its economic activities. In this model, the international price is fixed in advance based on the small country assumption. The international prices of all commodities except for energy are fixed at the initial year. The international energy price is given according to the scenarios based on the AIM/Emissions model. Moreover the total value of imports and exports in the future is given according to the scenario, and individual imports and exports are calculated endogenously.

11.2.2 Mathematical formulation of AIM/Material

For simplicity, the model of the following three sectors as shown in Table 3 is assumed. Sectors 1 and 2 are ordinary economic sectors and sector 3 is a waste management sector. Commodities 1 and 2 are ordinary commodities and commodity 3 is a waste management service.

Market for produced commodities and production factors

The market equilibrium for produced commodities can be represented by the following equations.

Table 3. Social accounting matrix in the hypothesized economy

			Production sector			Final Consumption	Investment			Supply	Price	Endowment
			Sector 1	Sector 2	Sector 3		Sector 1	Sector 2	Sector 3			
Input	Intermediate inputs	Commodity 1	X_{11}	X_{12}	X_{13}	C_1	I_{11}	I_{12}	I_{13}	$Y_{11}+Y_{21}+Y_{31}$	P_1	
		Commodity 2	X_{21}	X_{22}	X_{23}	C_2	I_{21}	I_{22}	I_{23}	$Y_{12}+Y_{22}+Y_{32}$	P_2	
		Commodity 3	X_{31}	X_{32}	X_{33}	C_3				Y_{33}	P_3	
		Capital	K_1	K_2	K_3						P_K	K^*
		Labor	L_1	L_2	L_3						P_L	L^*
		Final disposal			W_3						P_W	W^*
Output		Commodity 1	Y_{11}	Y_{21}	Y_{31}						P_1	
		Commodity 2	Y_{12}	Y_{22}	Y_{32}						P_2	
		Commodity 3			Y_{33}						P_3	

$$P_i \left\{ \sum_{j=1}^{3} Y_{ji} - (\sum_{j=1}^{3} X_{ij} + C_i + \sum_{j=1}^{3} I_{ij}) \right\} = 0, \quad P_i \geq 0, \text{ and}$$

$$\sum_{j=1}^{3} Y_{ji} - (\sum_{j=1}^{3} X_{ij} + C_i + \sum_{j=1}^{3} I_{ij}) \geq 0$$

These equations indicate the relationship between the commodity price P_i and the demand and supply of that commodity. The excess demand for each commodity i does not exist. In the case of a market equilibrium, that commodity has a positive price, P_i. On the other hand, in the case of a disequilibrium state, that is to say, in the case of excess supply, the price of this commodity i becomes 0. In the example of Table 3, there is no demand for the waste management service produced by sector 3 as investment goods. In addition, Y_{31} and Y_{32}, which are commodity 1 and 2 produced by sector 3, can be regarded as recycled goods.

The capital market and labor market also can be explained by the following equations.

$$P_K \left\{ K^* - \sum_{j=1}^{3} K_j \right\} = 0, \quad P_K \geq 0, \text{ and } K^* - \sum_{j=1}^{3} K_j \geq 0$$

$$P_L \left\{ L^* - \sum_{j=1}^{3} L_j \right\} = 0, \quad P_L \geq 0, \text{ and } L^* - \sum_{j=1}^{3} L_j \geq 0$$

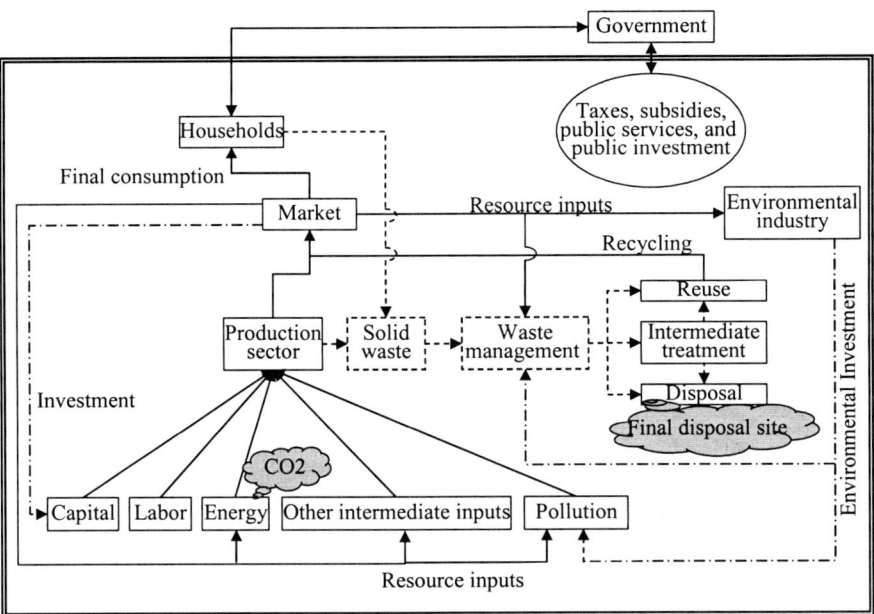

Note:
- - - ▶ shows the flow of solid waste. In the model, municipal waste management sector and industrial waste management sector are set by the waste category. As for the waste category, see Table 2.
— · ▶ shows the flow of the capital investment and environmental investment. Although the capital is endowed in household, for simplicity, the investment is input from market or environmental industry directly in this figure.

Fig. 1. Overview of AIM/Material

In the capital market and the labor market, total demand for these is less than the endowment of capital and labor, respectively. In a state of equilibrium for these production factors, these prices become positive values, but in a state of disequilibrium, that is to say, the supply of these factors is more than the demand, the prices become 0.

As for the quantity of disposable waste, W, sector 3 disposes of the residual materials generated through waste treatment. The maximum quantity of this material for final disposal is W^*. Sector 3 cannot dispose of more than W^* of the residual materials. When the quantity for disposal is equal to W^*, the marginal cost for dumping, P_W, becomes a positive value. The situation in which the quantity for final disposal is less than W^* is regarded as an excess supply, and the price of final disposal is 0. That is to say, the quantity of materials for final waste disposal can be expressed by the following equation in the same way as for capital and labor.

$$P_W \left\{ W^* - \sum\nolimits_{j=1}^{3} W_j \right\} = 0, \ P_W \geq 0, \text{ and } W^* - \sum\nolimits_{j=1}^{3} W_j \geq 0$$

In the example of Table 3, the values of W_1 and W_2 are 0, since only sector 3 generates materials for final waste disposal. If the self management of solid wastes and the dumping of the residual materials from this self management are considered, the values of W_1 and W_2 are effective.

Production sectors

The revenues and expenditures in each production sector can be expressed by the following equation.

$$\sum\nolimits_{i=1}^{3} P_i X_{ij} + P_K K_j + P_L L_j + P_W W_j = \sum\nolimits_{i=1}^{3} P_i Y_{ij}$$

The left hand side of this equation shows the expenditure for buying intermediate goods and other production factors in order to produce commodities. The right hand side represents the total revenues from selling the produced commodities in each sector. This equation also shows that the revenues and expenditures are balanced in the sector j. In order to represent the relationship between inputs and outputs, the appropriate production function is defined. In this model, in order to maintain a materials balance besides a monetary balance, the values of the elasticity of substitution or distribution are assumed to be 0 or infinity, except for the aggregation of different type of materials, such as capital and labor.

Household sector

The household retains resources such as capital, K, labor, L, and the final disposal area for solid waste, W. The household sector supplies these resources to the production sectors, and receives the income as the counter values of these resources. The household consumes the goods to maximize its utility, subject to income constraints. The income of the household can be expressed as follows by using the rent for capital, P_K, the wages, P_L, and the marginal cost of the final disposal of solid waste, P_W.

$$H = P_K \sum\nolimits_{j=1}^{3} K_j + P_L \sum\nolimits_{j=1}^{3} L_j + P_W \sum\nolimits_{j=1}^{3} W_j$$

On the other hand, in order to calculate the quantity of demand for each commodity, the demand function has to be specified. Here, savings are included in the demand function as the demand for investment goods I_{ij} besides the final consumption C_i. The next equations show the expenditure of the household sector, and the change in capital stock. The total of new investment is added to the existing stock depleted at δ_j. The parameters for technological change, such as energy efficiency improvements, are calculated from the new technology assumption and the new equipment as a proportion of the total capital.

$$H = \sum\nolimits_{i=1}^{3} P_i (C_i + \sum\nolimits_{j=1}^{3} I_{ij})$$

$$K_{j,t+1} = (1 - \delta_j) K_{j,t} + \sum\nolimits_{i=1}^{3} I_{ij}$$

Flow of solid waste

In Table 3, the solid wastes can be managed only by the waste management sector. However, in reality, the quantity of solid wastes managed by each sector is relatively large. As a result, this model includes not only a waste management sector, but also the self management of solid wastes in each sector. The generated solid wastes in each sector are distributed between the self management sub-sector in the sector itself or the waste management sector. After the management of solid wastes, some of the solid wastes are recycled to the market, some of them are reduced through the incineration process, and the rest are dumped at the final disposal site. The endowment W^* represents the maximum quantity of the final disposal of solid wastes. W^* can be defined with respect to each waste type. Through the definition of this solid waste treatment activity, the appropriate solid waste treatment strategies can be estimated, such as the recycling policy and introduction of innovative technologies in solid waste management.

CO$_2$ emissions

The CO_2 emissions from each sector are calculated from the quantities of combusted fossil fuels and their emission factors. If the CO_2 emissions exceed an upper limit, such as the Kyoto Target, a carbon tax is imposed on the combusted fossil fuels, as with the P_W in Table 3.

Government

The government sector is also one of the final demand sectors, in the same way as the household sector. However, the roles of the government sector are quite different from those of the household sector. The roles of the government in this model are defined as follows:

- The government sector collects taxes, supplies public services and makes public investment.

- When an environmental burden is beyond the standard, the government sector imposes an environmental tax. The rate for this environmental tax is equivalent to P_w in Table 3.

11.3 Application of AIM/Material to Japan

11.3.1 Economic impact of the management of CO_2 and solid wastes

Currently Japanese society has many environmental problems that require immediate attention. For example, greenhouse gas emissions reduction, waste management, air pollution, and so on. While considering these environmental problems, economic growth should not be ignored. It is generally thought that environmental conservation and economic development are in conflict. However, in order to achieve a sustainable society, these two problems need to be solved simultaneously. AIM/Material indicates one solution.

In order to apply the AIM/Material model to Japan, a social accounting matrix is constructed based on input-output tables and other related databases for 1995. Table 4 represents the database used for this application. The simulation period is from 1995 to 2010. In order to estimate the economic impacts from the environmental constraints and mitigation due to environmental policies, the following scenarios were proposed.

1) Reference scenario
2) Environmental constraints scenario
3) Environmental constraints and countermeasures scenario

In scenario 2) and scenario 3), two environmental constraints were considered. One is the CO_2 emissions reduction target based on the Kyoto Protocol. The other

Table 4. Dataset for AIM/Material [Japan]

Economy	Management and Coordination Agency ed. (1999)
Waste	Ministry of Health and Welfare (1998a)
	Ministry of Health and Welfare (1998b)
Investment	Management and Coordination Agency ed. (1999)
	Department of National Accounts, Economic Research Institute, Economic Planning Agency (1999)
Environmental industry	The Japan Society of Industrial Machinery Manufacturers
Tax	Management and Coordination Agency ed. (1999)
	Ministry of Finance
	Research institute of international trade and industry (2000)
Energy	Ministry of international Trade and Industry (1996, 1997)
	Agency of Natural Resources and Energy, Ministry of International Trade and Industry (1999)

Note: The names of ministries are the names when these literatures were published.

is the reduction in the final disposal of solid waste based on the government target that the quantity of materials for final disposal as waste in 2010 should be half of that in 1996. The countermeasures in scenario 3) represent the increase in environmental investment, the introduction of efficient equipment for solid waste treatment, subsidies for power generation from waste, and the shift to green consumption, that is to say replacement of ordinary commodities by eco-products. Until 2000, in order to reproduce the actual values, the parameters for technological improvements, propensity to save and so on are adjusted. Moreover, in the reference scenario, these parameters are calibrated in order to follow the future economic development path represented in the existing estimations. Figure 2 represents the GDP trajectory in scenario 1).

Figure 3 shows the change in the GDP in scenario 2) and scenario 3) based on scenario 1). From a comparison between scenario 1) and scenario 2), the

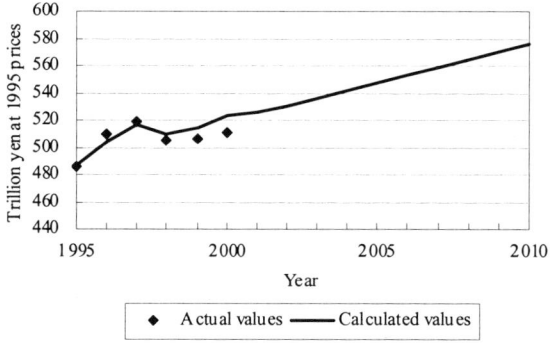

Fig. 2. GDP trajectory of scenario 1)

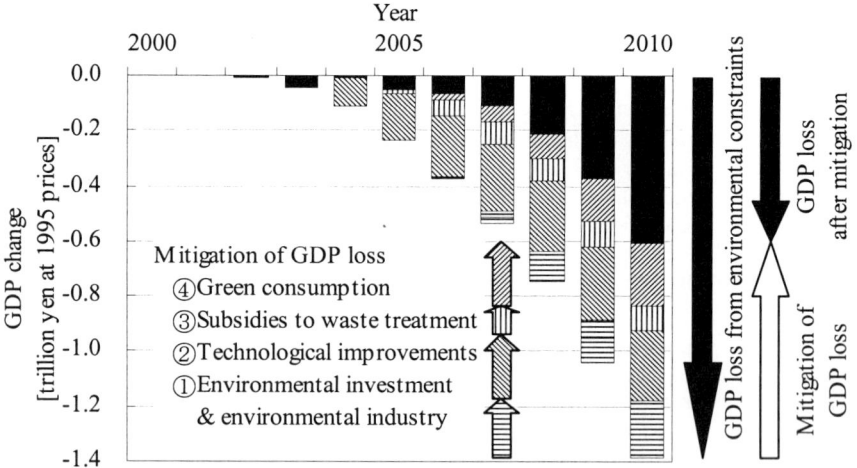

Fig. 3. GDP loss from environmental constraints and its mitigation

environmental constraints will have a negative impact on the economy. In 2010, the GDP loss will be 0.5 % of the GDP in scenario 1). On the other hand, introducing various countermeasures as shown in scenario 3) will make it possible to recover more than half of the economic losses suffered from the environmental constraints.

The reasons for the mitigation of the GDP losses are considered to be as follows. Firstly, the mitigation of environmental constraints will be achieved by these countermeasures. The second reason is the increase in environmental investment and green consumption will activate environmental industries and the production sectors that produce the eco-products. Moreover, the increase in environmental investment and eco-products has the potential to reduce the environmental burden. These advantages affect the mitigation of the GDP loss and economic activities.

11.3.2 Linkage of AIM/Material to the bottom-up model

AIM/Material is a top-down model. In order to solve the AIM/Material model, assumptions related to the efficiency of technology and social change are required. The solution of AIM/Material can reproduce consistent interaction among sectors. In order to assess the feasibility of the introduction of technology, the support of a bottom-up model, such as the AIM/Emissions model, is inevitable.

In order to meet the needs, a technology model on sewage sludge treatment has been constructed and linked to the AIM/Material model as a trial. Since the structure is too complex and the numbers of the variables are too large when the technology model is included in the AIM/Material, AIM/Material and the technology model are linked serially year by year. The concept for this linkage is shown in Fig. 4. In this linkage, the quantity of sewage sludge generation, the price of final disposal and materials recycling demand calculated in AIM/Material from one model are used for the assumptions in the technology model. The efficiency of sewage sludge treatment from the technology model is used for the

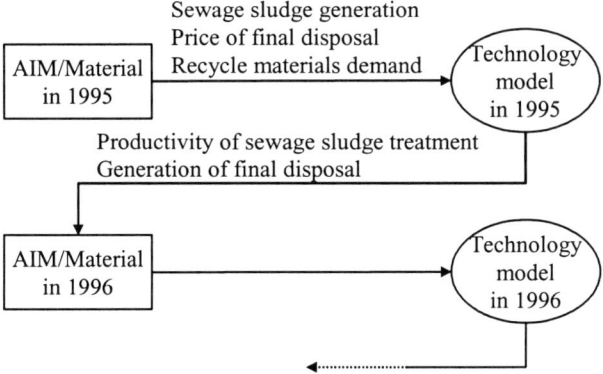

Fig. 4. Linkage of AIM/Material and technology model

assumption in AIM/Material. Figure 5 shows the changes in the technologies under different constraints in relation to the final disposal of solid wastes. In the case of the reference case without constraints on final disposal sites, only standard technologies will be introduced. On the other hand, the scenario with constraints on final disposal sites introduces new technology that is more efficient, although the cost is higher.

By linking these two types of models, a top-down model and a bottom-up model, more realistic countermeasures can be assessed. The AIM team is now trying to expand the technology model in order to analyze other fields.

11.3.3 Effects on CO_2 emissions from changes in the tax rates on fossil fuels

In Japan, taxes imposed on fossil fuels are very high at present. The current tax rates are tentative, and are planned to be reduced to almost half in 2003. Table 5 represents the values for the tax rates. These high tax rates have a role in reducing CO_2 emissions, just like a carbon tax. If the tax rates are now returned to the original (after 2003) values, the opposite effect of a carbon tax, that is to say, an increase in CO_2 emissions, may occur. By using AIM/Material and changing the taxation rates on fossil fuels, the changes in CO_2 emissions can be estimated.

Figure 6 shows the results for CO_2 emissions. CO_2 emissions in 2010 will increase by 6.5 MtC according to the change in the tax rates. These values are equivalent to 2.2 % of the emissions in 1990. That is to say, more than half of the amount of the carbon sink under the Bonn agreement, 3.9% of the emissions in 1990, are offset.

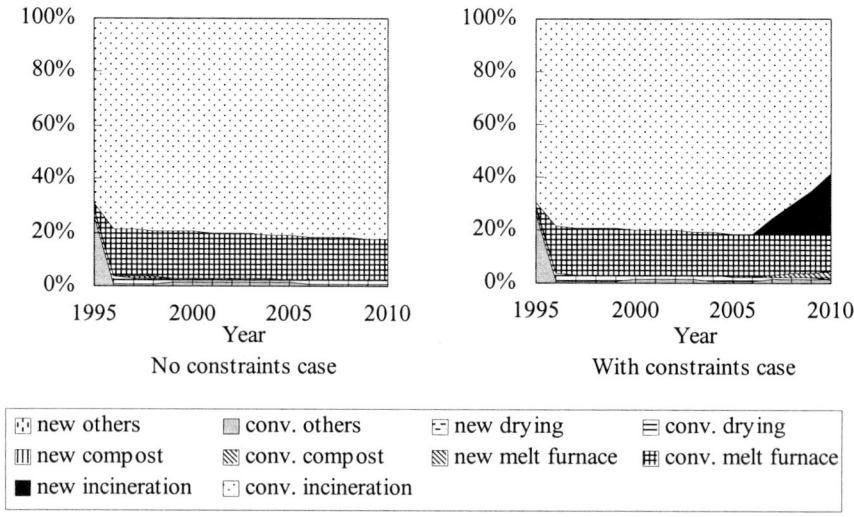

Note: "Conv." means "conventional."

Fig. 5. Changes of technology share in two scenarios

Table 5. Tax rate of gasoline and light oil in Japan

	Present (tentative) rate	Original (after 2003) rate
Gasoline	53.8yen/l	28.7yen/l
Light oil	32.1yen/l	15.0yen/l

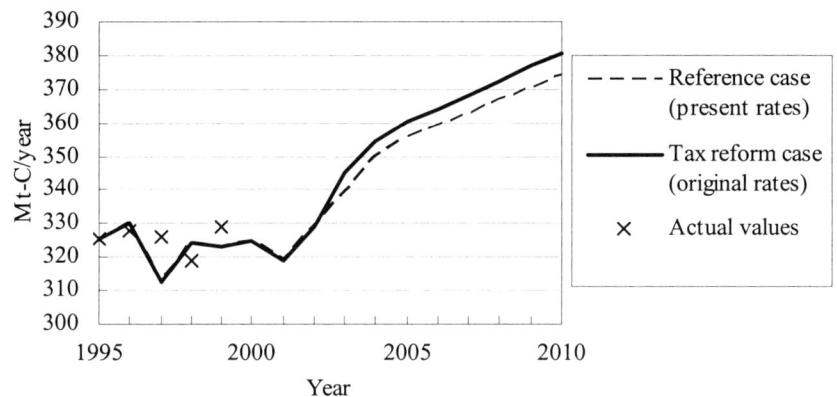

Fig. 6. CO_2 emissions change by tax reform

11.4 Application of AIM/Material to India

11.4.1 Modifications to the model

For the application of AIM/Material to India, some changes were incorporated taking into consideration the specific environmental problems of India as well as problems related to the availability of data that is as detailed as in the case of Japan. In India the problem of hazardous waste has a higher priority than dealing with other types of industrial solid waste. To take this into account, a significant modification in the India model is the separate treatment of hazardous, or toxic, waste and its disposal. A separate limit is set for the discharge of toxic waste into the environment. This limit is similar in character to the limit on the final disposal of waste, W^*, described in section 11.2.2.

Due to the limited data, the present version of AIM/Material [India] has 26 economic sectors and 24 commodities as shown in Table 6. The base year is taken as 1994 and the model is calculated year by year until 2010. Solid wastes are categorized into 15 types, including toxic waste, as shown in Table 7. Furthermore, solid wastes are classified as either industrial or municipal wastes. The industrial wastes are defined as wastes generated from production processes, and municipal wastes are defined as wastes from households and business activities.

11.4.2 Data sources

An input-output table, with data on the energy sector and data on solid wastes is required for the model. The input-output tables for 1993-94 available from the Central Statistical Organization of the Government of India were utilized (CSO 2000). Data on the energy sectors were taken from TEDDY 2000-2001 (TERI 2001). To estimate solid wastes from individual sectors, data on waste generation intensities for industries available from the documents of the Central Pollution Control Board of India (CPCB, various issues) and some other published sources was compiled. Typically, waste generation intensities are tabulated by waste type and production quantities in the base year and are applied to arrive at an estimation of the total waste generated in the sector. Usually the task is not straightforward as the waste generation intensities as well as waste types may differ within a single sector, depending on the technology in use. For example, in the case of the paper industry, large and medium-sized paper mills use wood as

Table 6. Classification of sectors and commodities in AIM/Material [India]

ID	Description	ID	Description
AGR	Agriculture, forestry, fishing	WTR	Water supply
MIN	Mining	SRV	Services
FOD	Food	MWM	Municipal waste management
TEX	Textiles	IWM	Industrial waste management
PLP	Paper and pulp	EMC	Environment industry
CHM	Chemicals	GOV	Government service
NMM	Non-metallic minerals	COL	Coal
BMT	Basic metals	OIL	Oil
FMT	Fabricated metals	GAS	Gas
MCH	Machinery	HYD[+]	Hydro power generation
ELM	Electrical machinery	THE[+]	Thermal power generation
TRE	Transport equipment	NUC[+]	Nuclear power generation
OTH	Other manufacturing	ELE*	Electricity
CNS	Construction		

[+] Only sector;   * Only commodity

Table 7. Classification of solid wastes in AIM/Material [India]

ID	Description	ID	Description
ASH	Ash	SCM[**]	Scrap metal
SLD	Sludge	WGC[**]	Waste glass
WOL	Waste oil	SLG	Slag
WPL[**]	Waste plastic	WCT	Construction waste
WPP[**]	Waste paper	DST	Dust
WWD	Waste wood	WZZ[**]	Other waste
WTX[**]	Waste textile	WWT	Toxic waste
WAP[**]	Animal and plant waste		

[**] Both industrial and municipal waste categories

well as agricultural residues for pulp making, while small industries use wastepaper as inputs, which results in a substantial difference in the quantity and quality of their waste generation. In the fertilizer industry, solid waste generation is 0.085 tons per ton of fertilizer production. Steel plants produce solid waste in the form of slag and dust. The amount of slag and dust produced is about 500 kg and 25 kg per ton of pig iron produced from blast furnaces. Pig iron is converted to steel through an open-hearth furnace or oxygen furnace, producing about 22 kg of dust per ton of steel. The manufacture of sulfuric acid generates solid waste in the form of sulfur sludge which amounts to 2.5 – 2.8 kg/ton of sulfuric acid. Construction activity generates solid wastes, which include sand, gravel, concrete, stone, bricks, wood, metal, glass, plastic, paper, etc. The management of construction and demolition waste is a major concern for town planners due to the increasing quantities of demolition rubble, a continuing shortage of dumping sites, increases in transportation and disposal costs and, above all, growing concern about pollution and environmental degradation. The Central Pollution Control Board has estimated that waste from the construction industry accounts for 25% of the total volume of solid waste generated in India. Such a high volume of waste puts enormous pressure on solid waste management systems. Municipal solid waste data is based on the Planning Commission report published in 1995.

Data on toxic waste is mainly drawn from the report of the High Powered Committee (HPC) established in 1997 by the Supreme Court of India to produce a comprehensive study related to hazardous wastes in the country. In its detailed final report in 2000, the committee clearly indicates the dubious nature of data that is reported by industry with respect to the hazardous wastes generated by them (High Powered Committee on the Management of Hazardous Wastes 2000). The data, made available at its behest and further refined by officials, records 4.4 million tonnes of hazardous waste being generated per year within the country. This is further classified under various toxic waste categories according to the Hazardous Waste Rules 1989 guidelines. For the present study, the categories of toxic sludge that contributed the largest proportion were considered, covering around sixty percent of the total volume of toxic waste. This was done in order to make it possible to easily attribute the generated waste to the relevant waste generating sectors. Dealing with all categories together poses a challenge of attribution, in the absence of other data.

11.4.3 Scenarios

Three scenarios were constructed for this application.

Scenario 1:   This is the reference or business-as-usual scenario and there are no interventions to limit waste disposal.

Reference case, Scenario 1, depicts an average annual GDP growth of 5.85% for the period 1994-2010 for India. Figure 3 shows the match between the actual statistics for the GDP and that simulated using the model in the overlapping periods. Total industrial waste generation is simulated to grow at 4.9% during the model period. Growth in the disposal of non-toxic waste is 3.5% per year while that for toxic waste disposal is 3%. The growth rate for the waste management service sectors (IWM and MWM) is 5% and the environment industry grows at 6.28%.

Scenario 2:   Toxic Constraints Scenario – In this scenario a limitation on the discharge of toxic wastes is imposed.

This scenario is not driven by a government target but represents the effect of a combination of many actions that are expected to be taken by state governments and pollution control boards. The aforementioned HPC on Hazardous Waste has made several recommendations regarding the siting of secure landfills, the first of which states that reliance on land-based disposal should be minimized or eliminated as land-based disposal is the least favored method of managing hazardous wastes. Open dumping contaminates drinking water from underground and surface supplies and this practice can lead to real hazards for human health and the environment. Corrective action is bound to be expensive, complex and time consuming. Hence, reliance on land-based disposal should be minimized or eliminated. Other actions include stopping indiscriminate disposal of toxic waste, shutting down illegal dumping sites, relocation of industries away from ecologically sensitive sites, strict adherence to environment impact assessment criteria for sites, and so on. Scenario 2 is constructed by limiting the amount of resources, or endowment, for toxic waste disposal. From the year 2002, the available endowment for toxic waste disposal is decreasing at a rate of 10 percent every year.

Scenario 3:   In this scenario countermeasures are introduced through increased environmental investment and labor efficiency improvements in the waste management sector.

Environmental investment is one way of boosting the capacity for the treatment

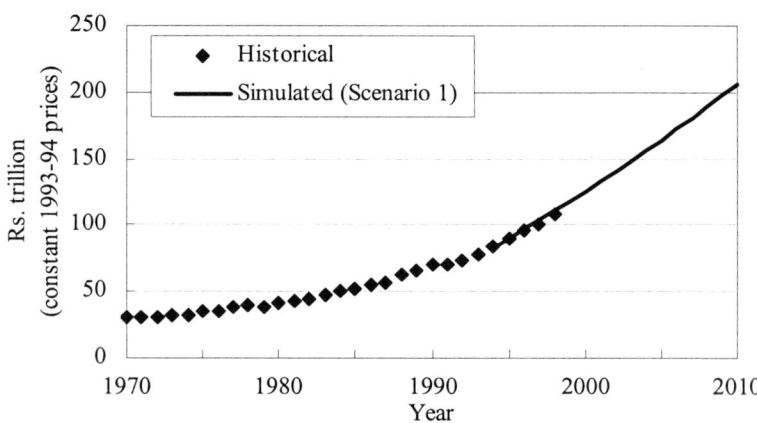

Fig. 7. Historical GDP and GDP in Scenario 1

and management of waste, thereby relieving the constraints posed in Scenario 2. At the same time, as mentioned earlier, it is clear that much of the mitigation can be achieved at a low marginal cost through improvements in waste management systems, improvements in organizational efficiency and other low-tech measures, such as greater segregation of wastes. A 5% increase in environment investment accompanied by labor efficiency improvements of 10% are taken into consideration in Scenario 3 from 2005.

11.4.4 Results

When faced with a constraint in Scenario 2, the following effects are observed in the economy. Toxic waste disposal is restricted to 1.29 Mt by 2010, a reduction of more than 60% over Scenario 1 for the same year, while disposal of other (non-toxic) waste decreases marginally over Scenario 1 (Fig. 8). Reduced toxic waste disposal is accompanied by restraint on the part of the toxic waste producing sectors as well as other sectors due to cross-linkages between them. By comparing the sectoral GDP in Scenario 2 in various years with that in Scenario 1, the overall GDP loss in 2010 can be observed to be around 2% over Scenario 1.

In Scenario 3, however, some GDP loss is recovered as the capacity to manage waste is enhanced (Fig. 9). The output for most of the sectors is also recovered in Scenario 3 (Fig. 10). Only 8 out of 26 sectors show a slight decrease in their GDP over Scenario 2. These sectors are Agriculture (AGR), Food (FOD), Water services (WTR), Government (GOV), Gas (GAS), and the two waste management sectors - MWM and IWM. None of these sectors generates hazardous waste, but due to the increased output of other sectors there is some level of demand switching. The reason for the decrease in the output of the waste management sectors is that there is a reduction in demand for their services since the cost of direct disposal is lower in this scenario. Among other sectors, the environmental

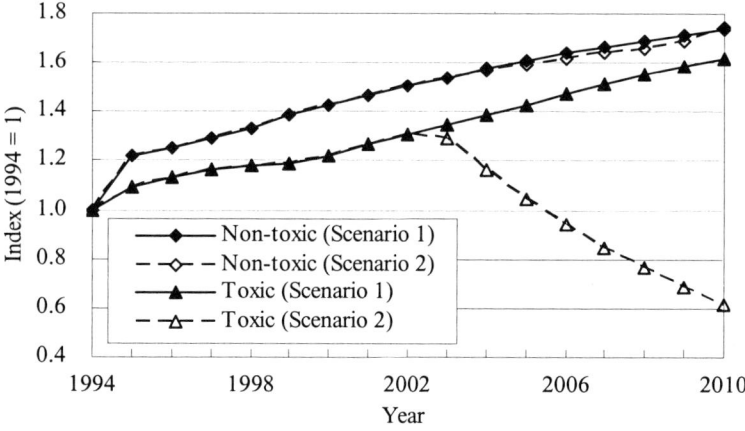

Fig. 8. Trajectories of final disposal of waste

industry (EMC) output increases substantially and its growth rate in this scenario is projected to be more than 20% and, compared with other equipment industries (MCH and ELM), to experience lower growth. (EMC is not shown in Fig. 10 because of the scale.) This result indicates the importance of this industry for pollution control.

Simulation results for toxic disposal constraints and mitigation do not indicate any significant effect on CO_2 emissions mitigation. Effects on waste recycling were not studied in detail since the information provided for the model is limited.

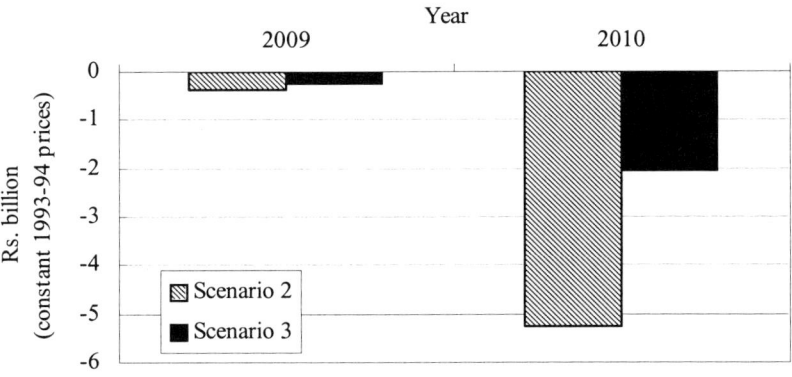

Fig. 9. GDP change in scenario 2 and scenario 3

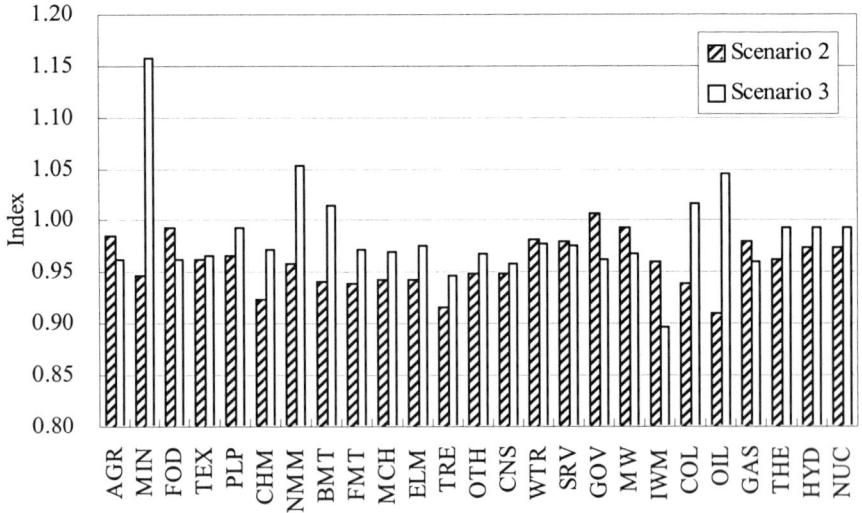

Fig. 10. Economic sectors showing change in output in 2010 over Scenario 1

11.5 Conclusion

As shown in the application of AIM/Material to Japan and India, various countermeasures can be assessed from the viewpoint of both a reduction of the environmental burden and the mitigation of economic losses. AIM/Material is a significant tool for assessing the policies designed to achieve sustainable development not only in the developed countries, but also in the developing countries. Although this is the top-down model, the solutions with the reality based on technology can be presented by linking it to the bottom-up model. Now the AIM team is trying to include other environmental problems, such as waste water treatment, and to apply it to other countries in the Asia-Pacific region.

References

Agency of Natural Resources and Energy, Ministry of International Trade and Industry ed. (1998) Energy balance table in Japan in 1997. International Trade and Industry Research Publishing (in Japanese)

Central Pollution Control Board (CPCB), Comprehensive industry documents. various issues

Central Statistical Organisation (CSO) (2000) Input output tables 1993-94. CSO

Department of National Accounts, Economic Research Institute, Economic Planning Agency (1999) Gross capital stock of private enterprises. (in Japanese)

Management and Coordination Agency ed. (1999) 1995 input-output tables. National Federation of Statistical Associations (in Japanese)

Masui T, Matsuoka Y, Morita T (2000) Development of applied general equilibrium model integrated environment and economy. Environmental Systems Research 28: 467-475 (in Japanese)

Masui T, Tsuchida K, Matsuoka Y, Morita T (2001) Integration of computable general equilibrium model and end-use model for sewage sludge management. Environmental Systems Research 29: 237-244 (in Japanese)

Ministry of Finance, Ministry of finance statistics monthly. various issues (in Japanese)

Ministry of Health and Welfare (1998a) Waste management in Japan 1995. (in Japanese)

Ministry of Health and Welfare (1998b) Industrial waste generation and treatment in 1995. (in Japanese)

Ministry of International Trade and Industry (1996) Yearbook of the current survey of energy consumption in manufacturing. International Trade and Industry Statistics Association (in Japanese)

Ministry of International Trade and Industry (1997) The statistical survey of energy consumption in commerce and manufacturing. International Trade and Industry Statistics Association (in Japanese)

Ministry of the Environment, The basic law for establishing the recycling-based society. http://www.env.go.jp/recycle/circul/kihonho/low-e.html

Planning Commission (1995) Urban solid waste management in India, Report of High Power Committee. Government of India

Report of the High Powered Committee on Management of Hazardous Wastes (2000) Supreme court of India, writ petition No.657/95. Research Foundation for Science, Technology and Natural Resource Policy v/s Union of India and ors., Government of India

Research institute of international trade and industry ed. (2000) Handbook of industrial tax in 2000. Research institute of international trade and industry (in Japanese)

Tata Energy Research Institute (2001) TERI energy data directory and yearbook (TEDDY) 2000-2001. Tata Energy Research Institute, New Delhi

The Japan Society of Industrial Machinery Manufacturers, Production of environmental equipment, every year (in Japanese)

12. Impact and Adaptation Assessment on a National Scale: Case Studies on Water in China

Kiyoshi Takahashi[1], Songcai You[2], Jiulin Sun[2], Zehui Li[2], Toshihiko Masui[1], Tsuneyuki Morita[1], Yuzuru Matsuoka[3], and Hideo Harasawa[1]

Summary. In order to assess vulnerability to climate change, not only the degree of climate change impact but also the adaptive capacity of the region with regard to the impact needs to be considered. To achieve such an assessment, a national-scale impact assessment model is being developed. This paper introduces two case studies related to water problems in China. One of them is a scenario for an assessment of future water withdrawal and consumption in China taking into consideration alternative future socio-economic development patterns. This indicates that the adoption of appropriate policies and efficient technologies can ensure that future water consumption is maintained at the current level, even though water withdrawal will increase under all socio-economic scenarios, which reflects the expected increase in industrial withdrawal. In the other case study, an evaluation is made of the adaptation policies in China for mitigating current flood damage due to climate variability as well as the future increase in flood damage due to climate change is evaluated. Investment in infrastructure for flood prevention in accordance with the expected future climate change can also reduce current flood damage and is found to be robust and low-regret adaptation strategy.

12.1 Introduction and Framework of AIM/Impact [Country]

As shown in Chap. 3 (see the chapter by Harasawa *et al.*), an AIM/Impact model has been developed mainly for the purpose of assessing climate change impacts on several sectors (water, agriculture, ecosystems, and human health) on a global scale and it has been used to elucidate the regions that would be significantly affected by anticipated climate change. However, in order to assess the vulnerability of a region to climate change, not only the degree of the impacts of climate change, but also the adaptive capacity of the region to cope with these impacts needs to be considered. To analyze concrete adaptation measures for mitigating climate change impacts, more detailed analysis is required on an appropriate spatial scale corresponding to that of the stakeholders (e.g. governments, local communities, individuals) with regard to the adaptation measures. Moreover, the role of each country in reporting the results of impact assessment to international processes,

---

[1] National Institute for Environmental Studies, Tsukuba 305-8506, Japan
[2] Institute of Geographical Sciences and Natural Resources Research, P.O. Box 9717, Beijing 100101, China
[3] Kyoto University, Kyoto 606-8501, Japan

such as the National Communication for the UNFCCC, increases the importance of national scale assessments carried out by each country itself. Within the AIM/Impact project, there are increasing requests from collaborative research teams in the Asian countries to pay more attention to assessments on the national scale. Taking into account this situation, the development of a national-scale assessment model of impact and adaptation has started, mainly based on the models that have been developed for AIM/Impacts on the global scale. This has been named AIM/Impact [Country].

AIM/Impact [Country] is a package of impact models, tools, databases, and visualization modules with a user interface to enable these elements to be operated in an integrated way. AIM/Impacts [Country] is used to assess climate change impacts on various sectors on a national scale. Fig. 1 shows the framework of AIM/Impact [Country], which illustrates the elements of the package (data, models, tools, parameters, etc.) and their relationships and dependencies. Impact assessment models for several sectors form the central part of the package (the upper right rectangle bounded by a dotted line in Fig. 1). Tools and databases included in the package support the development of national-scale input data for the impact assessment models (the upper left rectangle bounded by a dotted line in Fig. 1) and the spatial and numerical analysis of the output results of an assessment (the bottom rectangle bounded by a dotted line in Fig. 1). Although ready-made standardized data and model parameters are provided with the models and tools for countries with data limitations, these can be replaced by the users according to the specific conditions of each country. A visualization tool with simple functions to graphically display the spatial data and results is also included in the package, and

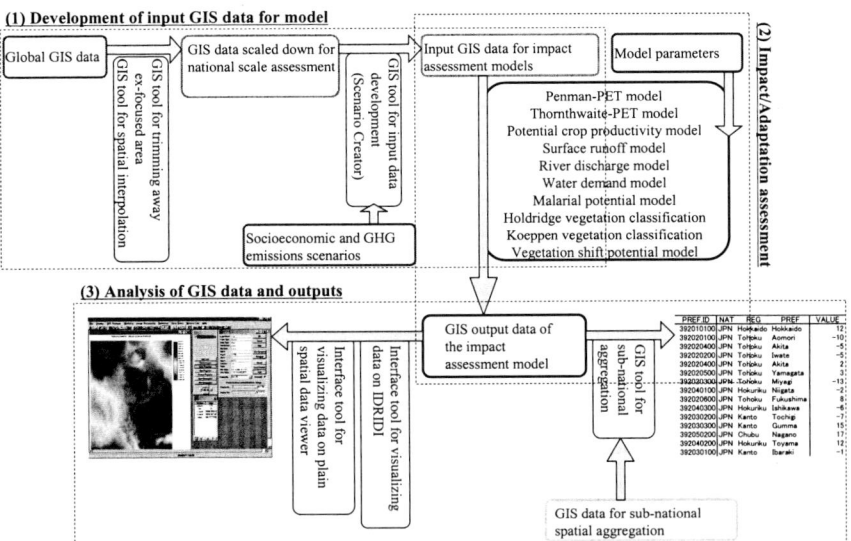

Fig. 1. Framework of AIM/Impact [Country] (see color plates)

users can quickly view the spatial information without purchasing expensive GIS software.

While the main focus of AIM/Impact has been the assessment of the direct physical impact of climate change, AIM/Impact [Country] is designed to more explicitly consider socioeconomic aspects of the problem and adaptation, besides the physical impacts. Therefore, in this chapter, two recent case studies related to water problems in China are introduced, whose models are planned to be included into AIM/Impacts [Country] in a more generalized form. Section 2 describes an assessment of future water withdrawal and consumption in China taking into consideration alternative future socio-economic development patterns. Section 3 introduces an assessment of adaptation policies in China for mitigating current flood damage due to climatic variability as well as future increased flood damage due to climate change.

12.2 Projection of Water Demand through 2032

12.2.1 Water withdrawal and consumption under alternative socioeconomic development scenarios

According to the dynamic changes related to human activities, such as the increase in the population, industrialization, irrigation development, shortages of water resources and consequent environmental problems are becoming more obvious in many regions of the world. If the rapid development scenario is taken, which assumes a significantly higher rate of population increase and industrialization in developing countries, more severe and more frequent water shortages are expected to occur. More extensive water shortages can be a factor limiting future socioeconomic development. There is even speculation that regional conflicts or wars may occur as a result of competition for limited water resources in the 21st century.

While such a pessimistic view of the prospects for water resource problems is being emphasized, future water demand depends substantially on future socioeconomic conditions, such as population changes and levels of economic development. Since future socioeconomic conditions are still uncertain, diverse sets of future development scenarios that could be potentially be chosen need to be analyzed. In order to assess the impact of climate change on water resources, the estimation of water demand taking into consideration socioeconomic conditions is as important as the estimation of future water supply under altered climatic conditions.

In this study, an assessment model of water withdrawal and water consumption on a national scale was developed and then used to estimate water withdrawal and consumption in China under four alternative future development scenarios for the period from 1995 to 2032. In the model, historical trends in the irrigated area and the population supplied with water as well as assumed future trends in relation to various socioeconomic factors, such as an increase in the population and economic development, are taken into account to estimate future withdrawal and consumption. In many studies for the projection of water demand made so far, the amount of water withdrawal has been taken as the index of water demand. However, for the purpose of evaluating the quantitative balance between water supply and demand in order to investigate any shortfall in water supply, water consumption is a better index than water withdrawal. Here, water withdrawal means the gross amount of water taken from water resources for human activities, on the other hand, water consumption means the net amount of water consumed mainly through evaporation before the water is recycled to form part of the available water resources. The ratio of water consumption to water withdrawal is called the water consumption ratio, while the ratio of water returned to water resources to water withdrawal is called the return flow rate. In the industrial and domestic sectors, most of the withdrawal is discharged back to replenish water resources (the water consumption ratio is low), thus the amount of water consumed is relatively small even if water withdrawal increases rapidly in future. On the other hand, in the agricultural sector, most withdrawal is consumed through evaporation and transpiration (the water consumption ratio is high).

12.2.2 Water demand model

Figure 2 shows the estimated flow of water withdrawal and water consumption. Water withdrawal and consumption in three sectors (industry, agriculture, domestic) are estimated separately. For the base year (1995), sectoral water withdrawal data compiled by the World Resource Institute (2001) is used. Firstly, based on the scenario for socioeconomic factors (urban/rural population and economic growth) and technological factors (water use efficiency improvements in each sector), future water withdrawal in each sector is estimated. Then, considering the scenario for the water consumption ratio, future water consumption is calculated. The basic

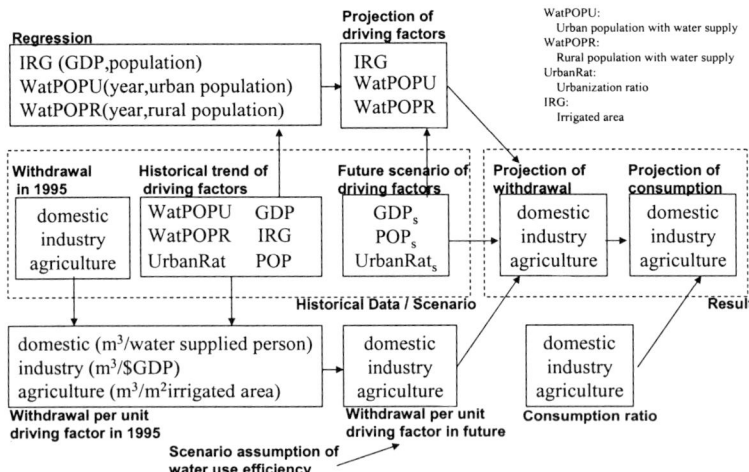

Fig. 2. Overview of water demand model

driving force for changes in industrial water withdrawal is economic growth in the model. It is assumed that industrial water withdrawal will change in proportion to changes in the GDP. The amount of irrigated area is taken as the driving force for any change in agricultural water withdrawal. Agricultural water withdrawal is assumed to change in proportion to any change in the irrigated area. The future irrigated area is estimated from a multi regression with the logarithm of population and the logarithm of per capita GDP as the explanatory variables. The population supplied with water is taken as the driving force for domestic water withdrawal. The future population supplied with water is estimated taking into consideration the historical trend of the ratio of the population supplied with water and the future population scenario. In order to take into account technology improvements, sectoral withdrawals based on the basic driving forces are then multiplied by a factor for water use efficiency improvements, which is assumed taking into consideration the form of socioeconomic development. The sectoral water withdrawals calculated in this way are summed up to provide the total water withdrawal for the country. Some portion of the withdrawal is consumed mainly through evaporation, while the other portion is recycled to replenish the water resource. In our model, the ratio of the consumed volume to the withdrawal (consumption ratio) is assumed to be 0.16, 0.8, and 0.3 for the industrial, agricultural and domestic sectors, respectively, according to information from Shiklomanov (1998).

12.2.3 Scenarios

Water withdrawal and consumption in the three sectors under the alternative four sets of assumptions regarding socioeconomic changes are estimated (Table 1). These four scenarios of socioeconomic changes are identical to the scenarios adopted in the global environment outlook 3 (GEO3) formulated by UNEP (2001). The comparative level of population change, economic development, and water-use efficiency improvements for the four scenarios are shown in Table 1. Among these, the scenarios for population change and economic development are deduced from the scenario of GEO3.

The scenario for water use efficiency improvements was set by the authors by maintaining consistency with the background sequence of the GEO3 scenarios. In the PR scenario, it is assumed that currently available efficient water use technologies will be introduced from the first half of the projected period (1995 – 2015) with the support of environmental policies, and that innovative efficient water use technologies will become available in the second half of the period of the projection (2015 – 2032). In the GT scenario, water use efficiency improvements in the first half will be at the same level as in the MF scenario. However, efficient technologies will be available in the second half of the period. In the MF scenario, technology improvements are slower to emerge than in the PR scenario in the first half, and are slower to emerge than in the PR and the GT scenarios in the second half of the period. In the FW scenario, water use technology improvements are much slower to emerge than in the other scenarios for the whole period of the projections.

Table 1. Socioeconomic development scenarios analyzed in the study

	Market Forces (MF)	Fortress World (FW)	Policy Reforms (PR)	Great Transition (GT)
Population growth rate	Medium	High	Medium	Medium
Economic growth rate	High	Low	High	Medium
Water use efficiency improvements	Medium	Slow	Rapid	Medium – Rapid

12.2.4 Analysis

Figure 3a - 3c shows the water consumption for the three sectors. A common trend for all the scenarios is that industrial consumption will increase even taking into consideration water use efficiency improvements. Reflecting the expected rapid economic growth and industrialization, the growth of industrial consumption will be the greatest in the MF scenario. In the PR scenario and the GT scenario, the growth of industrial consumption will be slightly lower than that in the MF scenario, while it is rather low in the FW scenario due to slow economic growth. As for agricultural consumption, this will not increase significantly even in the FW scenario with a high increase in the population. It will decrease in the other scenarios with a lower increase in the population and higher water use efficiency improvements. These trends reflect the recent historical situation, which indicates that irrigation in China has reached saturation point. Domestic consumption will increase only in the FW scenario due to its high increase in population and low rate of efficiency improvements.

Figure 4a and 4b show the total water consumption and total withdrawal in China, respectively. By comparing these two figures, it can be found that they have a different order with regard to the scenarios. In viewing the results of withdrawal, the MF scenario involves the largest increase. This is because industrial withdrawal will increase much more rapidly in the MF scenario than in the other scenarios. On the other hand, the reason of the second highest total withdrawal in the FW scenario is not an increase in industrial withdrawals, but an increase in agricultural and domestic withdrawals reflecting rapid population growth and slow efficiency improvements. However, when viewing the results of consumption, the FW scenario will show the greatest increase, and the MF scenario does not show a significant increase. Moreover, water consumption will decrease slightly in the PR scenario and the GT scenario. This difference of order between consumption and withdrawal is caused by the high water consumption ratio in the agricultural sector. This result indicates that it is very important to differentiate water consumption and withdrawal when assessing water demand.

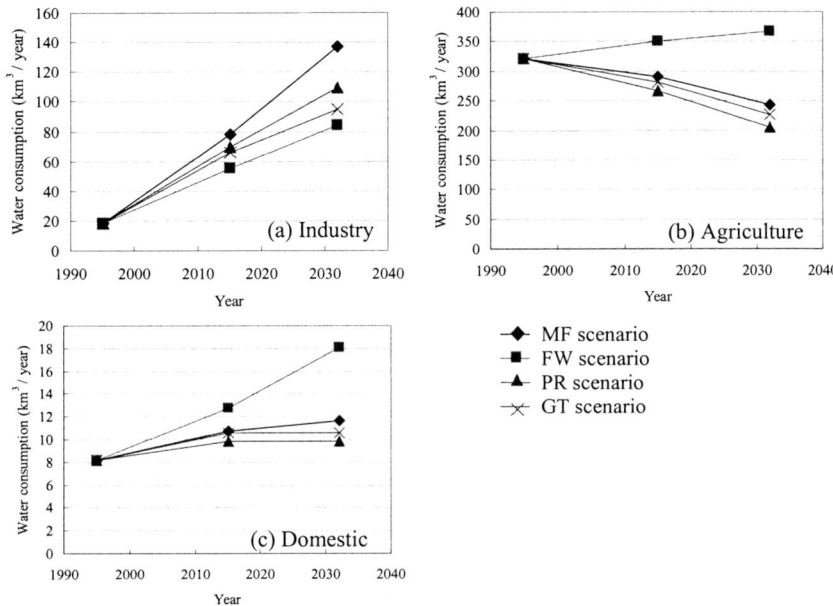

Fig. 3a-3c. Water consumption in the three sectors (km^3/year)

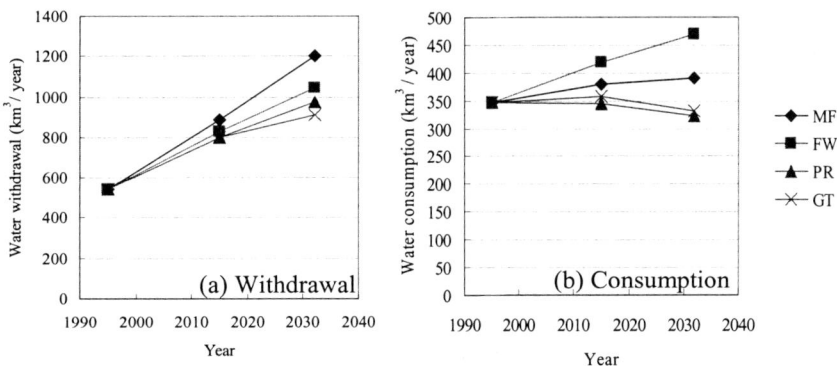

Fig. 4a and 4b. Total water withdrawal and consumption

12.2.5 Findings and future improvements

In this case study, water withdrawal and consumption in China under four alternative future development scenarios were estimated for the period from 1995 to 2032. Although withdrawal will increase in all the scenarios, reflecting the significant increase in industrial withdrawal, water consumption can be kept at the current level if appropriate policies are adopted to accelerate the introduction of efficient technologies. Increasing industrial water consumption, which is unavoidable under rapid industrialization, can be accommodated through efficiency improvements in the agricultural sector.

This study requires improvement with regard to some points. Firstly, in this study, only the quantity of water withdrawn from water resources and consumed through human activities has been analyzed. However, water quality has not been considered in any form. Even if a sufficient quantity of water is available for future human activities, the actual quality of the water may not be appropriate for some uses. The aspect of water quality should be included in the assessment. Secondly, the results of the assessment depend substantially on the assumptions with regard to water use efficiency improvements as well as on the socioeconomic scenarios for the driving forces. In this study, the assumptions regarding efficiency improvements are set exogenously considering the historical trend. However, these assumptions need to be related to the availability of concrete technology options for more efficient water use. Detailed databases on technology options with information on costs and level of efficiency need to be developed to add greater reality to the assessment.

12.3 Adaptation Policies in China for Mitigating Flood Damage

12.3.1 Importance of long-term adaptation policy

Adaptation to climate change impacts is considered to be an important and efficient strategy. Nevertheless, the efficiency of each adaptation strategy has not been sufficiently analyzed to make it possible to propose a detailed action plan, though adaptation options have been compiled for each sector. The limitations are mainly derived from the following features of climate change impact and adaptation studies: (a) the climate change impacts are still uncertain; (b) the mechanisms of adaptation are too complex to be evaluated. In spite of these difficulties, each country should actively seek to elucidate what can and should be done in the early 21st century to mitigate the negative impacts of climate change, as there are long time lags in relation to the introduction of capital stock and new technologies in human activities in response to changing economic conditions.

China is expected to experience significant impacts on the hydrological cycle from climate change. However, it is very difficult for the Chinese government to introduce a long term adaptation policy since China places the priority on short- to medium-term policies. In order for China to introduce long term policies, the benefits of the integration of long term adaptation policies with short term policies need to be evaluated. A macroeconomic model is introduced to evaluate the effects of investment in flood mitigation infrastructure in China considering climatic variability and climate change.

12.3.2 Model structure

There are some references to the multi-sector models such as by Duraiappah (1993) and Masui *et al.* (2000). In this study, the focus is a dynamic optimization model with two economic sectors: the agricultural sector and the non-agricultural sector. Two climate discount factors are considered in the model: damage from climatic variability and damage from climate change. The solution of this model is highly normative, since the production and distribution of goods are determined so as to maximize the temporary sum of the discounted utility derived from final consumption. The set of model equations is presented below.

Objective function

The fundamental assumption is that policies should be designed to maximize the generalized level of consumption now and in the future. This approach rests on the view that more consumption is preferred over less, and in addition that increments of consumption become less valuable as consumption levels increase (Nordhaus 1994). In technical terms, these assumptions are embodied by maximizing a social welfare function that is the discounted sum of the utility of per capita consumption. The objective function to be maximized is:

$$\text{Max } U = \sum_t \left[\left(\prod_i C_i(t)^{v_i} \right) \times (1+\rho)^{-t} \right] \tag{1}$$

Where U is the flow of utility, C_i is the flow of consumption from sector i per capita at year t, and ρ is the pure rate of social time preference that allows for distinguishing the relative emphasis on different generations, a value of 0.03 per year is applied by Nordhaus (1994). Also set to 0.03. v is the consumption share of each sector product, the sum of v_i (i=1,2) is equal to 1. The planning horizon spans 105 years, from 1995 to 2100, with each time period in the model specification representing a one-year interval.

Production function

Total intermediate input and the production factor are aggregated to obtain total output using the Leontief production function. The total outputs of the agricultural and non-agricultural sectors are expressed as follows:

$$Ya(t) = \min\left(\frac{Maa(t)}{a2a}, \frac{Mna(t)}{n2a}, \frac{YGDPa(t)}{a2ao} \right) \tag{2}$$

$$Yn(t) = \min\left(\frac{Mnn(t)}{n2n}, \frac{Man(t)}{a2n}, \frac{YGDPn(t)}{n2no}\right) \tag{3}$$

Where, *Ya(t)* and *Yn(t)* are the total outputs of the agricultural and non-agricultural sectors, respectively; *Maa(t)* and *Man(t)* are intermediate inputs from the agricultural sector to both the agricultural and non-agricultural sectors, respectively; *Mnn(t)* and *Mna(t)* are intermediate inputs from the non-agricultural sector to both the non-agricultural and agricultural sectors, respectively. *YGDPa(t)* and *YGDPn(t)* represent the productions of the agricultural and non-agricultural sectors; *a2a*, *n2a*, *n2n*, and *a2n* are input coefficients and are equal to 0.16064, 0.24199, 0.59804, 0.05394 respectively; a2ao and n2no are production factors and are equal to 0.59737 and 0.34803, respectively. They are aggregated from six sectors (SSB 1999).

Agricultural production is expressed as follows:

$$YGDPa(t) = Aa(t) \times Ka(t)^{\beta} \times La(t)^{\gamma} \times F(t)^{\lambda} \tag{4}$$

Where *Aa(t)* is the total factor of productivity; *Ka(t)* is the capital stock, *La(t)* is the labor input, *F(t)* is the land input, β is the elasticity of the capital input, γ is the elasticity of labor, λ is the elasticity of land. Based on the historical data from 1972 to 1995, β, λ and γ are determined to be 0.55, 0.20 and 0.25 in agriculture, respectively, during the period 1991-1995 (Zhu and Liu 1998).

The non-agricultural sector production is expressed as a function of technology, capital and labor.

$$YGDPn(t) = An(t) \times Kn(t)^{1-\alpha} \times Ln(t)^{\alpha} \tag{5}$$

Where α is the elasticity of output with respect to capital, *An(t)* is the total factor of productivity, *Kn(t)* is the capital stock, *Ln(t)* is the labor input. The non-agricultural capital elasticity (α) is set to 0.3 in this study according to the latest study by Jiang *et al.* (1998).

It is assumed that land is primarily used for agricultural activities, and that the total area remains unchanged throughout the simulated period, which is the goal of the government in guaranteeing the supply of food for its huge population. Since the rapid development of the non-agricultural sector accounts for some land, keeping the total area unchanged involves reclaiming some abandoned areas that in general have contributed to a decline in land quality. According to a study (Gao *et al.* 1998), land quality decreased 2.53% from 1985 to 1995 at mean annual rate of 0.25%. It could be expected that the occupation of cultivated land will approach zero when the population growth reaches zero and the urbanization process basically stops. So it is assumed that land quality will decrease 0.25% annually until 2030 and 0.15% per year from 2031 to 2050 and 0% after 2050.

National accounts and capital constraints

The distribution relation of goods is of the standard input-output style. It states that the total sectoral output must be equal to the sum of consumption demand, investment, intermediate requirements, and adaptation investments, taking into consideration negative climate change impacts. It is assumed that the non-agricultural sector is the primary sector that contributes to capital formation for itself as well as for the agricultural sector. Imports and exports are not considered independently in goods distribution.

$$Ya(t) = Ca(t) + Ia(t) + Iav(t) + Iac(t) + Maa(t) + Man(t) \qquad (6)$$

$$Yn(t) = Cn(t) + In(t) + Inv(t) + Inc(t)$$
$$+ Mnn(t) + Mna(t) + Ina(t) \qquad (7)$$

Where, $Ia(t)$, $Iav(t)$ and $Iac(t)$ are contributions of the agricultural sector to agricultural capital stock, investment for flood control, and extra investment for projected flood damage from climate change. $In(t)$, $Inv(t)$ and $Inc(t)$ are contributions of the non-agricultural sector to non-agricultural capital stock, investment for flood control, and extra investment for projected flood damage from climate change. $Ca(t)$ and $Cn(t)$ represent the consumption of agricultural and non-agricultural goods, respectively; Ina is the contribution of the non-agricultural sector to the formation of agricultural capital stock.

$$Ka(t) = (1 - \delta)Ka(t-1) + Ia(t-1) + Ina(t-1) - Dak(t-1) \qquad (8)$$

$$Kn(t) = (1 - \delta)Kn(t-1) + In(t-1) - Dnk(t-1) \qquad (9)$$

$$INF(t) = (1 - \delta)INF(t-1) + Iav(t-1) + Inv(t-1) \qquad (10)$$

$$INFc(t) = (1 - \delta)INFc(t-1) + Iac(t-1) + Inc(t-1) \qquad (11)$$

Where: δ is the depreciation rate of the capital stock and flood mitigation infrastructure; Ka and Kn are capital stocks of the agricultural and non-agricultural sectors, respectively; INF, $INFc$ are the infrastructure stock that is designed to mitigate flooding from climatic variability and projected climate change, respectively, Dak and Dnk are the flooding damage to capital stocks of the agricultural and non-agricultural sectors.

The respective capital stocks for the agricultural sector and for all sectors in 1995 were 395.91 billion Yuan (Zhu and Liu 1998) and 5663.3 billion Yuan (Jiang et al. 1998) at 1980 prices. Both of them are converted, based on the overall retail price index (SSB 1996), to 1300.21 billion Yuan and 17298.74 billion Yuan at 1995 prices. The stock of the infrastructure for flood mitigation was 26.045 billion Yuan in 1995 at 1995 prices, too, calculated cumulatively from 1956. The depreciation rate of the capital stock and infrastructure in China varies from year to year, with a tendency towards a gradual increase. The depreciation rate for the fixed assets of stated-owned enterprises increased from 4.1% in 1980 to 5.5% in 1992

(SSB 1995). It is set at 5% in this study.

Damage function

It is quite difficult to establish the damage function between flood mitigation infrastructure stock per capita and flood damage to the agricultural and non-agricultural sectors, respectively, since there exists only very limited reported flood damage, and even that is in an aggregated form. There are two types of direct flood damage: damage to land/crops in the agricultural sector, and damage to capital stock in both sectors. The first type damages crop yields by inundation, often leading to a loss of the harvest, or damages land as a result of erosion or fluvial and alluvial sedimentation. In the latter case, floods not only destroy the sown crops, but also make future cultivation impossible. In this study, it is assumed that flood damage to land has no impact on future use.

Based on the total direct flood damage and the statistical data for the sown crops covered and affected by flood (SSB 1981), the following damage functions are established.

$$Dnkv(t) = 10^{nk0} \times \left[INF(t-t')/P(t-t') \right]^{nk1} \tag{12}$$

$$Dakv(t) = 10^{ak0} \times \left[INF(t-t')/P(t-t') \right]^{ak1} \tag{13}$$

$$Dalv(t) = 10^{al0} \times \left[INF(t-t')/P(t-t') \right]^{al1} \tag{14}$$

Where, $Dnkv$, $Dakv$, $Dalv$ represent damage to capital stocks of the non-agricultural and agricultural sectors, and land, respectively. P is population. t' is the time lag during which investment takes effect and equals 5; $nk0$, $nk1$, $ak0$, $ak1$, $al0$, $al1$ equal to 1.513, -0.918, 0.794, -0.771, 0.984, -0.355, respectively.

By assuming that equations (12)~(14) are still applicable in the future if climate change occurs and that the marginal cost of reducing unit flood damage from climate change is the same as that from climatic variability, the damage functions of climate change can be derived from equations (12)~(14).

$$Dnkc(t) = 10^{nk0} \times \left[\frac{INFc(t-t')}{P(t-t')} + \left(\frac{Dc(t)}{10^{nk0}} \right)^{1/nk1} \right]^{nk1} \tag{15}$$

$$Dakc(t) = 10^{ak0} \times \left[\frac{INFc(t-t')}{P(t-t')} + \left(\frac{Dc(t)}{10^{ak0}} \right)^{1/ak1} \right]^{ak1} \tag{16}$$

$$Dalc(t) = 10^{al0} \times \left[\frac{INFc(t-t')}{P(t-t')} + \left(\frac{Dc(t)}{10^{al0}} \right)^{1/al1} \right]^{al1} \tag{17}$$

$$Dnk(t) = (Dnkv(t-1) + Dnkc(t-1)) \times YGDPn(t-1) \qquad (18)$$

$$Dak(t) = (Dakv(t-1) + Dakc(t-1)) \times YGDPa(t-1) \qquad (19)$$

$$F(t) = F0 \times [1 - Q(t)] \times [1 - Dalv(t) - Dalc(t)] \qquad (20)$$

F is the effective land input considering flood damage and the reduction of land quality. The equation for climate change damage, $Dc(t)$, is expressed as

$$Dc(t) = D_{ref} T(t)^2 / 6.25 \qquad (21)$$

Where $T(t)$ is the temperature increase in year t. Damage caused by flooding under climate change with a 2.5°C increase in temperature is assumed to be D_{ref}, the quadratic term of the temperature reflects the assumption that the damage is quadratic, along with the temperature increase (Nordhaus 1994). Based on a study of physical impacts on surface runoff, and the intensity and frequency of floods under projected climate change (paper in preparation), it is assumed that the flood damage under a future climate in the year 2100 will double, and the damage from climate change is the same as the current maximum damage of 3.9% in 1994.

Growth rate of technology and the total productivity factor

The growth rate of technology during the period 1991-1995 was 2.5% for the agricultural sector (Zhu and Liu 1998) and 3.1% for all sectors, estimated based on a study (Jiang et al. 1998). The figure 3.1% is taken as the approximate growth rate of technology in the non-agricultural sector due to the approximately 20% share of the total GDP held by the agricultural sector. So the initial values for the technology growth rate for the agricultural and the non-agricultural sectors are 2.5% and 3.1%, respectively. Due to a lower technological level and a lower contribution of technology growth to economic growth (about 30%) (Ministry of Agriculture 1996; Jiang et al. 1998), it is expected that technology growth will continue to speed up within the next several decades, and then level off and afterwards decline annually. So it is assumed that the rate of change in the technology growth rate is to be 1%, 0% and -1%, annually for the periods 1996-2030, 2031-2050 and 2051-2100, respectively. The growth rates of technology for the agricultural and the non-agricultural sectors at year t are estimated as follows (Nordhaus 1994):

$$GTa(t) = GTa(t-1) \times [1 + \varphi a(t)] \qquad (22)$$

$$GTn(t) = GTn(t-1) \times [1 + \varphi n(t)] \qquad (23)$$

where $\phi a(t)$ and $\phi n(t)$ are the rates of change in technology growth for the agricultural and the non-agricultural sectors. GTa and GTn are the growth rates of technology for both sectors. Then the total productivity factors are calculated as,

$$Aa(t) = Aa0 \times \exp[GTa(t)] \qquad (24)$$

$$An(t) = An0 \times \exp[GTn(t)] \tag{25}$$

The values of the total productivity factors in the initial year, *Aa0* and *An0*, can be calculated, based on equations (4) and (5).

Population and labor

Labor is assumed to be proportional to population. According to a report (Li *et al.* 2000), demographers at the China Renmin University put forward three population scenarios for the next 100 years (High, Medium and Low). The projected population in 2100 is 1.5 billion, 1.0 billion, and 0.8 billion under High, Medium and Low scenarios, respectively. The common assumption for the three scenarios is that population increases from 1.21 billion in 1995 to near 1.4 billion by 2010, 1.6 billion by 2030 and then levels off after 2030 until 2035. The population growth rate in 1995 is 1.0605%, and the decrease in the rate of population growth from 1996 to 2030 is estimated to be 1.6%.

In 1995, 52.2% of the labor force was employed in the agricultural sector and 47.8 % in the non-agricultural sector (SSB 1996). As the share of the agricultural sector declines along with economic development, labor migrates to the cities and enters the formal or informal labor market. In this research, a simplified migration scenario is assumed in which 70% and 80% of labor will be employed in the non-agricultural sector by 2050 and 2100 respectively.

12.3.3 Scenarios of adaptation and climate change

Four scenarios combining climate change and investment in the flood mitigation infrastructure were assumed for simulation, based on the assumption that policy makers will optimize investment in flood prevention infrastructure to reduce the cost of damage caused by floods from the current climatic variability, whether climate change occurs or not (Table 2).

1. CnAn: policymakers do not arrange adaptation investment for the projected climate change and climate change does not occur.
2. CyAn: policymakers believe climate change will not occur and thus no adaptation investment is implemented to mitigate flood damage from climate change, but unfortunately, climate change occurs.
3. CyAy: policymakers believe climate change will occur and thus adaptation investment is arranged to mitigate flood damage from climate change and climate change occurs.
4. CnAy: policymakers believe that climate change will occur and adaptation investment is arranged to mitigate flood damage from climate change, the amount of investment is assumed to be the same as that in CyAy, but climate change does not occur.

Table 2. Scenarios considering policy options and the probability of climate change occurring

| | Invest | Climate change | |
		Yes	No
Policy options	Yes	CyAy	CnAy
	No	CyAn	CnAn

12.3.4 Analysis

The model first is run on the following assumptions: population growth follows the medium scenario (1.0 billion in 2100); the non-agricultural sector employs 70% of the labor force by 2050 and 80% by 2100; the marginal adaptation costs of adapting to climate change are assumed to be the same as those for current climatic variability. Taking CnAn as a base scenario, flood damage to cultivated land (Fig. 5) from climatic variability and climate change is expected to increase to 1.13% in 2100 even if investment takes climate change into consideration and climate changes do occur (scenario CyAy); the highest level of damage is about 1.58% around 2050. The damage gradually increases to 3.11% by the end of the century when there is no investment in flood prevention infrastructure to combat projected climate change (scenario CyAn). When climate change does not occur while adaptive investment is being implemented (scenario CnAy), flood damage is lower than the base scenario (scenario CnAn) as adaptation investment adds extra adaptive capacity, thus mitigating the damage from climatic variability. Damage to the agricultural capital stock and non-agricultural capital stock (Figs. 6 and 7) under these four scenarios shows a similar pattern to that of agricultural land. In-

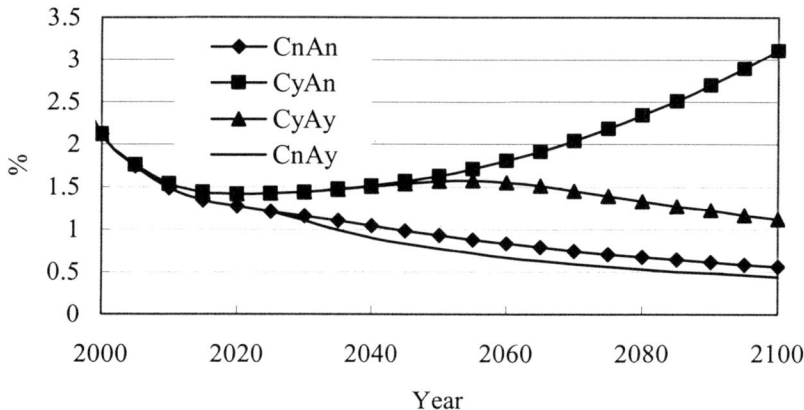

Fig. 5. Flood damage to cultivated land

vestments ignoring climate change will cause damage starting around 2020 when climate change occurs.

Utility per capita decreased by 0.83% when no adaptation investment is implemented while climate change occurs, but the reduction can be reduced to 0.1% if adaptation investment is implemented (Fig. 8).

Changes in utility per capita were accumulated to analyse the effects of long term adaptation policy and short term policy. Two criteria ("maximax" and "maximin") for decision making under uncertainty were applied in the study to select the best option in response to uncertain climate change. The maximax criterion is based on the assumption of an optimistic decision maker. The best outcomes expected under each decision alternative considering uncertainty are compared, and the alternative that represents the best of the best outcomes is selected. The maximin criterion is based on the assumption of a conservative (pessimistic) decision maker. Worst outcomes expected under each decision alternative are compared, and the alternative that represents the best of the worst outcomes is selected.

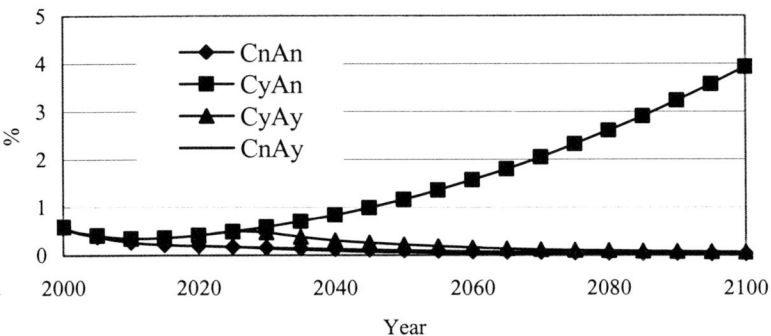

Fig. 6. Damage to capital stock in the agricultural sector

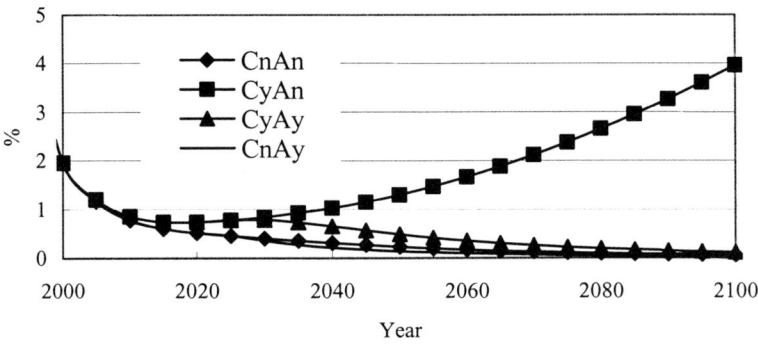

Fig. 7. Damage to capital stock in the non-agricultural sector

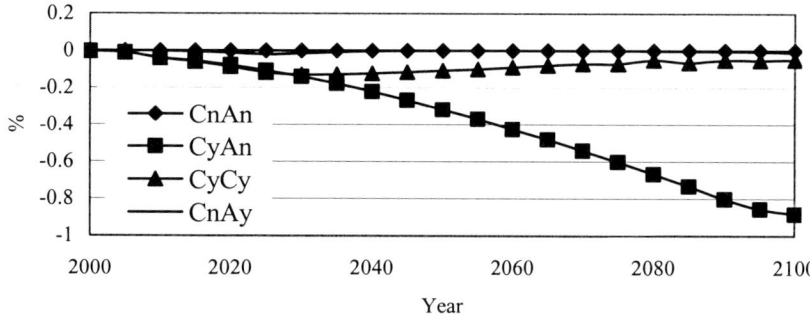

Fig. 8. Changes in the utility compared with the baseline (CnAn)

Table 3. Decision making based on changes in utility

	Invest	Climate change		Best option	
		Yes	No	Maximax	Maximin
1995-2100	Yes	-563.2	-17.6		✓
	No	-4158.7	0.0	✓	
1995-2020	Yes	-22.8	-1.8		
	No	-21.4	0.0	✓	✓

The result (Table 3) shows the best option in the short term is not to adapt to climate change. In the long term, the best option is to adapt to climate change, if policy makers prefer risk aversion decisions (maximin criterion).

12.3.5 Findings from the simulation

In this case study, a dynamic macroeconomic model dealing with investment optimization considering climate variability and climate change is created and a simulation extended for 105 years from 1995 to 2100 in China. A feature of this model is that it considers two sectors, two discount factors from climatic variability and climate change, as well as labor migration and capital flows existing between the two sectors. The following results have been obtained by using this model:

1. Investments that are optimized ignoring climate change will cause severe damage starting about 2020 and reaching its peak by 2100 when climate change occurs. Optimizing investment considering both climatic variability and climate change can effectively mitigate flood damage from climate change.
2. Investment against projected climate change is the best option from a century-

long perspective in a situation of uncertainty in relation to climate change. In the short term, the best option is not to invest.

12.4 Conclusion

The importance of impact assessment on a regional or national scale has been increasing. In order to evaluate regional impacts and adaptation measures, in addition to spatially detailed data, new assessment tools that consider diverse alternative socioeconomic development paths and effective adaptation strategies are expected to be developed. Moreover, there is also a need for tools to build more facilities for local actors, who are better able to comprehend the significant problems and the feasible solutions, to evaluate impact and adaptation strategies by themselves. AIM/Impacts [Country] is expected to meet these needs.

In this report, the results of case studies on China were presented, where collaborative research with local organizations has continued for a long time. In the next stage, while the AIM/Impact [Country] model continues to be developed and refined, the area that is subject to national scale assessment will expand from China to the South and Southeast Asian countries, where the most drastic socioeconomic developmental transformation and serious climate change impacts are expected.

References

Duraiappah AK (1993) Global warming and economic development, a holistic approach to international policy co-operation and co-ordination. Kluwer Academic Publishers.

Gao ZQ, Liu JY, Zhuang DF (1998) The dynamic changes of the gravity center of the farmland area and the quality of the farmland ecological background in China. Journal of Natural Resources 13(1): 92-96 (in Chinese)

Jiang JL, Ruan GF, Ren L (1998) Analysis on effect of technology progress on economic development in China. In: Estimation of technology progress contribution to economic development: theory and practice. China Planning Press (in Chinese)

Kong SH (1999) Effects of transfer of surplus rural labor on advance of industrial structure. China Economics Publishing House (in Chinese)

Li Y, Shen ZH, Guo N (2000) One birth or two births? Life Weekly 10: 31-44 (in Chinese)

Masui T, Morita T, Kyogoku J (2000) Analysis of recycling activities using multi-sectoral economic model with material flow. European Journal of Operational Research 122: 405-415

Ministry of Agriculture (1996) China agricultural development report-96. China Agricultural Press (in Chinese)

Nordhaus WD (1994) Managing the global commons - the economics of climate change. The MIT Press

Shi QQ and Qin BT (1998) Study on effect of technology progress on economic development. In: Estimation of technology progress contribution to economic development:

theory and practice. China Planning Press (in Chinese)

Shiklomanov IA (1998) Assessment of water resources and water availability in the world, Comprehensive assessment of the freshwater resources of the world, Stockholm Environment Institute

SSB (1981) China Statistical Yearbook-1981. China Statistical Publishing House (in Chinese)

SSB (1995) China Statistical Yearbook-1995. China Statistical Publishing House (in Chinese)

SSB (1996) China Statistical Yearbook-1996. China Statistical Publishing House (in Chinese)

SSB (1999) China Statistical Yearbook-1998. China Statistical Publishing House (in Chinese)

UNEP (2002) GEO-3, Past, present and future perspectives. Earthscan

WRI (2001) World resources 2000-2001: People and ecosystems: The fraying web of life. World Resource Institute

Zhu DJ (1999) Study on labor migration between rural and urban under market of three sectors. China Economics Publishing House (in Chinese)

Zhu XG, Liu YF (1998) Estimation of the contribution of technology growth to agricultural development in China. In: Estimation of the contribution of technology progress to economic development: theory and practice. China Planning Press (in Chinese)

13. AIM/Trend: Policy Interface

Junichi Fujino[1], Shigekazu Matsui[2], Yuzuru Matsuoka[3], and Mikiko Kainuma[1]

Summary. The model for assessing the future environmental loads based on the past socio-economic trends has been developed. By using this model, the environmental trends until 2032 in the Asia-Pacific 42 countries have been estimated. This AIM/Trend model has characteristics of econometric model and communication tool to enhance discussions and search for countermeasures by envisaging environmental conditions in a consistent way. Results of AIM/Trend model show that Asia-Pacific countries have a wide range of economy, energy, and environment conditions. It makes the country-level study in this region more important to deal with the diversity of Asia-Pacific countries.

13.1 Introduction

The Asia-Pacific region is expected to lead global economy due to its rapid increase of population and economic growth in the earlier of 21st century. On the other hand, it is said this economic development will cause significant damage on regional and global environment without any appropriate measures.

Though the study on its economy, energy, and environment is becoming more important, the variety of Asia-Pacific culture, society, and natural resources in a country-level makes it difficult to understand policy and future projection. It is necessary to develop the model that evaluates the effects of each country's policy to country-level, regional, and global environment. It is also important that each country's policymakers who know well the country can use the model easily. Our team has been developing the model for assessing the future environmental loads based on the past socio-economic trends and future scenarios. By using this model, the environmental trends until 2032 in the Asia-Pacific countries have been estimated (Fig.1).

AIM/Trend model is built as a simple econometric model and written in ATPL (AIM/Trend Program Language) using VBA in Microsoft Excel. The user can change the model structure and make sensitivity analysis. Results of AIM/Trend model are used as Asia-Pacific scenario in UNEP/GEO-3 (UNEP 2002).

---

[1] National Institute for Environmental Studies, Tsukuba 305-8506, Japan
[2] Fuji Research Institute Corporation, Tokyo 101-8403 Japan
[3] Kyoto University, Kyoto 606-8501, Japan

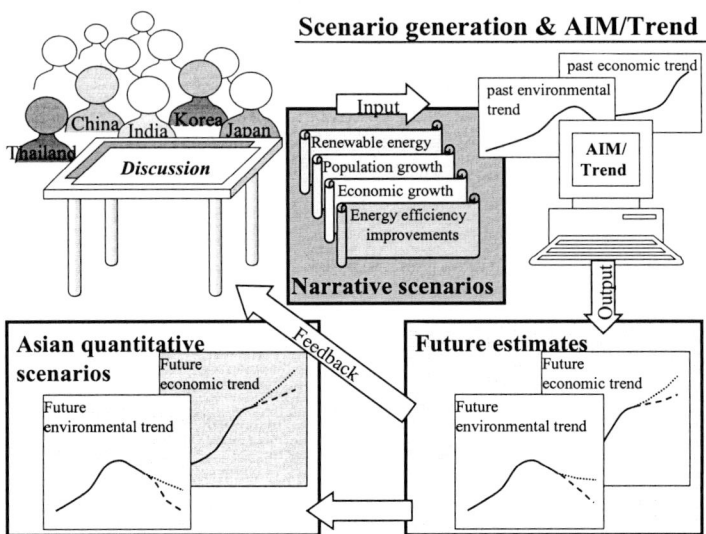

Fig. 1. Concept of AIM/Trend model (see color plates)

13.2 Structure of AIM/Trend Model

13.2.1 Framework

AIM/Trend model is an econometric model to estimates future conditions of economy, energy, and environment (Fujino 2002). It calculates the relationships between each parameter by regression method and extrapolates those relationships for the future projection. It makes simulations of energy supply and demand, GHG emissions, waste emission, water supply and demand, and so on by setting basic data –population GDP, GDP per capita, GDP share, etc. – as driving forces (Fig.2). In this chapter, the module of energy supply and demand is mainly focused.

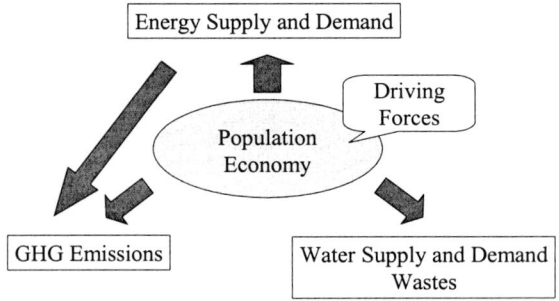

Fig. 2. Framework of AIM/Trend model

Target countries

This model covers 42 countries in the Asia-Pacific region shown in Table 1. They are divided into two categories: Model A and Model B. Model A uses detailed energy data, IEA energy statistics data. Model B uses simple energy data, UN energy data. The countries can be grouped in five regions: South Asia, Southeast Asia, East Asia, Central Asia, and ANZ and South Pacific.

Table 1. Target countries

	\multicolumn{3}{c}{Detailed Data (Model A)}		\multicolumn{3}{c}{Simple Data (Model B)}
	Code	Country	Group		Code	Country	Group
1	BGD	Bangladesh	SA	1	AFG	Afghanistan	SA
2	IND	India	SA	2	BTN	Bhutan	SA
3	IRN	Iran	SA	3	MDV	Maldives	SA
4	LKA	Sri Lanka	SA	4	BRN	Brunei	SEA
5	NPL	Nepal	SA	5	KHM	Cambodia	SEA
6	PAK	Pakistan	SA	6	LAO	Laos	SEA
7	IDN	Indonesia	SEA	7	MNG	Mongolia	EA
8	MMR	Myanmar	SEA	8	FJI	Fiji	SP
9	MYS	Malaysia	SEA	9	KIR	Kiribati	SP
10	PHL	Philippines	SEA	10	NRU	Nauru	SP
11	SGP	Singapore	SEA	11	PLW	Palau	SP
12	THA	Thailand	SEA	12	PNG	Papua New Guinea	SP
13	VNM	Vietnam	SEA	13	PYF	French Polynesia	SP
14	CHN	China	EA	14	SLB	Solomon Islands	SP
15	JPN	Japan	EA	15	TON	Tonga	SP
16	KOR	Korea, Rep	EA	16	VUT	Vanuatu	SP
17	PRK	Korea, Dem	EA	17	WSM	Samoa	SP
18	TWN	Taiwan	EA				
19	KAZ	Kazakhstan	CA				
20	KGZ	Kyrgyz Republic	CA		Code	Group	
21	TJK	Tajikistan	CA		SA	South Asia	
22	TKM	Turkmenistan	CA		SEA	Southeast Asia	
23	UZB	Uzbekistan	CA		EA	East Asia	
24	AUS	Australia	SP		CA	Central Asia	
25	NZL	New Zealand	SP		SP	ANZ and South Pacific	

Note: Though it has IEA energy statistics data, Brunei is treated in Model B because of its special structure of energy supply and demand.

Simulation period

The Johannesburg summit was held at August 2002. The target year is set as 2032, 30 years after the summit.

Target indices

Following indices are selected and will be evaluated by AIM/Trend model (sign "*" means the element under consideration):

- Population: population, rate of urbanization
- Economy: GDP (growth rate, per capita), GDP share (agriculture, industry, service, PFC (private final consumption), car numbers
- Energy: primary energy supply by fuel, final energy demand by fuel and sector, energy plant, economic intensity, carbon intensity
- GHG emissions: CO_2, SO_x, NO_x, CH_4, N_2O, CO
- Waste: waste emission, landfill*, recovery of waste*
- Water: withdrawal, consumption (agriculture, industry, domestic), population in water stress*
- Food and Agriculture*: average daily consumption, vegetable food consumption, animal food consumption, fraction of meat from feedlots, fish production, crop production, feed production, nitrogen fertilizer consumption
- Land use*: crop land, irrigated cropland, potential cultivable land, mature forest, growing forest, pasture, protected, other land
- Human Health*: SPM (PM10, PM2.5)
- Biodiversity*: species, degree of threat to biological diversity, area of habitat remaining

13.2.2 Model structure

AIM/Trend model consists of Modal A with detailed energy data and Model B with simple energy data. The model structures of Model A and Model B are explained in following sections.

Detailed data model (Model A)

Model A covers 25 countries that have the IEA energy statistics. The calculation flow of Model A focused on energy is shown in Fig. 3. Model A is built up by following order: final energy demand, energy conversion, and primary energy.

(1) Energy classification

Following 10 energies are treated in this Model A: COL (coal), OIL (crude oil and petroleum products), GAS (gas), CRW (combustible renewables and waste), NUC

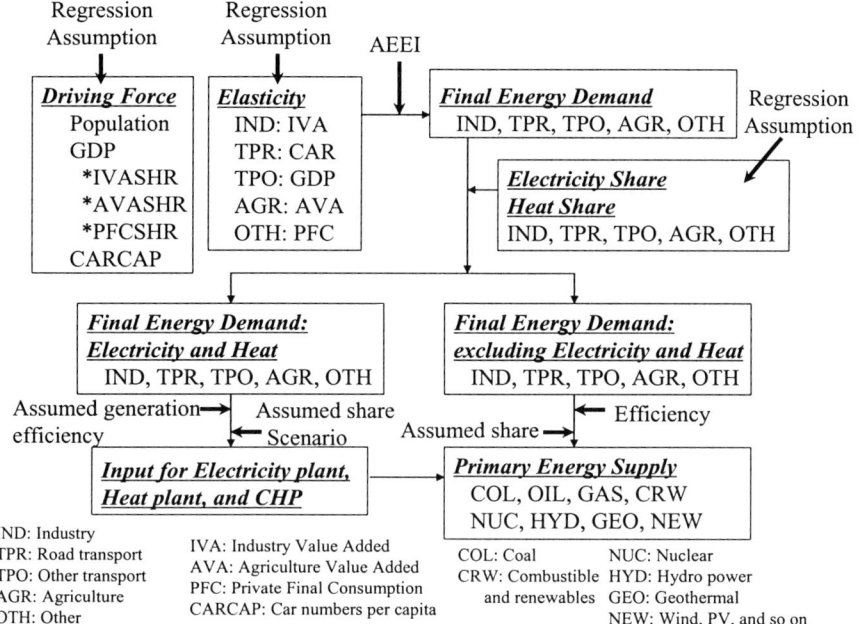

Fig. 3. Calculation flow of detailed data model (Model A)

(nuclear), HYD (hydro power), GEO (geothermal), NEW (wind, PV, and so on), HET (heat), and ELE (electricity).

(2) Energy sector classification

Energy conversion sector has following 5 sub-sectors: ELP (power generation), HTP (heat supply system), CHP (combined heat and power plant), DST (distribution of energy), and TFM (transformation). The final energy demand sector has following 6 sectors: IND (industry), TPR (transport on road), TPO (other transport), AGR (agriculture), OTH (other), and NEU (non-energy use).

(3) Estimation of final energy demand

Final energy demand except NEU is described as the function of driving force (DRV). Driving force of IND, TPR, TPO, AGR, and OTH is basically defined as industrial value-added (IVA), numbers of car (CAR), GDP, agricultural value-added (AVA) and private final consumption expenditure (PFC), respectively. Elasticity between each final energy demand and driving force is calculated by regression analysis using historical data. If these data are not available, GDP can be used for the driving force. Following equation is assumed:

$$TFE_i(t) = A_i(t) \times TFE_i(t_0) \times \{DRV_i(t)/DRV_i(t_0)\}^{ELS_i} \qquad (1)$$

$$A_i(t) = (1 - AEEI_i(t)/100)^{(t-t_0)}$$

where,

$TFE_i(t)$: total final energy demand for sector i, time period t

$DRV_i(t)$: driving force for sector i, time period t

ELS_i : elasticity for sector i

$AEEI_i(t)$: autonomous energy efficiency improvement for sector i, time period t

i : final energy demand sector, i = {IND,TPR,TPO,AGR, OTH}

t : simulation time period, t_0: initial time period

NEU (Non-energy use demand) is estimated from the following equations. It is supposed that NEU consists of oil by historical data.

$$NEU(t) = FE_{IND,OIL}(t)/FE_{IND,OIL}(t_0) \times NEU(t_0) \qquad (2)$$

where,

$FE_{i,e}(t)$: final energy demand for sector i, energy e, time period t

e : energy, e = {OIL, COL, GAS, CRW, NUC, HYD, GEO, NEW, ELC, HET}

(4) Estimation of driving force

To estimate the trajectory of driving force (IND, TPR, TPO, AGR, and OTH), IVASHR (GDP share of IVA), CARCPT (car numbers per capita), PFCSHR (GDP share of PFC), and AVASHR (GDP share of AVA) are estimated from regression analysis using GDP per capita for independent variables. To estimate IVASHR in other way SVASHR (GDP share of SVA (service value added) is also estimated. The relationships between AVASHR, SVASHR and GDP per capita using South Asia, East Asia, and South East Asia data from 1980 to 1995 of HYDE/RIVM are shown in Figs. 4 and 5 (Klein, *et al.* 1995).

(5) Share of energy in final energy demand

Final energy demand consists of electricity, heat, and others.
a) Share of electricity: The share of electricity in each final energy sector is estimated by using the regression analysis using driving force for independent variable.
b) Share of heat: The share of heat is fixed at that in the latest data existed year in all sectors.
c) Share of other energy: The share of fossil fuel energy is estimated by using the regression analysis using driving force for independent variable.

(6) Fuel share and energy efficiency in energy conversion sector

CHP and HTP are only used in specific countries in the Asia-Pacific region. Share of fuel input into HTP and share of fossil fuel input into CHP are assumed to be

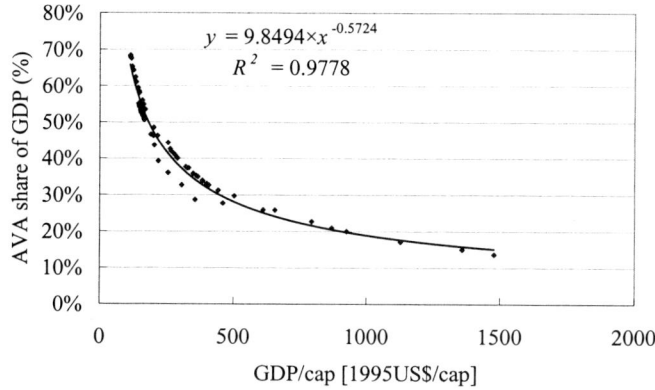

Fig. 4. Relationship between GDP per capita and AVA share of GDP

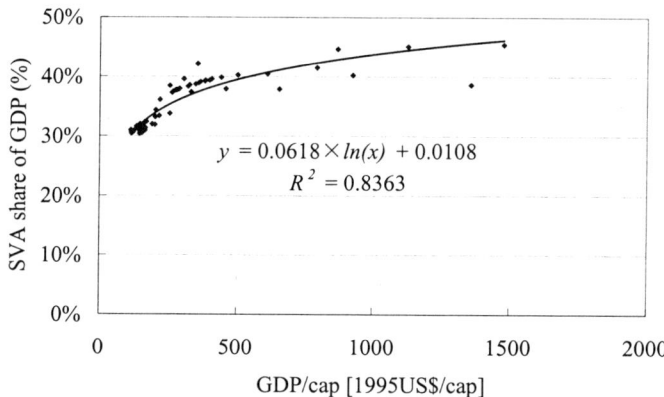

Fig. 5. Relationship between GDP per capita and SVA share of GDP

constant. Share of fossil fuel input into ELP is calculated by regression analysis. Non-fossil fuel input into CHP and ELP are depended on scenario assumption. The electricity generation efficiency of ELP and CHP with COL, OIL, GAS, and CRW is assumed by using exogenous energy efficiency improvement parameter. The generation efficiency of NUC, HYD, GEO, and NEW is fixed as IEA's definitions (NUC: 0.33, HYD: 1.0, GEO. 0.1, NEW: 0.1). The heat generation efficiency is assumed to be fixed as that in initial simulation period.

(7) Total primary energy supply

Primary energy supply is calculated with final energy demand, energy conversion process, and distribution loss. Distribution loss of fossil fuel, electricity, and heat is assumed to be constant as that in initial simulation period.

(8) GHG (CO_2, NO_X, SO_X, CH_4, N_2O, CO) emissions

Energy related GHG emissions are calculated by simulation result and assumed emission factor. NOx and SOx emissions are assumed to be reduced according to increase of GDP per capita, known as Kuznets curve.

Simple data model (Model B)

For the several countries, that are not available of IEA energy statistics data, Model B is constructed for estimation of environmental loads in the future.

UN Energy Statistical Yearbook 2000 (UN 2000) is used as energy data. Energies are categorized as Liquids, Solids, Gas, Electricity and Traditional fuelwood. Liquids, Solids, and Gas correspond to OIL, COL, and GAS in Model A respectively. Electricity (ELC) consists of supply from geothermal, hydro, nuclear, solar, tide, wind, wave, import and export. Traditional fuelwood (TRF) corresponds to CRW in Model A. Each energy supply is assumed to be decided by GDP and AEEI.

This can be given by the following equation:

$$PE_e(t) = A_e(t) \times PE_e(t_0) \times \{GDP(t)/GDP(t_0)\}$$
$$A_e(t) = (1 - AEEI_e(t)/100)^{(t-t_0)}$$

(3)

$PE_e(t)$: primary energy supply for energy e, time period t

$GDP(t)$: GDP, time period t

$AEEI_e(t)$: autonomous energy efficiency improvement for energy e, time period t

e : energy, e ={OIL, COL, GAS, ELC, TRF}

t : simulation time period, t_0: initial time period

Model data

Following data are mainly used:

- Population: historical data (WB 2000), projection (UN 1998)
- GDP: historical data (WB 1999), projection (IMF 2001) (EIA/DOE 2001), IPCC/SRES scenario (IPCC 2000)
- Energy: for Model A (IEA 2001), for Model B (UN 2000)
- Emission factor: (IPCC 1996)

13.2.3 Model interface

AIM/Trend model is executed with ATPL (AIM/Trend Program Language) that is built with VBA of Microsoft Excel. Major commands consist of load, save, future parameter setting, future projection, format and regression. An illustration of AIM/Trend detailed data model (Model A) interface is shown in Fig. 6. Users can choose country and select cases for load and save. Clickable buttons such as Projection All, and Pam Set All are designed to perform simulations effectively. They are written in ATPL and users can write or change programs for their own purpose.

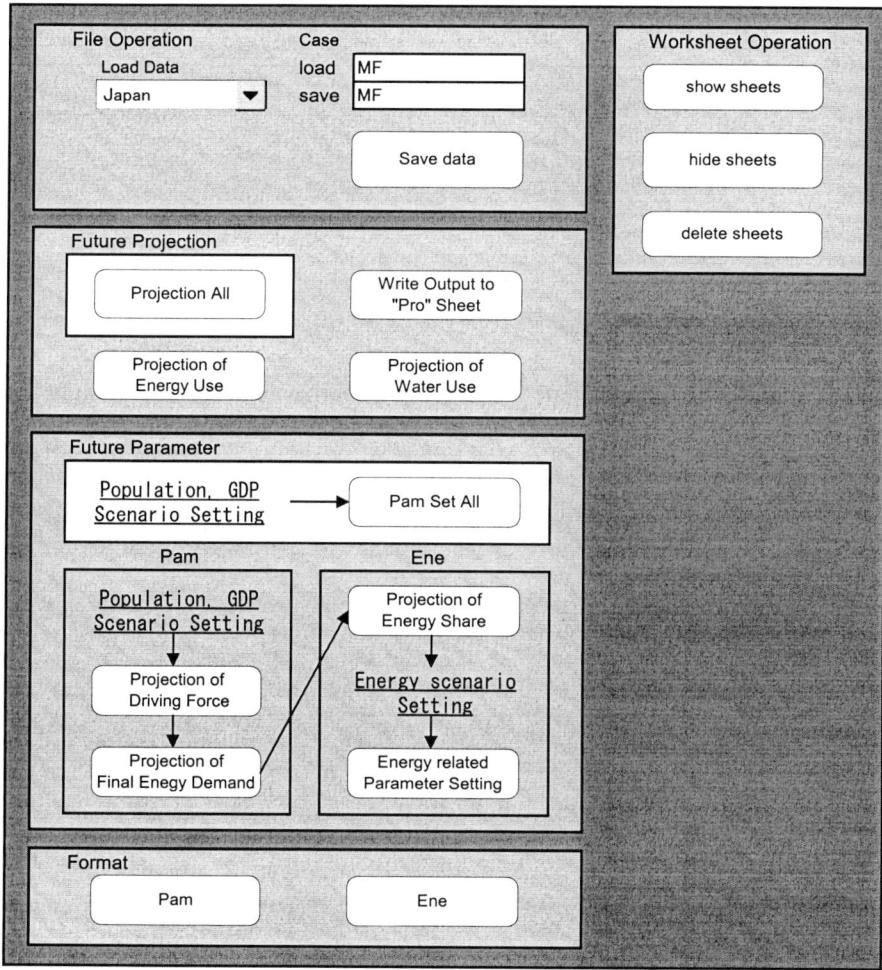

Fig. 6. AIM/Trend interface for detailed data model (Model A)

13.3 Middle-term Economy, Energy, Environment Scenario Simulation in the Asia-Pacific Region

13.3.1 Scenario study

UNEP published GEO-3: Global Environment Outlook 3 (UNEP 2002) before Johannesburg Summit. It focused on future scenarios from 2002 to 2032. The results of AIM/Trend model were adopted as the future projections in the Asia-Pacific region. GEO-3 prepared following 4 scenarios: Market First scenario, Policy First scenario, Security First scenario, and Sustainability First scenario. The Market First scenario envisages a world in which market-driven developments coverage on the values and expectations that prevail in industrialized countries; In the Policy First world, strong actions are undertaken by governments in an attempt to reach specific social and environmental goals; The Security First scenario assumes a world of great disparities, where inequality and conflict prevail, brought about by socio-economic and environmental stresses; and Sustainability First pictures a world in which a new development paradigm emerges in response to the challenge of sustainability, supported by new, more equitable values and institutions.

To address GEO-3 scenarios with AIM/Trend model, several parameters (GDP, population, AEEI, energy conversion efficiency, non-fossil energy supply, GHG emission control and so on) are set for each scenario.

13.3.2 Overview results

Figure 7 shows the historical data and future projection of major economy, energy, and environment indices in China. GEO3/Sustainablity First scenario is used to estimate future conditions of environmental loads.

Population will reach 1,500 million around 2030, though rate of increase in population will be decreasing annually. GDP will increase steadily and GDP per capita will become $5,000 in 2030. Population and GDP assumptions are given as scenario.

GDP share, which is driving force of final energy demand, is projected by regression. According to growth of GDP per capita, AVA share will decrease and SVA share will increase. This is a common phenomenon observed in developed countries. Though IVA share increased in the 4th quarter of 20th century, it will be replaced by SVA share in 21st century.

Total final energy demand will increase due to the growth of driving forces. It mainly consists of industry sector and other sectors. Total primary energy supply will also increase and it will reach at around 2,500MTOE in 2030. equivalent to 30% of 1999 total primary energy supply in the world. It is mainly formed from coal, oil, and CRW. Though CO_2 emissions will become close to 2 Gt-C in 2032, its rate of increase will slow down because of introduction of nuclear and expansion of natural gas. CO_2 intensity will also decrease after 2015. Energy intensity decreased after 1970 dramatically. This decreasing trend is projected to continue in future.

Though SO_2 emissions will increase after 2000, but it will start to decrease after 2015. NO_x emissions will also begin to decline around 2020. It is well-known as Kuznets effect that air polluting sources such as SO_2 and NO_x will be reduced according to economic growth. Kuznets effect is more pronounced and is projected to occur earlier for SO_2 emissions than for NO_x emissions.

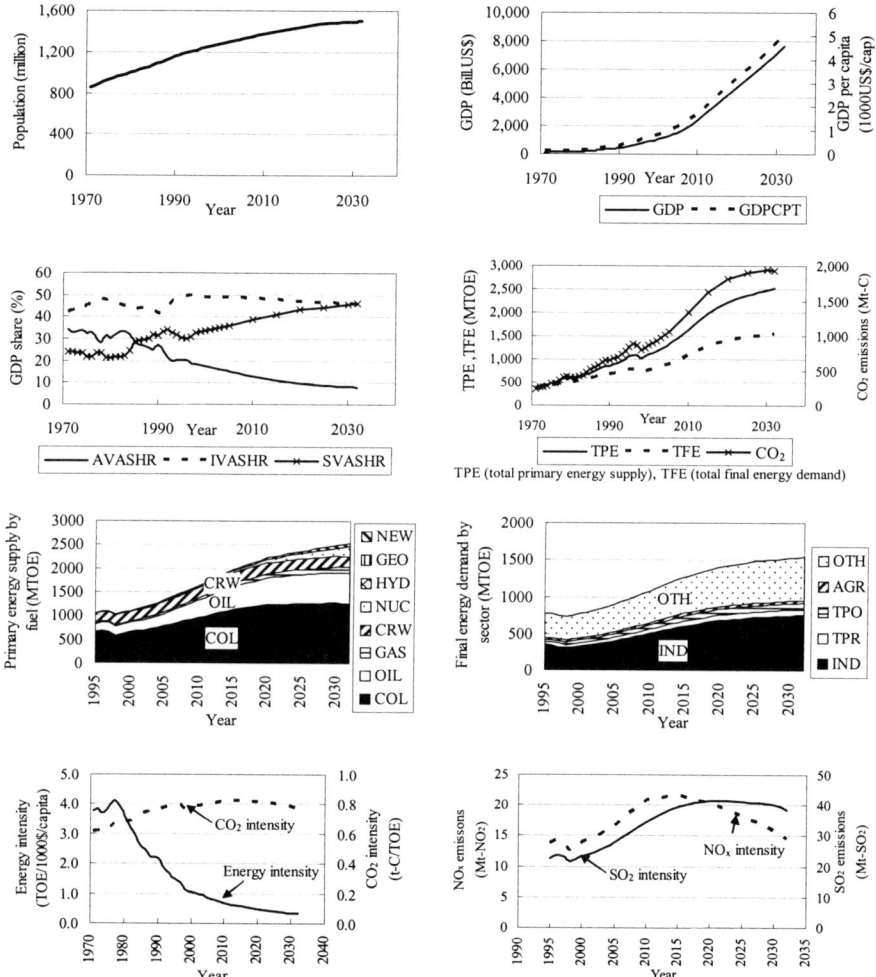

Fig. 7. Projection of each index for China

Figure 8 shows the projection of CO_2 emissions per capita for each country in GEO3/Sustainability First scenario. Countries are arranged in order of CO_2 emissions per capita in 2032. China and India, which have larger population in Asia-Pacific countries, will be located in the middle position in 2032. Singapore and Brunei, which have special energy structure, will stay at larger CO_2 per capita emissions through 2032.

Asia-Pacific countries exhibit a wide range of CO_2 per capita emissions. This highlights the importance of country-level study in this region.

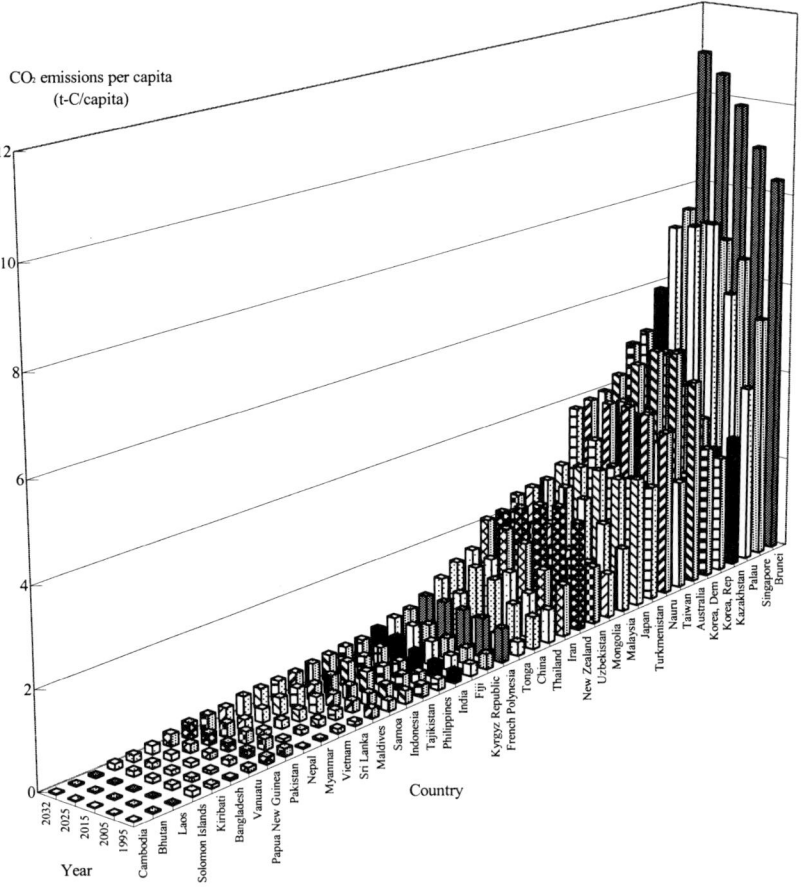

Fig. 8. Projection of CO_2 emissions per capita for each country

13.3.3 Energy related GHG emissions results

Energy related CO$_2$ emissions

Figure 9 shows the energy related CO$_2$ emissions in the Asia-Pacific region in the four scenarios. Compared to Policy First and Sustainability First, emissions of CO$_2$ increase more rapidly in Markets First circumstances because of high economic growth. In Policy First, advanced technologies are introduced to reduce CO$_2$ emissions. Because a Sustainability First society shifts from conventional to sustainable lifestyles, CO$_2$ emissions are somewhat mitigated. On the other hand a Security First society holds on to technologies with low energy efficiency. CO$_2$ emissions increase most rapidly in this scenario everywhere except in Central Asia where low economic activities mitigate CO$_2$ emissions vis-à-vis Markets First.

Figure 10 shows change in energy related CO$_2$ emissions by 2032 relative to 2002 in sub-regions of the Asia-Pacific region. China in the East Asia sub-region, India in South Asia, and Kazakhstan in Central Asia experience maximum change in CO$_2$ emissions over the 30 year period while the ANZ and South Pacific sub-region including Australia and New Zealand experiences least change in all the scenarios. An interesting observation is that the ANZ and South Pacific sub-region and Japan show a decline in CO$_2$ emissions in the Policy First scenario. In all sub-region but Central Asia, CO$_2$ emissions are higher in the Security First scenario than in the Markets First scenario for the reason mentioned in the previous paragraph.

Energy related SO$_2$ emissions

As with the CO$_2$ emission factor, SO$_2$ emission factor is set for each country. Kuznets curve, which postulates an inverse U relationship between income level and pollution, are assumed to deal with the change of emission reduction rate according to economic growth. In modeling this phenomenon, thresholds are set as the function of GDP per capita and the emission reduction rate changes at these thresholds.

SO$_2$ emissions will increase most rapidly in the Security First scenario because little money is invested to reduce SO$_2$ emission in a low economic growth world (Fig.11). On the other hand, in the other scenarios, the increase of SO$_2$ emissions will be slow as measures are taken to avoid severe air pollution. This is especially true in Policy First and Sustainability First where SO$_2$ emissions will be controlled more strictly. Due to the Kuznets effect, a downward trend is observed in Policy First and Sustainability First scenarios from around 2015 while it is observed in Markets First from around 2025. It is not observed in Security First scenario. Results for changes in SO$_2$ emissions between 2002 and 2032 (Fig.12) provide some interesting observations. In the ANZ and South Pacific, SO$_2$ emissions are reduced by 40 to 50 percent between 2002 and 2032 in all scenarios. In modeling the scenarios, it is assumed that all the industrialized countries would reduce their SO$_2$ emissions regardless of the scenario. Australia dominates emissions in the ANZ and South Pacific and therefore the whole area appears to be decreasing its

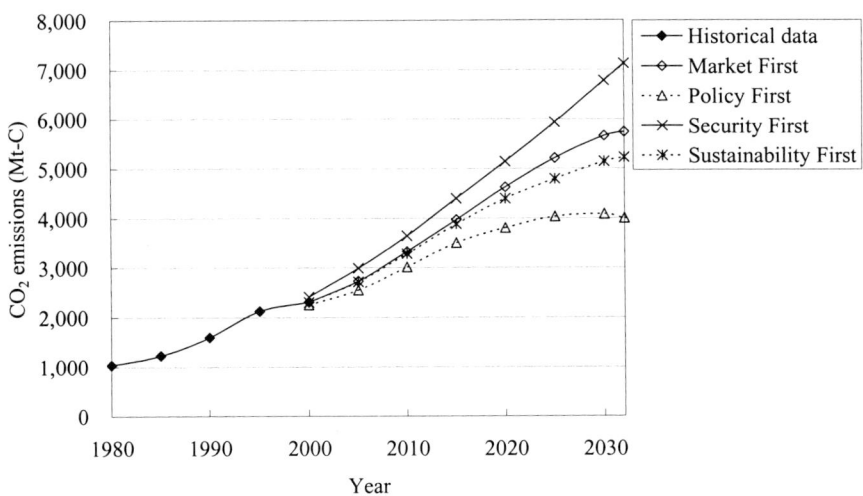

Fig. 9. Energy related CO_2 emissions in the Asia-Pacific region (see color plates)

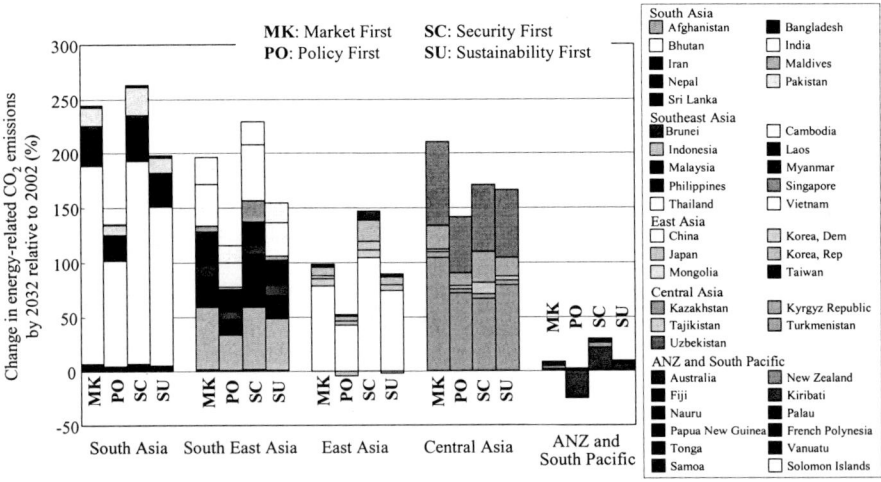

Fig. 10. Change in energy related CO_2 emissions in sub-regions of the Asia-Pacific region (see color plates)

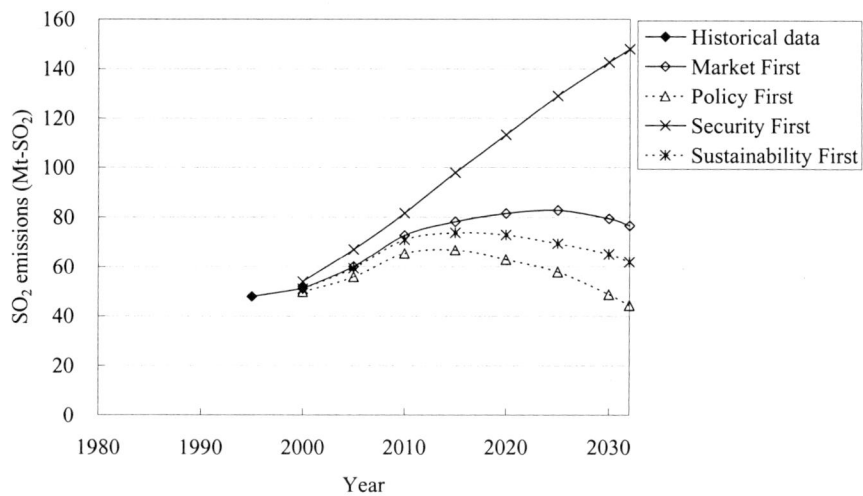

Fig. 11. Energy related SO$_2$ emissions in the Asia-Pacific region (see color plates)

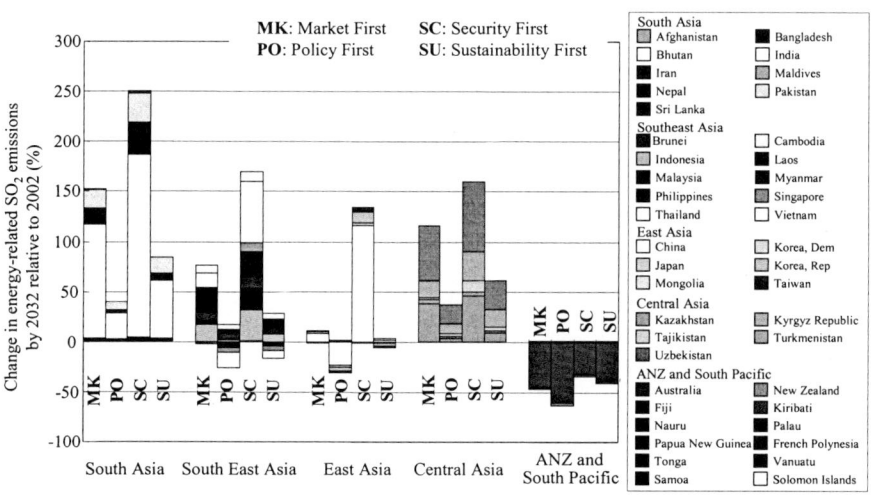

Fig. 12. Change in energy related SO$_2$ emissions in sub-regions of the Asia-Pacific region (see color plates)

SO_2 emissions. In the East Asia sub-region, China's emissions reduce at a slower rate than those of Japan. As a result, the sub-region on the whole increases its emissions in all scenarios except Policy First scenario. Emissions also reduce for some countries in Southeast Asia and East Asia sub-regions under Policy First and Sustainability First scenarios. In all the regions except ANZ and South Pacific emissions increase most rapidly in Security First scenario because of relatively low investment in emissions reduction, especially in India and China where the emissions more than double over the 30 year period.

13.4 Concluding Remarks

In this chapter, the framework of AIM/Trend model and its results are presented. This model has been developed to assess the future environmental loads based on the past socio-economic trends and scenarios though 2032 in the Asia-Pacific 42 countries of the Asia-Pacific region. It is built as a simple econometric model and written in ATPL using VBA in Microsoft Excel. The user can change the model structure and make sensitivity analysis easily. Results of AIM/Trend model show that the Asia-Pacific region is going to experience a rapid growth in CO_2 emissions in all UNEP/GEO-3 scenarios, and Asia-Pacific countries have a wide range of economy, energy, and environment conditions. It makes the country-level study in this region more important to deal with the diversity of Asia-Pacific countries. In future, AIM/Trend model will be improved to make people interested in immediate actions to control the incidence of increased environmental loads on air, land as well as water.

References

EIA/DOE (2001) International energy outlook 2001. EIA's National Energy Information Center, Washington, D.C.
Fujino J (2001) Development of AIM-Trend model as a communication tool to enhance discussions about prospect of energy and environment in each Asia-Pacific country. In: Proceeding of IFAC Workshop on modeling and control in environmental issues, Yokohama, pp109-114
IEA (2001) Energy statistics 2001 CD-ROM. IEA, Paris
IMF (2001) World economic outlook May 2001. IMF, Washington, D.C.
IPCC (1996) Revised IPCC guidelines for national greenhouse gas inventories: reference manual, vol 3. Hadley Centre, Bracknell
IPCC (2000) Special report on emission scenarios. Cambridge University, Cambrige Press
Klein, Goldewijk CGM, Battjes, JJ (1995) The IMAGE 2 hundred year (1890-1990) data base of the global environment (HYDE). RIVM rapport 481507008
UN (1998) World population prospects : The 1998 revision. UN publication, New York
UN (2000) Energy statistics yearbook 1997. UN publication, New York
UNEP (2002) Global environment outlook 3 2002. Earthscan Publications, London
WB (2000) World development indicators 2000 CD-ROM. WB, Washington, D.C.

14. AIM/Common Database: A Tool for AIM Family Linkage

Go Hibino[1], Yuzuru Matsuoka[2], and Mikiko Kainuma[3]

Summary. Several teams have been developing AIM models in different countries to analyze their own policy options to mitigate greenhouse gas emissions. Although they need to collect their own data, there are a lot of data that can be shared. The AIM/Common Database, consisting of common formats and code system, has therefore been developed as a tool for AIM Family linkage. There are several advantages by using this database. First, as the formats and code system are common, one team can use data collected by the other teams without special data conversion. Thus the duplication of data collection and conversion can be greatly avoided. It also helps in setting up country specific data by referring to the data provided by other teams. Second, outputs of one model can be used by another model without having to check the data formats. It helps to share the relevant outputs of different models in AIM family. Third, as AIM/Common Database can provide a user-friendly interface based on the common data format, it facilitates users to access input and output data stored in the database.

14.1 Background

The AIM Family consists of a number of models, including AIM/Enduse, AIM/Trend, AIM/CGE, AIM/Material, AIM/Climate and AIM/Impact. Each model has been developed in various countries such as China, India, South Korea and Japan. As many teams in different countries are involved in model development, AIM/Common Database plays a very important role in the AIM Family by linking the different models and modelers in different locations.

It is necessary to collect and process a large amount of statistical data in the model development and application stages. There is a significant overlap in the data required by different models and modeling teams. To ensure efficient implementation of this work, there is a need to develop standard database system containing common formats and codes.

Within the AIM Family, the simulation results of one model are often utilized as input by other models. Additionally, data and results of one modeling team are often useful to other teams as well. Moreover, users of the simulation results are not limited to the AIM development teams, but also include researchers, government officials, NGOs, etc., who use them as a reference in their decision-making.

---

[1] Fuji Research Institute Corporation, Tokyo 101-8443, Japan
[2] Kyoto University, Kyoto 606-8501, Japan
[3] National Institute for Environmental Studies, Tsukuba 305-8506, Japan

With this large number of widely dispersed users, data sharing must be carried out under a unified code system and unified file format. Otherwise, each user would have to learn the data structure for each model independently, which would impair efficiency.

The AIM/Common Database was developed in order to solve these problems and construct linkages for the AIM Family.

14.2 Outline of AIM/Database

Figure 1 shows an outline of AIM/Database. AIM/Database contains various types of datasets including statistics, outputs by AIM Family modelers, outputs by other modelers, and estimates by international organizations and governments.

AIM/Database consists of the AIM/Primary Database and the AIM/Common Database. The AIM/Primary Database is made up of observed value and estimated value by international organizations, national governments and research institutions. These file format is the original state, so is not standardized. There are magnetic media including statistics edited by international organizations, the first processed statistics for the AIM/Common Database, worksheets input manually from book type statistics, and so on.

On the other hand, the AIM/Common Database is used directly by other models in the AIM Family. The code systems and data table formats are standardized so as to allow easy use of the data. The AIM/Common Database mainly stores datasets related to observed values quoted from statistics and estimated values obtained by the models and organizations with a specific format. The statistical data are stored automatically by a module from the AIM/Primary Database. The module makes it easy to update the statistical data. As modelers use datasets in the AIM/Common Database, the module developed by each modeler imports data from the database directly. The modelers do not necessarily have to use the statistics for all input data in the database. As for the output of the model, the modelers are required to store data in the AIM/Common Database in a specific format in view of data sharing. Both manual operation and the use of modules developed by modelers are acceptable as export methods. Users other than AIM development teams, such as administrators, other modelers, NGOs, and so on can therefore efficiently use the datasets though the interface of the database without knowing the detailed data structure of each model.

At the operational level, AIM/Database is a decentralized system. Specific databases and interfaces are maintained by different users in a decentralized way. However, in order for the work related to data gathering and data exchange to proceed smoothly, certain guidelines have been established regarding the development and management of the AIM/Common Database.

The AIM/Common Database is distributed via a Web site, and the AIM project teams in individual countries download the database files from this Web site. The Web site has an area where access is restricted to the development teams only, and an area accessible to the public. The development data are published only in the

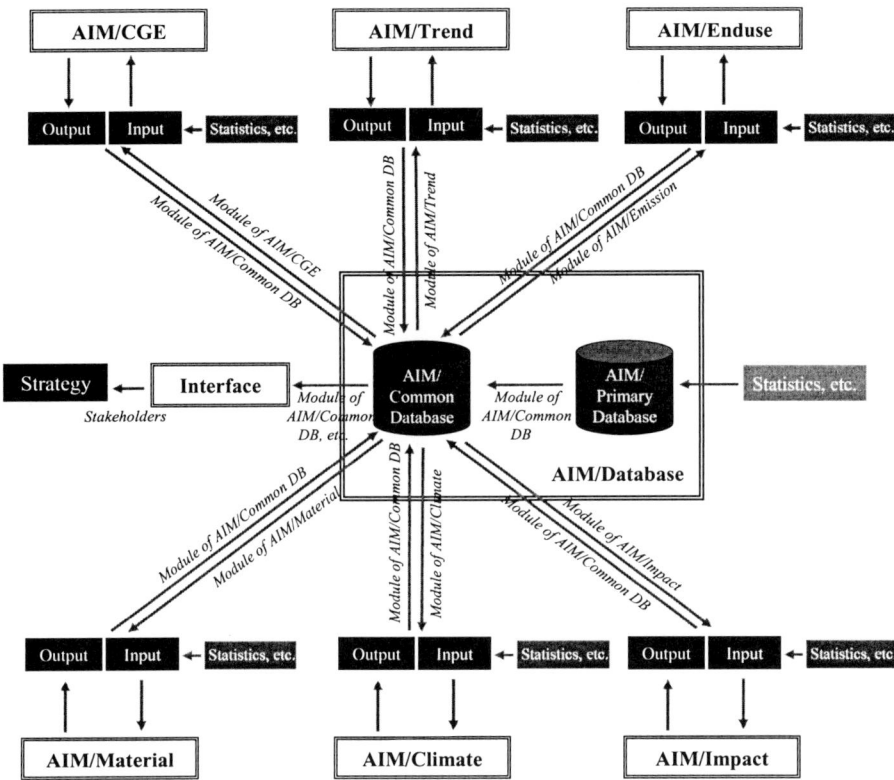

Fig. 1. Outline of AIM/Database

area restricted to the development teams. Users other than AIM development teams, such as researchers, policy makers, and NGOs, can obtain the output of the AIM Family from the site through the area accessible to the public.

14.3 AIM/Primary Database

The AIM/Primary Database is made up of observed value and estimated value by international organizations, national governments and research institutions. These file format is the original state, so is not standardized. The formats vary among the different materials; e.g., magnetic media, books, Web sites, copies of theses, and so on. Magnetic media include not only media such as CD-ROMs and so on issued by statistics institutions, research institutes, etc., but also worksheet format files of data from printed matter input by AIM development teams, and files of data contained in CD-ROMs, etc. that have been converted or processed into a format allowing them to be easily imported into the AIM/Common Database.

With regard to the components of the AIM/Primary Database, lists describing the items such as Media, Language, Data format, Abstract, Data collection period, Published year, Price, etc. have been prepared in order to maintain and manage the components.

14.4 AIM/Common Database

14.4.1 Structure of AIM/Common Database

The AIM/Common Database contains a large number of database files in Microsoft Access format. One of these is a database file that manages the code system of the AIM/Common Database, Code.mdb. The others are database files that mainly manage the statistical datasets and output of AIM Family models. These datasets are stored in the data table of the database file. Table 1 shows the data table in Energy_IEA.mdb. The table stores the country-wise, sector-wise, and energy-wise energy flows quoted in IEA (2001a, 2001b). The data table format in the other database files is almost the same as this data table. The codes ITEM, COMMODITY, SECTOR, COUNTRY, and REFERENCE in the fields correspond to the codes managed in Code.mdb.

Database file to manage code system, Code.mdb

Code.mdb is a database file that manages the code system of the AIM/Common Database. The database file contains tables listing the codes for items, commodities, sectors, countries, districts, and references. Each table is explained in turn below.

Code_Item. Code_Item is the code table for items (Table 2). An item shows what value the data indicate. An item code with the same unit is used in all the tables. The unit should always be listed in the NAME field. Input of the item is indispensable in the data tables.

Code_Commodity. Code_Commodity is the code table for commodities (Table 3). The code is used as the data table stores commodity-wise data; e.g., industrial commodity production, energy consumption (energy-wise), or agricultural commodity production. When the commodity code is not required, it can be omitted in the data table.

Table 1. Data table in Energy_IEA.mdb

ITEM	COMMODITY	SECTOR	COUNTRY	YEAR	VALUE	REFERENCE
EN_FLW	EN_COL	TPES	AUS	1961	16107	IEA_99
EN_FLW	EN_COL	TPES	AUS	1962	16394	IEA_99
....	....	....	....	......	....	....

Table 2. Code_Item code table in Code.mdb

ID	CODE	NAME	NOTE
101	EN_FLW	Energy flow (kTOE)	
102	EN_ELE	Electricity output (GWh)	
...	......	......	......
201	PP_TTL	Female population (thousand)	
202	PP_MAL	Male population (thousand)	
...	......	......	......
401	IN_PRD	Industrial commodity production (tons)	
402	IN_PRD_L	Industrial commodity production (meters)	
...	......	......	......
501	NA_GDP_DF	GDP deflator	
502	NA_GDP_CD	GDP (current million US$)	
503	NA_GDP_CN	GDP (current million LCU)	
...	......	......	......

Table 3. Code_Commodity code table in Code.mdb

ID	CODE	NAME	NOTE
...	......	......	......
203	EN_OLK	Kerosene	
204	EN_OLG	Gasoline	
205	EN_OLD	Diesel oil	
...	......	......	......
401	MN_IRN	Iron ore	
402	MN_LMS	Limestone	
...	......	......	....

Table 4. Code_Sector code table in Code.mdb

ID	CODE	NAME	NOTE
...	......	......	......
201	IN_STL	Iron and steel	
202	IN_CHM	Chemical and petrochemicals	
203	IN_NMT	Non-metallic minerals	
...	......	......	......
401	OT_AGR	Agriculture	
402	OT_CMM	Commercial and public services	
403	OT_RSD	Residential	
...	......	......	....

Code_Sector. Code_Sector is the code table for sectors (Table 4). The code is used as the data table stores sector-wise data; e.g., energy consumption or value added. When the sector code is not required, it can be omitted in the data table.

Code_Country. Code_Country is the code table for countries (Table 5). The relationships between the codes and major organizations' codes are shown in the table. The definition of a country, such as changes resulting from separation or amalgamation, is listed in the NOTE field. Input of the country is indispensable in the data tables.

Code_District. Code_District is the code table for districts in a country (Table 6). The code is used as the data table stores district-wise data. When the district code is not required, it can be omitted in the data table.

Code_Reference. Code_Reference is the code table for references (Table 7). The data sources of values in the data table are listed. Input of the reference is indispensable in the data tables.

Table 5. Code_Country code table in Code.mdb

ID	CODE	NAME	UN	WB	ISO1	ISO2	....	NOTE
4	AFG	Afghanistan	AFG	AFG	AFG	4		
8	ALB	Albania	ALB	ALB	ALB	8		
12	DZA	Algeria	DZA	DZA	DZA	12		
16	ASM	American Samoa	ASM	ASM	ASM	16		
20	AND	Andorra	AND	ADO	AND	20		
24	AGO	Angola	AGO	AGO	AGO	24		
28	ATG	Antigua and Barbuda	ATG	ATG	ATG	28		
31	AZE	Azerbaijan	AZE	AZE	AZE	31		
32	ARG	Argentina	ARG	ARG	ARG	32		
36	AUS	Australia	AUS	AUS	AUS	36		
40	AUT	Austria	AUT	AUT	AUT	40		
......	......		......	......	......	......		

* UN: Country code used in UN statistics (2001), WB: Country code in World Bank (2002), ISO1: ISO3166 code A3, ISO2: ISO3166 code number

Table 6. Code_District code table in Code.mdb

ID	COUNTRY	CODE	NAME	NOTE
1	IND	101	Adiladbad	
2	IND	102	Anantapur	
3	IND	104	Cuddapah	
......	......	......	......	......

Table 7. Code_Reference code table in Code.mdb

ID	CODE	NAME	NOTE
101	IEA_99	IEA Energy Balance 1999	IEA Energy Balances of OECD Countries 1960-1999 (CD-ROM), IEA Energy Balances of Non-OECD Countries 1971-1999 (CD-ROM)
200	UN_STY_45	UN Statistical Yearbook Forty-fifth issue	UN Statistical Yearbook Forty-fifth issue (CD-ROM)
201	UN_CMM_01	UN Industrial Commodity Production 1999	UN Industrial Commodity Production Statistics Database 1950-1999 (CD-ROM)
202	UN_POP_01	UN World Population Prospects 2000	UN World Population Prospects 2000 (CD-ROM)
......	......	......	......

Database files to manage datasets

The database files other than Code.mdb mainly manage statistical datasets and output of AIM Family models. The structure of the files is shown in Fig. 2. The two upper tables manage datasets, while two tables, one form, and one module play the role of importing data from the AIM/Primary Database and the output of other models. The tables, form, and module are described below.

Data table. This table consists of data imported from the statistical data in the AIM/Primary Database and the output of models using the unified code system. The table format is standardized and is composed of the Item, Commodity, Sector, Country, District, Year, Value, and Reference fields.

Code table. This table shows explanations concerning the codes used in the Item, Commodity, Sector, Country, District, and Reference fields in the Data Table. It is linked with the table in Code.mdb.

Import file table. This table is used when the module imports statistical data from the Primary Database or the output of models to the Common Database. The table indicates from which item and which file in the Primary Database to import the data and to which item in the Common Database.

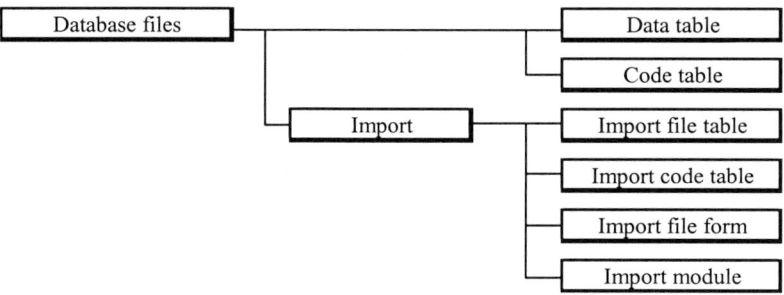

Fig. 2. Structure of database files

Import code table. This table is used when the module imports statistical data from the Primary Database or the output of models to the Common Database. The table indicates which code in the Common Database corresponds to the code in the Primary Database.

Import form. This form is used when the module imports statistical data from the Primary Database or the output of models to the Common Database. It is the interface for importing data from the Primary Database and users can execute the module on the form.

Import module. This module is used when the module imports statistical data from the Primary Database or the output of models to the Common Database. It is written with Access VBA.

14.5 Development and Management of AIM/Common Database

Maintenance and management of the AIM/Common Database is not carried out by a specific team in a centralized way, but in a decentralized way by the model development teams of each country. However, in order for the work related to data gathering and data exchange to proceed smoothly, certain guidelines have been established within the AIM development team regarding the development and management of the AIM/Common Database. These guidelines are introduced below.

14.5.1 Planning phase

The guidelines on planning phase of AIM/Common Database are described below.

- AIM/Common Database is to be developed with using MS Access 2000 or later.
- Data table is to be made up of Item, Commodity, Sector, Country, District, Year, Value and Reference fields. Data in Item, Country, Year, Value, and Reference are indispensable.
- The code system (Code.mdb) of the AIM/Common Database is to be common to each development team in each country, and the codes in the data files must conform to this system.

14.5.2 Accumulation phase

The guidelines on accumulation phase of AIM/Common Database are described below.

- Coordination and management of the code system is to be carried out by AIM/Common Database coordinators (one to two persons for the overall AIM project).
- When creating a database file for the AIM/Common Database, if a new data source is used, the relevant information is to be added to the data source lists of the Primary Database following to the prescribed format.
- When a database file of the AIM/Common Database is created or updated, the document is to be prepared following the prescribed format for the contents of the file.
- When a database file of the AIM/Common Database is created or updated, notification to that effect is to be given on the AIM Development Team Web Site.
- When the creation of new data is desired, notification to that effect is to be given on the AIM Development Team Web Site.

14.5.3 Communication phase with AIM family models

The guidelines on communication phase with AIM family models via AIM/Common Database described below.

- Users within the AIM development teams need to have basic knowledge of MS Access (database structure, how to view data, copying to MS Excel, etc.).
- A precondition for the development of AIM Family models is that they conform to the data format and code system of the AIM/Common Database.
- The output of AIM Family models is to be provided following the data format and code system of the AIM/Common Database so as to enable smooth exchanges of data between different teams.
- Since the database files of the AIM/Common Database are in Microsoft Access format, modelers need to learn how to extract data files in MS Access.

14.5.4 Dissemination phase

The guidelines on dissemination phase of AIM/Common Database are described below.

- AIM/Common Database is to be disseminated for external uses via Web site.
- For external users of data accessible to the public, an interface is to be provided that enables use of the data without knowledge of MS Access. Therefore, modelers of AIM Family are to provide the aggregated results in the html format for external users.
- Concerning detailed data, the files including simulation results are to be uploaded in MS Access format. External users download the file and use it on their end. Modelers are to provide not only data table but also pivot table and pivot chart in the file. The table and chart make it easy for external users without knowledge of MS Access to use data.

14.6 An Example: How to Interact with AIM Family Models

Linkage between AIM/Enduse and AIM/Trend through the AIM/Common Database is described here as an example of AIM Family linkage.

The AIM/Enduse model is the tool to estimate future greenhouse gases emission with using detailed, sector-by-sector information about future driving forces given externally. For example, the production volume of each industrial product in the industrial sector and the demand volume of each application (e.g., heating and air conditioning) in the home sector must be given. As most of the estimates provided by governments and international research institutes are macro indicators such as economic growth rates and population, each analyst is often required to individually estimate the driving forces on a segment-by-segment basis. Analyses previously conducted in Japan, China, India, and Korea, based on end-use models, also used estimated segment-by-segment future driving forces, although they were based on the GDP, population, and other macro indicators predicted and planned by governments.

The prediction of these future values is based on a very simple methodology. First, regression equations are derived from the actual driving force volumes and GDP measurements. Then, the future driving force volumes are estimated using the regression values and future GDP measurements predicted by the governments or research institutes. Finally, the values obtained are converted into the form that the model can use for simulation. Despite the simplicity of the methodology, this series of actions is still not systematically documented, which makes it difficult to update the estimated values and transfer the process of estimate to others.

Under the circumstances, new capabilities were added to the AIM Family models and databases to construct a system that can systemize and automate the above series of actions. In addition, using the AIM/Trend Program Language (ATPL)

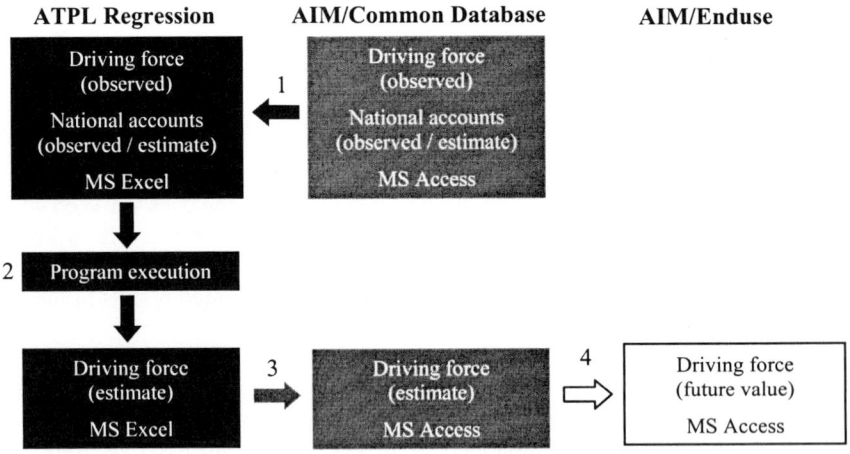

Fig. 3. Work flow: Linkage between AIM/Enduse and AIM/Trend through AIM/Common Database

developed by AIM Project Team (2001), ATPL Regression was constructed as a new program that derives future driving force volumes from the regression equations for driving forces and GDP measurements (Fig. 3 and Fig. 4).

1. The Import module of ATPL Regression imports the information about the actual driving force volumes and actual/estimated future value added figures contained in the AIM/Common Database into an Excel file.
2. The Regression/Projection module of ATPL Regression derives regression equations from the actual driving force volumes and actual value added figures. Then, the derived regression equations and future value added figures are used to estimate the future driving force volumes. The computation results are output as Excel data.
3. The Import module of AIM/Common Database imports the computation results provided by AIM/Trend into the AIM/Common Database as MS Access data.

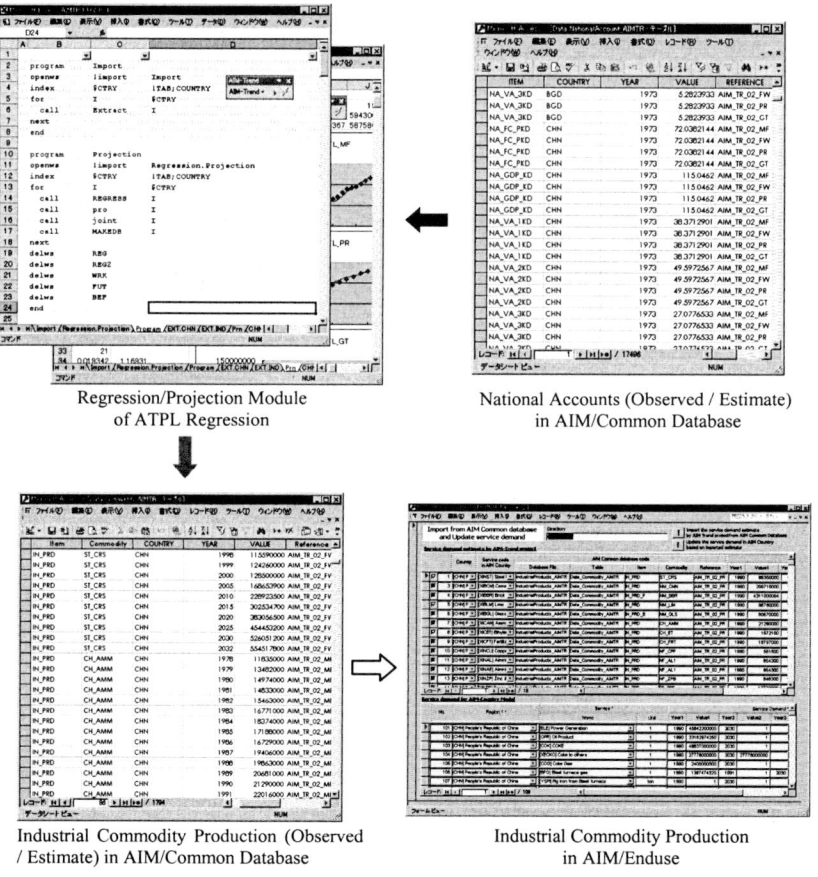

Regression/Projection Module
of ATPL Regression

National Accounts (Observed / Estimate)
in AIM/Common Database

Industrial Commodity Production (Observed
/ Estimate) in AIM/Common Database

Industrial Commodity Production
in AIM/Enduse

Fig. 4. Examples of model and database displays: Linkage between AIM/Enduse and AIM/Trend through AIM/Common Database

4. The Import module of AIM/Enduse imports the future driving force volumes contained in the AIM/Common Database into AIM/Enduse as MS Access data.

14.7 Concluding Remarks

There are several AIM modeling teams in Asian countries such as China, India, South Korea and Japan. Although they need to collect their own data, there are a lot of data that can be shared. If an efficient database system is developed to facilitate this sharing, data collection work can be reduced. The AIM/Common Database, consisting of common formats and code system, has therefore been developed as a tool for AIM Family linkage. Maintenance and management of the AIM/Common Database is not carried out by a specific team in a centralized way, but in a decentralized way by the model development teams of each country based on certain guidelines.

The following effects have been realized through application of the AIM/Common Database.

• A team is able to easily utilize statistical data gathered or processed by another teams, and overlapping of work is greatly reduced.
• Results of other models can be easily utilized as input data or reference data for a model, without having to check the data formats of the other models.
• The output of models is in a common format, allowing it to be readily used by stakeholders and promoting the efficient utilization of simulation results.

The AIM/Common Database should play an important role to link the models in AIM Family that are used for integrated assessment. In order to maintain the efficiency, it is essential for each modeling team to observe the guidelines for its proper maintenance and management.

References

AIM Project Team (2001) AIM Trend model user's manual. AIM Project Team, AIM/Trend Model distribution CD

IEA (2001a) Energy balances of OECD countries 1960-1999. International Energy Agency, CD-ROM

IEA (2001b) Energy balances of non-OECD countries 1970-1999. International Energy Agency, CD-ROM

UN (2001) Statistical yearbook. United Nations, CD-ROM

World Bank (2002) World development indicators. World Bank, CD-ROM

IV. Manual

A Guide to AIM/Enduse Model

Go Hibino[1], Rahul Pandey[2], Yuzuru Matsuoka[3], and Mikiko Kainuma[4]

1. AIM/Enduse: An Overview

1.1 Introduction

AIM/Enduse is a technology selection framework for analysis of country-level policies related to greenhouse gas emissions mitigation and local air pollution control. It can also assist in energy policy analysis. It simulates flows of energy and materials in an economy, from supply of primary energy and materials, through conversion and supply of secondary energy and materials, to satisfaction of enduse services. AIM/Enduse models these flows of energy and materials through detailed representation of technologies.

This guide is intended to enable the user to easily learn and work with AIM/Enduse model. This chapter provides an overview of the model. Theoretical formulation is explained in chapter 2. Methodology for data preparation, as explained in chapter 3, is written to guide the user in the process of collecting and estimating data required for the model. Chapters 4 and 5 illustrate application of the model to India and Japan. Thirteen appendices provide useful information like terms of reference, GAMS code, description of AIM/Enduse database system, list of international publications of relevant data, standard numbers for efficiency and cost of selected technologies, standard numbers for calorific value, price and emission coefficients of fuels, and classifications and data for AIM-India and AIM-Japan.

1.2 Concept of the Model

AIM/Enduse is a bottom-up model of technology selection within a country's energy-economy-environment system. Energy and material flows through technology systems in an economy, and consequent emissions, are modeled elaborately. Selection of technologies takes place in a linear optimization framework where system cost is minimized under several constraints like satisfaction of service demands, availability of energy and material supplies, and other system constraints.

---

[1] Fuji Research Institute Corporation, Tokyo 101-8443, Japan
[2] National Institute for Environmental Studies, Tsukuba 305-8506, Japan (on leave from Indian Institute of Management, Lucknow 226013, India)
[3] Kyoto University, Kyoto 606-8501, Japan
[4] National Institute for Environmental Studies, Tsukuba 305-8506, Japan

System cost includes fixed costs and operating costs of technologies, energy costs, and other costs like taxes or subsidies. The model can perform calculations simultaneously for multiple years. Various scenarios including policy countermeasures can be analyzed in AIM/Enduse. Key features of the model are described below (refer to Appendix A for definitions of concepts and terms used in the model and this manual).

1.2.1 Reference energy system

An integrated reference energy system is captured in the model. Energy and material flows through various technologies – from primary energy extraction, through energy conversion and supply, to energy-enduse or final service conversion and delivery – can be modeled in detail for a given year. Model's structure is flexible to permit modeling of energy and material flows and technology linkages to any degree of scope.

As an aggregated option, a user can model energy and material flows beginning from energy conversion technologies producing secondary energy from primary energy, and ending at energy enduse technologies producing final services. Alternatively, as a disaggregated option, a user can model these flows beginning from primary energy extraction technologies, passing through primary energy transportation technologies, energy conversion technologies, secondary energy transportation and distribution technologies, and ending at final service conversion and delivery technologies. In the disaggregated option, each technology can be further represented as a combination of multiple devices, as explained in section 1.2.2.

Energy, materials and services can be represented in two ways in the model. An energy or material that is supplied externally to the model is an external energy. Similarly, a service that is demanded externally to the model is an external or final service. Any energy or material that is produced and consumed within the model is an internal service or energy (as shown in Figs. 1.2). Flow of an internal service or energy is balanced within the model. This feature of AIM/Enduse allows extremely disaggregated modeling of reference energy system as described in previous paragraph.

Service demands are the main external drivers that trigger technology selection decisions based on information on costs. Energy-mix and material-mix are derived from the technology-mix. Finally, emissions of CO_2, SO_2, and NO_x are derived from the information on emission characteristics of energy, materials, and technologies.

1.2.2 Technology representation and selection

AIM/Enduse permits detailed representation of technologies where, on one hand, a device and a combination of emission removal processes can be coupled together (see Figs. 1.1 and 1.2), and on the other hand, multiple devices can be linked in a complex network of sequential and parallel relationships that are close to reality (see Fig. 1.3). Thus, in AIM/Enduse, concepts of device and removal process are integral to the concept of technology, and need to be carefully understood by the user.

A device may be defined as a distinct equipment or machine that is used in real life. Alternatively, a device may be defined as an aggregate concept comprising more than one distinct equipment or machines of real life (see Appendix A for definition and Appendices J and L for examples in case of AIM-India and AIM-Japan). A device can have multiple energy or material inputs and multiple outputs. An example of device is smelting process in aluminum-making technology that produces molten aluminum (internal service) from alumina (internal energy produced as internal service from Bayer's process) by utilizing electricity (internal energy produced as internal service from power plants) and fuel oil (internal energy produced as internal service from oil refineries). All devices are linked by a complex network of flows of energy and materials within a reference energy system of an economy.

Additionally, attachments or retrofits of specific combinations of emission removal processes to regular devices can be represented in the model (see Figs. 1.1, 1.2 and 1.3). A combination of emission removal process, when attached to a device, removes a particular pollutant gas emission during the device's operation. AIM/Enduse permits attachment of SO_2 and NO_x emission removal processes at three possible stages in a device's operation – pre-combustion stage, in-situ combustion stage, and post-combustion stage. A combination of removal process is a coupling of removal processes at these three stages. Coupling such a combination with a device results in a combination of device and removal process, as shown in Fig. 1.2. Coupling of a device with combination of removal process is optional in AIM/Enduse. Ash-removal process introduced before a boiler, limestone addition process during combustion of coal, and desulfurization of discharged flue gas, are examples of pre-combustion, in-situ combustion, and post-combustion removal processes respectively in a coal-fired power plant. Selection of a combination of removal process (defined as exchange of one combination by another) for attachment to a device is decided based on its costs and emission removal performance. For instance, under strict SO_2 mitigation target, the model may select SO_2 removal processes as attachments to high SO_2 emitting devices like high-sulfur coal-fired power plants. Thus AIM/Enduse allows detailed technological assessment for both regular devices and emission removal processes.

Two sets of decisions are made every year for technology selection in AIM/Enduse – level of capacity recruitment and level of operation of each device in a given year. Service demands determine the level of operating capacities required in a given year. Decision regarding recruitment-mix of devices in a given year is made based on their annualized capital costs. Decision regarding operating-mix of devices in a given year is made based on their annual running cost which comprises both energy costs and non-energy costs of operating the devices.

1.2.3 Stock transfers and multiple-year simulations

The model can be run simultaneously for multiple years as it simulates retirement of devices and transfer of un-retired stock of devices across successive years. Installed capacity of a device in a given year is represented by its stock in the model (see definition in Appendix A). Total stock of a device available in a given year is the sum of un-retired stock transferred from previous year and new stock recruited in the given year. Un-retired stock transferred from previous year is the total stock available in previous year minus the stock that retired at the end of previous year. A stock of a device retires after its life elapses from the year of its recruitment.

1.2.4 Outputs

Outputs of the model include recruited stocks of devices, operating quantities of devices, use of energy-types, level of flow of internal energy or services, emission quantities of CO_2, SO_2 and NO_x, system cost including recruitment costs, operating costs of devices and costs of exchanging removal processes, in each year. These outputs enable evaluation of a particular policy intervention or countermeasure on multiple criteria.

1.2.5 Scenario analyses and countermeasures

AIM/Enduse model permits analysis of various countermeasures. These are categorized as follows:
- Enduse stage countermeasures like efficiency improvement or better use and management of devices.
- Emission tax on fuels for each of the following gases – CO_2, SO_2 and NO_x.
- Energy tax on a (polluting) fuel to discourage its use.
- Regulatory constraint on quantity of emission of a gas in selected group of sectors in an economy.
- Regulatory constraint on quantity of use of an energy-type in selected group of sectors in an economy.
- Subsidy on capital cost or operating cost of a device to promote its selection.
- Subsidy to promote attachment of an emission removal process to a device.
- Regulatory constraint on use of a clean or efficient device or its combination with an emission removal process.

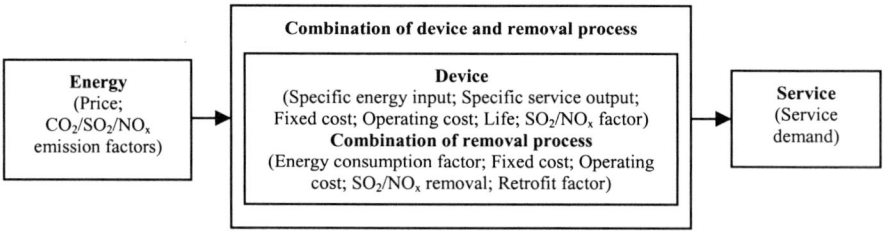

Fig. 1.1. Energy-Technology-Service linkage AIM/Enduse model

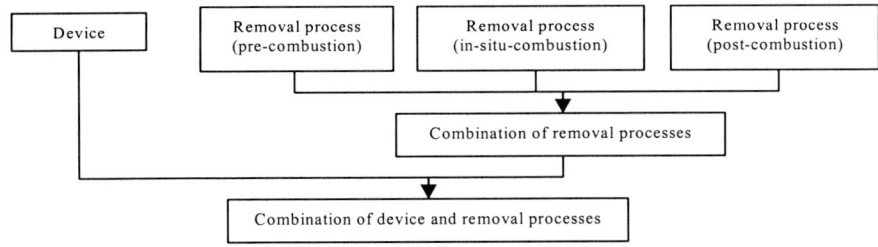

Fig. 1.2. Combination of device and removal processes in AIM/Enduse

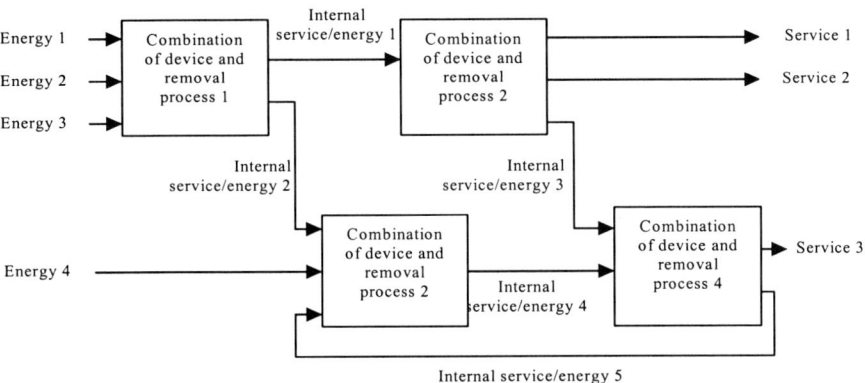

Fig. 1.3. Generalized technology representation in AIM/Enduse

1.3 AIM/Enduse Software

AIM/Enduse software comprises an integration of Optimization system (GAMS), Database system (MS-Access), and Geographical information system (IDRISI) (see Fig. 1.4). The mathematical formulation (see Chapter 2) is written and solved in GAMS (see Appendix B). AIM/Enduse database system, developed using MS-Access, is the interface for GAMS program. It can supply input data for GAMS program file AIM-CMB.gms and display results of simulation. It also provides a user-friendly interface to the user for input of data, and design and analysis of scenarios or countermeasures (see Figs. 1.4 and 1.5). Interface with IDRISI (a GIS and image processing software) permits geographical disaggregation and spatial representation of input data and output results. This manual does not include description of IDRISI data requirement and operation.

Users input data in AIM/Enduse through MS-Access tables or forms that correspond to numbered databases in Fig. 1.5. The figure shows flow of information in the database system. Combinations of indices indicated inside boxes for input data and output of GAMS program highlight both extensive input data requirement and flexibility of output results in the model. For example, combination (L, J, K, M, Y) in box 'Device Specification' means that required data for devices need to be specified for every valid combination of device, service-type, energy-kind, gas, and year. Similarly, combination (I, K, L, P, Y) in box 'Energy Consumption' means that energy use output can be specified for every valid combination of sector, energy-kind, combination of device and removal process, and year.

Appendix C describes the input data tables and each data item in detail. It also describes the procedure for export of input data to GAMS file, running GAMS program, import of output data from GAMS, and display of simulation results. Simulation results can be viewed through a flexible interface that allows both tabular and graphical views of any combination of output variables. For instance, year-wise results for energy use can be viewed by region, sector, or device; year-wise emission quantity of each gas can be viewed by region, sector, energy-type, or device; recruitment costs, operating costs, costs of exchanging removal processes, taxes, and total system costs can be viewed by region, sector, or device; etc. Users are advised to read Appendix C carefully to gain familiarity with the model interface and its implementation.

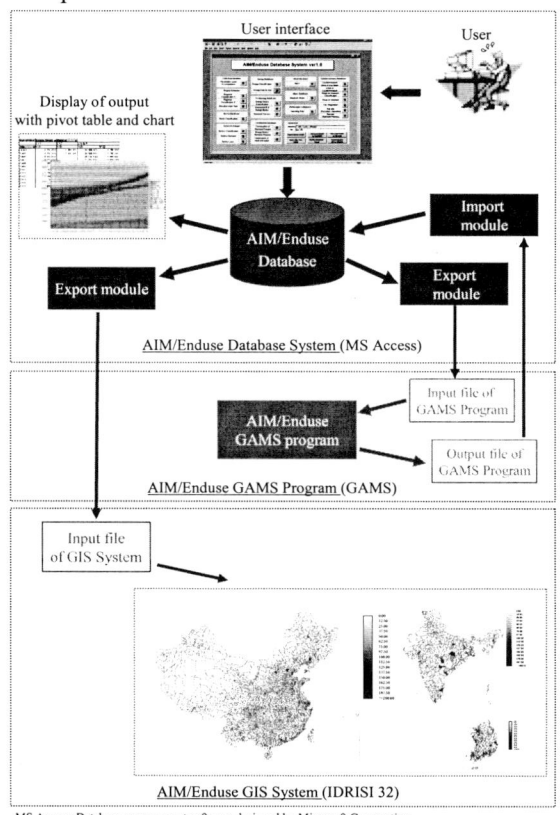

MS Access: Database management software designed by Microsoft Corporation
GAMS: General algebraic modeling system designed by GAMS Development Corporation
IDRISI 32: Geographical information system and image processing software designed
        by Clark Labs, Clark University

Fig. 1.4. AIM/Enduse database system, optimization system and GIS

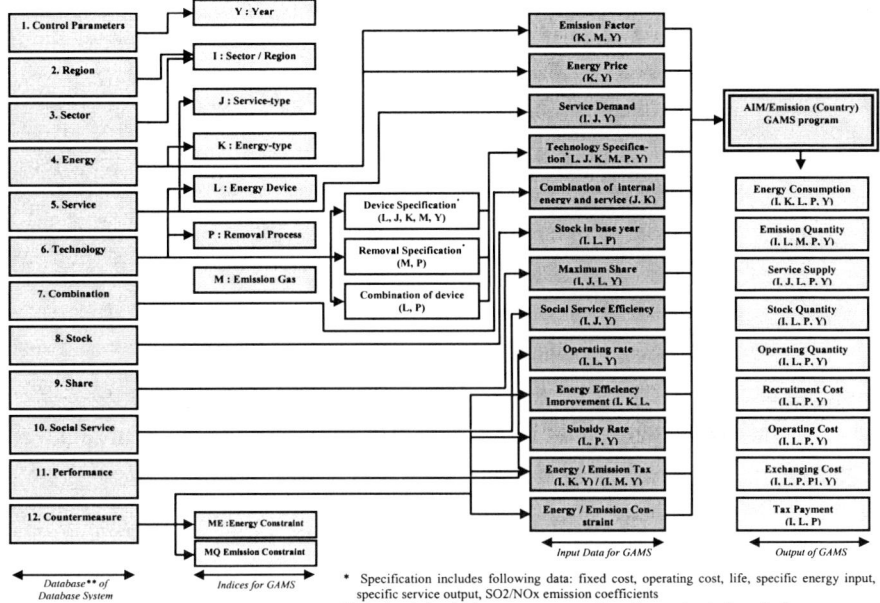

Fig. 1.5. Information flow in AIM/Enduse database system

To summarize, salient features of AIM/Enduse are as follows:

- Technology selection using linear programming framework
- Representation of technologies as
  - complex network of energy and material flows through multiple devices, and
  - emission removal processes as options for retrofit attachments to devices
- Service demands as external drivers
- Technology selection based on annualized capital cost and running cost of technologies including energy costs in a given year
- Retirement of technological stock at the end of its life, and transfer of non-retired stock across successive years
- Estimation of energy use for combustion and non-combustion operations of technologies
- Estimation of quantity of CO_2, SO_2 and NO_x emissions from fuel combustion as well as non-combustion operations of technologies
- Integrated software comprising GAMS, MS-Access, and IDRISI
- User-friendly database system and interface to facilitate easy input of data, design of scenarios or countermeasures, and analysis of results.

2. Theoretical Formulation of AIM/Enduse

This section explains the linear programming formulation of AIM/Enduse model. The code written in GAMS is included in Appendix B. The formulation described in this section is valid for a given year. Since the formulation allows transfer of relevant information like stock of devices and improvement in performance of devices across successive years, the model can be run for any number of years. Thus, if the simulation period selected by the user is 30 years, the model makes the calculations successively for each of the 30 years.

2.1 Indices and Sets

i	Sector
j	Service kind
k	Energy kind
l	Device
m	Gas (emission) kind
p	Gas (emission) removal process
W_j	Set of combinations of device and removal process (l, p) that can satisfy service kind j
z	Index denoting group of sectors categorized for the purpose of emission control
R_z	Set of sectors belonging to group z
G_i	Group of sectors selected for representing energy constraint
e	Index denoting group of sectors categorized for purpose of energy supply
R_e	Set of sectors belonging to group e
$V_{j,k}$	Set of internal service kinds j corresponding to internal energy kinds k (i.e. all k belonging to this set must be supplied by all j belonging to this set)

2.2 Expression for Emission Quantity Estimation

Emission quantity of a gas is estimated adding up quantity of emissions from all devices. Emission from a device is estimated by multiplying operating quantity of the device with emission quantity per unit of device.

$$Q_i^m = \sum_j \sum_{(l,p)\in W_j} \left(X_{l,p,i} \cdot e_{l,p,i}^m \right) \tag{2.1}$$

$$e_{l,p,i}^m = \left(f_{0,l}^m + \sum_k f_{k,l}^m \cdot \left(1 - \xi_{k,l,i}\right) \cdot E_{k,l,p,i} \cdot U_{k,l} \right) \cdot d_{l,p,i}^m \tag{2.2}$$

Where,

Q_i^m: Emission of gas m in sector i

$e_{l,p,i}^m$:   Emission of gas m from an operating unit of combination of device l with removal process p in sector i

$X_{l,p,i}$:   Operating quantity of combination of device l with removal process p in sector i

$E_{k,l,p,i}$:   Energy use of energy kind k per operating unit of combination of device l with removal process p in sector i (same as specific energy input)

$f_{0,l}^m$:   Emission of gas m from operations other than energy combustion of a unit of device l (same as gas m's emission coefficient of device l)

$f_{k,l}^m$:   Emission of gas m from combustion of energy kind k by a unit energy use of device l

$\xi_{k,l,i}$:   Energy saving ratio due to efficiency improvement in use of energy kind k by device l in sector i

$U_{k,l}$:   Proportion of energy kind k used in device l for combustion operations, or burning rate (Note: 1- $U_{k,l}$ or proportion of k used for non-combustion operations in device l is taken as input in database system)

$d_{l,p,i}^m$:   Emission rate (1- removal ratio) of gas m from combination of device l with removal process p in sector i

2.3 Expressions for Constraints

2.3.1 Emission constraints

Emission of gas m in sector i must not exceed allowable maximum emission limit in sector set R_z.

$$\sum_{i \in R_z} Q_i^m \le \hat{Q}_z^m \tag{2.3}$$

Where,

$\hat{Q}_z^m$:   Allowable maximum limit on emission of gas m in group z

2.3.2 Service demand constraints

For a given service, its demand must be met by the quantity of service output supplied by all devices.

$$D_{j,i} \le \left(1 + \Psi_{j,i}\right) \cdot \sum_{(l,p) \in W_j} A_{l,j,i} \cdot X_{l,p,i} \tag{2.4}$$

Where,

$A_{l,j,i}$:   Supply output of service j per operating unit of device l in sector i (same as specific service output)

$\Psi_{j,i}$:   A measure of service efficiency of service type j in sector i (Note: Negative of $\Psi_{j,i}$, a measure of loss of service j, is taken as input in database system; Negative of $\Psi_{j,i}$ is the loss incurred during delivery of service j, for example transmission and distribution loss of electricity supply)

$D_{j,i}$:   Service demand quantity of service type j in sector i

2.3.3 Device share ratio constraints

For a given service, ratio of service output of a device to total service output of all devices must not exceed its upper limit or maximum share.

$$\theta_{l,j,i} \cdot \sum_{(l',p')\in W_j} A_{l',j,i} \cdot X_{l',p',i} \geq A_{l,j,i} \cdot \sum_p X_{l,p,i} \tag{2.5}$$

Where,

$\theta_{l,j}$:   Maximum share of device l in service j

2.3.4 Operating capacity constraints

Operating quantity of a combination of device l with removal process p must not exceed its stock net of operating rate.

$$X_{l,p,i} \leq (1+\Lambda_{l,i}) \cdot S_{l,p,i} \tag{2.6}$$

Where,

$1+\Lambda_{l,i}$:   Operating rate of device l in sector i (Note: $1+\Lambda_{l,i}$ is taken as input in database system)

$S_{l,p,i}$:   Stock of combination of device l with removal process p in sector i

2.3.5 Stock exchange constraints

Every device has a life. Stock of a device recruited in a given year will retire at the end of its life, with its quantity reducing linearly during its lifetime. Thus, out of total stock of a combination of device l with removal process p that was available in the previous year, a certain fraction (inverse of life of device) retires and the balance stock is passed on to the current year. Certain stock of previous year's combination of device l with removal process p can be replaced (or exchanged) in the current year by its combination with another removal process p_l. However, the stock that is replaced in the current year cannot exceed the stock that is passed on from the previous year.

$$\overline{S}_{l,p,i} \cdot \left(1 - \frac{1}{\overline{T}_{l,i}}\right) \geq \sum_{p_1} M_{l,p \to p_1,i} \tag{2.7}$$

Where,

$\overline{S}_{l,p,i}$:   Stock of combination of device l with removal process p in sector i in the previous year

$M_{l,p \to p_1,i}$:Previous year's stock of combination of device l with removal process p that is replaced in the current year by its combination with removal process $p1$

$\overline{T}_{l,i}$:   Life of device l in sector i (this is the average life of stock of device l in previous year)

2.3.6 Energy supply constraints

Total quantity of supply of energy kind k cannot exceed its allowable maximum supply quantity in a sector.

$$\sum_{i \in G_i} \sum_{j} \left(\sum_{(l,p) \in W_j} (1 - \xi_{k,l,i}) \cdot E_{k,l,p,i} \cdot X_{l,p,i} \right) \leq \hat{E}_{k,G_i} \tag{2.8}$$

Where,

$\hat{E}_{k,G_i}$:   Allowable maximum supply quantity of energy kind k

2.4 Internal Service and Internal Energy Balance

All internal energy kinds must be supplied as internal services from within the model (from energy conversion and supply sectors).

$$\sum_{i \in R_e} \left\{ \sum_{j \in V_{j,k}} \left((1 + \Psi_{j,i}) \sum_{(l,p) \in W_j} A_{l,j,i} \cdot X_{l,p,i} \right) \right\} = \sum_{i \in R_e} \left\{ \sum_{k \in V_{j,k}} \left(\sum_{(l,p) \in W_j} (1 - \xi_{k,l,i}) \cdot E_{k,l,p,i} \cdot X_{l,p,i} \right) \right\} \tag{2.9}$$

2.5 Stock Balance

Stock of combination of device l and removal process p in the current year is equal to the sum of the stock of that combination transferred from previous year, the quantity of that combination recruited in current year, and the net stock of other combinations of device l that are exchanged by its combination with removal process p in current year.

$$S_{l,p,i} = \overline{S}_{l,p,i}\left(1-\frac{1}{\overline{T}_{l,i}}\right)+r_{l,p,i}+\sum_{p_1}\left(M_{l,p_1\rightarrow p,i}-M_{l,p\rightarrow p_1,i}\right) \qquad (2.10)$$

Where,

$r_{l,p,i}$:   Quantity of combination of device l with the removal equipment p recruited in current year in sector i

The performance of a device can also change over time. Average performance of combination of device l with removal process p on a given parameter in current year is estimated from the weighted average of performances of its stock passed on from previous year, its quantity recruited in current year, and the net stock of this combination that is obtained from exchanges with other combinations of device l in current year. Expressions (2.11), (2.12), (2.13), and (2.14) estimate the average performance of combination of device l with removal process p on different performance-parameters. Note: expression (2.14) is not used in current version on the model.

$$d_{l,p,i}^{m} \cdot S_{l,p,i} = \overline{d}_{l,p,i}^{m} \cdot \left\{\overline{S}_{l,p,i}\left(1-\frac{1}{\overline{T}_{l,i}}\right)-\sum_{p_1}M_{l,p\rightarrow p_1,i}\right\}+\overset{\circ}{d}_{l,p,i}^{m}\left(r_{l,p,i}+\sum_{p_1}M_{l,p_1\rightarrow p,i}\right) \qquad (2.11)$$

$$E_{k,l,p,i}\cdot S_{l,p,i} = \overline{E}_{k,l,p,i}\cdot\left\{\overline{S}_{l,p,i}\left(1-\frac{1}{\overline{T}_{l,i}}\right)-\sum_{p_1}M_{l,p\rightarrow p_1,i}\right\}$$

$$+\overset{\circ}{E}_{k,l,p,i}\cdot r_{l,p,i}+\sum_{p_1}\left(\overline{E}_{k,l,p_1,i}+\Delta_{k,l,p_1\rightarrow p}^{E}\right)\cdot M_{l,p_1\rightarrow p,i} \qquad (2.12)$$

$$A_{l,j,i}\cdot\sum_{p}S_{l,p,i} = \overline{A}_{l,j,i}\cdot\left\{\sum_{p}\overline{S}_{l,p,i}\left(1-\frac{1}{\overline{T}_{l,i}}\right)-\sum_{p_1}M_{l,p\rightarrow p_1,i}\right\}$$

$$+\overset{\circ}{A}_{l,j,i}\cdot\sum_{p}r_{l,p,i}+\overline{A}_{l,j,i}\cdot\sum_{p_1}M_{l,p_1\rightarrow p,i} \qquad (2.13)$$

$$T_{l,i}\cdot\sum_{p}S_{l,p,i} = \overline{T}_{l,i}\cdot\sum_{p}\overline{S}_{l,p,i}\left(1-\frac{1}{\overline{T}_{l,i}}\right)+\overset{\circ}{T}_{l,i}\cdot\sum_{p}r_{l,p,i} \qquad (2.14)$$

Where,

$\overline{d}_{l,p,i}^{m}$:   Emission rate (1- removal ratio) of gas m from combination of device l with removal process p in sector i in the previous year

$\overset{\circ}{d}_{l,p,i}^{m}$:   Emission rate (1- removal ratio) of gas m from combination of device l with removal process p in sector i, for stock of that combination obtained in the current year from either recruitment or exchange with other combinations.

$\overline{E}_{k,l,p,i}$:   Energy use of energy kind k per operating unit (or specific energy input) of combination of device l with removal process p in sector i in the previous year

$\overset{\circ}{E}_{k,l,p,i}$:   Energy use of energy kind k per operating unit (or specific energy input) of combination of device l with removal process p in sector i in the previous year, for stock of that combination recruited in current year.

$\Delta_{k,l,p_1 \to p}^{E}$:   Energy efficiency change due to exchange of combination of device l with removal process $p1$ to its combination with removal process p

$\overline{A}_{l,j,i}$:   Supply output of service j per operating unit (or specific service output) of device l in sector i in the previous year

$\overset{\circ}{A}_{l,j,i}$:   Supply output of service j per operating unit (or specific service output) of device l in sector i, for stock of that combination recruited in the current year.

$T_{l,i}$:   Average life of stock of device l in sector i in the current year

$\overset{\circ}{T}_{l,i}$:   Life span of the recruited equipment l in sector i, for stock of that device recruited in the current year (Note: This parameter is assumed constant since equation (2.14) is not used in current version of the model).

Change in average performance of combination of device l with removal process p over time can be calculated by repeatedly calculating expressions (2.11), (2.12), (2.13), and (2.14) in every year.

2.6 Expressions for Cost

2.6.1 Annualized initial investment cost (or annualized fixed cost or annualized capital cost)

Annualized initial investment cost as shown in expression (2.15) is used for evaluating recruitment of devices in a given year.

$$\sum_i \sum_j \sum_{(l,p) \in W_j} \left(\overset{\circ}{C}_{l,p} \cdot r_{l,p,i} + \sum_{p_1} \overset{\circ}{C}^x{}_{l,p_1 \to p} \cdot M_{l,p_1 \to p,i} \right) \tag{2.15}$$

$$\overset{\circ}{C}_{l,p} = \overset{\circ}{B}_{l,p} \cdot (1 - SC_{l,p}) \cdot \frac{\alpha(1+\alpha)^{\overset{\circ}{T}_{l,i}}}{(1+\alpha)^{\overset{\circ}{T}_{l,i}} - 1} \tag{2.16}$$

Where,

$\overset{\circ}{C}_{l,p}$:   Annualized investment cost of a unit of combination of device l with removal process p

$\overset{\circ}{C}^x{}_{l,p_1 \to p}$: Annualized investment cost of exchanging a unit of combination (l, p_1) to (l, p)

$\overset{\circ}{B}_{l,p}$:   Initial investment cost or fixed cost of recruiting one unit of combination of device l with removal process p

α :   Discount rate

$SC_{l,p}$: Subsidy rate

$\overset{\circ}{B}_{l,p}$ is estimated by expressions (2.17) and (2.18).

$$\overset{\circ}{B}_{l,p} = \overset{\circ}{B}'_l + \overset{\circ}{b}''_p \cdot \sum_i \sum_k E_{k,l,p,i} \tag{2.17}$$

$$E_{k,l,p,i} = (1 + e_p) \cdot E'_{k,l,i} \tag{2.18}$$

Where,

$\overset{\circ}{B}'_l$:   Initial investment cost or fixed cost of recruiting one unit of energy device l.

$\overset{\circ}{b}''_p$:   Initial investment cost or fixed cost of removal process p per energy use of combination of device l with removal process p.

$E'_{k,l,i}$:   Energy use of energy kind k per operating unit of energy device l.

e_p :   Additional energy use rate of removal process p.

2.6.2 Running cost

Running cost in a given year comprises cost of energy used by devices, and cost of operation and maintenance of devices.

$$\sum_{(l,p)\in W_j} \left(g^0_{l,p,i} + \sum_k g_{k,i} \cdot \left(1 - \xi_{k,l,i}\right) \cdot E_{k,l,p,i} \right) \cdot X_{l,p,i} \tag{2.19}$$

Where,

$g^0_{l,p,i}$:   Operating cost per unit of combination of device l with removal process p in sector i

$g_{k,i}$:   Price of energy kind k in sector i

$g^0_{l,p,i}$ is estimated by expressions (2.18) and (2.20)

$$g^0_{l,p,i} = g^{0'}_{l,i} + g^{0''}_{p} \cdot \sum_k E_{k,l,p,i} \tag{2.20}$$

Where,

$g^{0'}_{l,i}$:   Operating cost per unit of energy device l in sector i

$g^{0''}_{p}$:   Operating cost per unit of removal process p per energy use of combination of device l with removal process p

2.7 Objective Function

Objective function is the total cost in a given year as shown in expression (2.21). This comprises total annualized fixed cost (only for recruitments in that year), total running cost, and total cost of emission tax in that year. Decisions for recruitment quantity and operational quantity for all feasible combinations of devices and removal processes in a given year are made based on the criterion of total cost.

$$
TC = \sum_i \left(\sum_{(l,p) \in W_j} \left\{ \overset{\circ}{C}_{l,p} \cdot r_{l,p,i} + \sum_{p_1} \overset{\circ}{C}{}^{x}{}_{l,p_1 \to p} \cdot M_{l,p_1 \to p,i} \right. \right.
$$

$$
\left. \left. + \left(g^{0}_{l,p,i} + \sum_k (g_{k,i} + \varepsilon_{k,i}) \cdot (1 - \xi_{k,l,i}) \cdot E_{k,l,p,i} \right) \cdot X_{l,p,i} \right\} + \sum_m \zeta^{m}_{i} \cdot Q^{m}_{i} \right) \to \min
$$

$$(2.21)$$

Where,

TC: Total cost

$\varepsilon_{k,i}$:   Tax on energy k in sector i

ζ^{m}_{i}:   Emission tax on gas m in sector i

3. Methodology for Data Preparation

3.1 Choice of Start Year, End Year, and Discount Rate

Following data are required to be entered in AIM/Enduse for the start year of calculation – Service demand (for each service kind), Specific service output, Specific energy input, Operating rate, Fixed cost, Operational cost, Stock, Share and
Emission coefficient (for each device), Price and Emission coefficient (for each
energy kind). Start year should be the latest year for which the above-mentioned
data are either directly available from published sources or easily estimated by the
user team using the methodology described in section 3.4 of this chapter. It is advised that user teams select 2000 as reference year. However, if required data are
not available for 2000, then reference year should be the latest year for which data
are available. This should be updated to 2000 once updated statistics are available.

The choice of end year of calculation should depend on the trade-off between i)
the need for long term perspective required for GHG mitigation policy analysis,
and ii) the level of confidence that the user team can place in long-term projections of service demands and characteristics of future technologies. Most energy
optimization models for analyzing technological options for GHG mitigation have
30-50 year horizon. For application of AIM/Enduse, a time horizon extending up
to at least 2032 is recommended, since it coincides with Rio+40.

Each year in AIM/Enduse can be treated as either calendar year or financial
year. This choice should depend on the definition of year adopted in the majority
of sector-level publications of data that the user team refers to from within a country.

For the choice of annual discount rate, the user team is advised to refer to its
definition in Appendix A.

3.2 Classification of Region, Energy, Sectors and Services

Region classification is not used in current version of AIM/Enduse model. Only
one region, i.e. the country for which the model is to be set up, is considered.

Classification of energy should be based on standard classification used in IEA
Energy Balance Tables or IPCC documents. Appendix H lists a standard classification of energy, calorific values, emission coefficients, and prices.

Choice of sectors and services should depend on i) importance of a sector in national energy system, ii) importance of a service in its sector's energy use, and iii)
availability of data for service demand and technologies used for satisfying a service. Choice of unit of service demand should also depend on the availability of
data. As far as possible, the unit of service demand should be such that it is a
measurable representation of the service being provided. Appendix D gives two
possible classifications of sectors and services. Exact classification adopted by the
user team should depend on its own judgment.

3.3 Classification and Definition of Technologies

3.3.1 Energy device

The extent of detail while classifying the devices should depend on the availability of technology level data from published sources. In AIM/Enduse, a technology is represented in terms of a single device or a sequence of multiple devices (refer to Appendix A for definitions of device and technology). Each device is identified by its energy and material inputs and service outputs. A device can have multiple inputs and multiple outputs. Appendices J and L show examples of technologies as represented in AIM-India and AIM-Japan respectively. Technologies of most industrial processes can be represented in terms of sequence of multiple devices. Such disaggregated representation of technologies allows the user to introduce improvement options at each distinct device or stage of operation.

3.3.2 Removal process

This classification can be based on standard removal processes available in industries for reducing SO_2 or NO_x emissions. SO_2 removal processes can be broadly categorized as coal washing (pre-combustion stage), limestone addition (in-situ combustion stage), and flue gas desulfurization (post-combustion stage) technologies. NO_x removal processes can be broadly categorized as removal technologies used for stationary sources of emission like power plants, industrial boilers and furnaces, and those used for mobile sources of emission like transport vehicles. Appendix G provides a partial list of standard removal processes and their characteristics.

3.4 Estimation of Data for Start Year

3.4.1 Estimation of service demands

Data for service demands in start year should be obtained from published sources. Reliable domestic sources of information (e.g. publications by ministries, government agencies, industry associations, independent research organizations) should be preferred over international sources. Appendix E provides a list of some reliable international publications of country level statistics.

Choice of units for service demands is important. While on one hand the unit of a service demand should represent a good measure of the service, on the other hand it should be convenient from the point of estimating data (see Appendix D for possible choices of units). In case the data for a service demand is not available in desired unit, then either it should be converted to desired unit or the unit of service demand should be changed. Reasonable assumptions need to be made for conversion of data to desired unit. Table 3.1 shows some examples of methodology for conversion of data.

Table 3.1. Examples of methodology for conversion of service demand data in desired unit

Ex. no.	Service	Desired unit A	Unit in which data is available B	Additional assumptions required	Expression for conversion to desired unit
1	Road passenger transport	Person-km	Vehicle-km	C = Average persons per vehicle	A = B * C
2			Bi = Vehicle-km in vehicle category i	Ci = Average persons per vehicle in category i	A = Σ (Bi * Ci)
3			Bi = No. of vehicles in category i	Ci = Average persons per vehicle in category i Di = Average yearly km traveled by a vehicle in category i	A = Σ (Bi * Ci * Di)
4	Residential cooking	Kgoe of useful energy service for cooking	Bi = Kgoe of energy used by stoves of type i	Ci = Average efficiency of stoves of type i	A = Σ (Bi * Ci)
5			No. of households	C = Kgoe of daily useful energy service required for cooking by a typical household	A = B * C
6	Residential lighting	Lumen-hr	Bi = KWh of electricity used by lamp of type i	Ci = Lumens of light delivered by lamp of type i Di = Watt rating of lamp of type i	A = Σ {Bi * (Ci / Di) * 1000}

Note: These are mere examples; Exact methodology for estimating a particular service demand will depend on the specific data that are available to the user team.

3.4.2 Estimation of data for devices

Following data are required by AIM/Enduse for each technology in the start year: Fixed cost, Operational cost, Life, Specific service output, Specific energy input, Stock, Share, Operating rate, and Emission coefficients (for SO_2 and NO_x). These data should be estimated based on a combination of following steps:

1. Obtaining data from published sources: Often, all the data required for AIM/Enduse can not be obtained directly from published sources.

2. Using standard assumptions about efficiency, cost, and other parameters: Such assumptions should be made if all the data can not be obtained from published sources. Appendices F, K and M provide some standard assumptions for selected technologies.

3. Final estimation using bottom-up accounting approach: Final estimation of data required by AIM/Enduse should be made using both the data obtained from published sources and the standard assumptions, and making sure that

fundamental relationships between different parameters are not violated. Table 3.2 shows some examples of this methodology.

Table 3.2. Examples of methodology for estimation of device data in start year

Ex. no.	Parameters for which data or standard assumption are available	Desired unit of device	Additional assumptions required	Estimation of parameters required by AIM/Enduse			
				Specific service output	Specific energy input	Stock	Share
1	A_i = Population (no. of physical units) of device i, for all devices satisfying a particular service B_i = Quantity of service output delivered per unit input of fuel k by device i	One physical unit of device	C_k = Calorific value of fuel k D_i = Average activity of one unit of device i in a year (unit of this may be hours, or service output delivered) E_i = Amount of fuel k used by a unit of activity of device i	$B_i * D_i * E_i$	$D_i * E_i * C_k$	A_i	$A_i * B_i * D_i * E_i / \sum (A_i * B_i * D_i * E_i)$
2	A_i = Service output delivered in a year by device i B_i = Quantity of service output delivered per unit input of fuel k by device i	One physical unit of device	C_k = Calorific value of fuel k D_i = Activity of one unit of device i in a year (its unit may be hours, or service output delivered) E_i = Amount of fuel k used by one unit of activity of device i	$B_i * D_i * E_i$	$D_i * E_i * C_k$	$A_i / (B_i * D_i * E_i)$	$A_i / \sum A_i$
3	A_i = Amount of fuel k used in a year by device i B_i = Quantity of service output delivered per unit input of fuel k by device i	One physical unit of device	C_k = Calorific value of fuel k D_i = Activity of one unit of device i in a year (its unit may be hours, or service output delivered) E_i = Amount of fuel k used by one unit of activity of device i	$B_i * D_i * E_i$	$D_i * E_i * C_k$	$A_i / (D_i * E_i)$	$A_i * B_i / \sum (A_i * B_i)$
4	A_i = Amount of fuel k used in a year by device i B_i = Quantity of service output per unit input of fuel k by device i	One unit of service output of device	C_k = Calorific value of fuel k	1	$(1 / B_i) * C_k$	$A_i * B_i$	$A_i * B_i / \sum (A_i * B_i)$

Table 3.2 Examples of methodology for estimation of device data in start year (continued)

Ex. no.	Parameters for which data or standard assumption are available	Desired unit of device	Additional assumptions required	Estimation of parameters required by AIM/Enduse			
				Specific service output	Specific energy input	Stock	Share
5	A = Total energy used in a year by all devices satisfying a particular service Bi = Quantity of service output delivered per unit input of fuel k by device i	One unit of service output of device	Ck = Calorific value of fuel k Di = Share of device i in total energy used A	1	(1 / Bi) * Ck	Bi * Di * Ai / Ck	(Bi * Di * Ai / Ck) / ΣΣ (Bi * Di * Ai / Ck)

Note:
i) These are average estimates for each classification of device. If the user selects aggregate classification of device that in reality comprises multiple types of devices with different characteristics, then average data for the aggregate classification must be entered in 'Stock' table (see section 8 in Appendix C) and data for most recent or efficient type of device in that classification must be entered in 'Device' table (see section 6.1 in Appendix C). Data for both aggregate classification of device and specific type of device can be estimated using methods described in this table.
ii) Estimates for Fixed cost, Operational cost, and Life of a technology can be directly obtained from published sources like manufacturers' manuals.
iii) These are mere examples; Exact methodology estimating a particular technology's parameters will depend on the specific data that are available to the user team.

3.4.3 Estimation of data for removal processes

Appendix G briefly outlines the methods used for estimation of characteristics of some standard removal processes used in power plants, industries and transport sector. It also provides estimates for these processes.

3.5 Projection of Service Demands

It must be realized that long-term demand projections may not be accurate. There-fore, it is recommended that in addition to selecting a forecasting methodology that incorporates most determinants of service demands, users must also consider multiple demand scenarios in order to assess robustness of model results to accuracy of demand projections. Several alternative approaches exist for projecting service demands over 30-50 years. We briefly describe some general approaches below.

3.5.1 Obtaining projections from authoritative sources

User team should first obtain projections of service demands from authoritative sources like governmental five-year plans or expert estimates. Often projections for all service demands cannot be obtained from such sources. Moreover, projections available are typically for near-term and not for 30-50 years. In some cases, assumptions about growth rates may be made based on near-term projections.

3.5.2 Using a quantitative method to project service demands

A wide spectrum of quantitative methods can be used for long term projection of service demands. Complexity and data requirement of methods vary widely in this spectrum. On one end, complex models like CGE (e.g. AIM/CGE) that model the structure of an economy, can be adopted. On the other end, statistical methods like linear regression, or non-linear regression using an exponential or polynomial function, can be adopted.

While using such methods the user may sometimes project a service demand as a function of 'drivers'. Table 3.3 lists examples of drivers for various services. Choice of method and drivers is often restricted by the availability of time-series data. Judgment of user team is crucial in selecting the method and drivers.

Examples of methodologies for projection of demands adopted in case of India and Japan are illustrated in chapters 4 and 5 respectively. It must, however, be noted that these are mere examples and not necessarily the most desirable approaches in the context of a particular country.

Table 3.4. List of possible drivers for service demands

Service	Driver
Services in agriculture sector	GDP; Gross physical output of agriculture; Gross monetary output of agriculture; Irrigated area
Services in transport sector	GDP; Gross monetary output of transport
Road transport services	Length of roads; Road vehicles per km road
Road passenger transport service	Private vehicles per capita
Rail transport services	Length of railway tracks; Rail traffic per km
Freight transport services	Revenue from freight transport
Services in residential sector	Private final consumption expenditure; Number of households; Income per household;
Services in urban residential sector	Number of urban households; Income per urban household
Services in rural residential sector	Number of rural households; Income per rural household
Services in commercial sector	GDP; Gross monetary output of commercial services
Services in restaurants / hotels	Value added / employment in restaurants / hotels
Services in hospitals / clinics	Value added and employment in hospitals / clinics
Services in corporate and government offices	Value added in corporate and government offices
Services in industrial sector	GDP; Industry value added
Services in a particular industry	Gross monetary output of a industry

3.6 Projection of Improvement in Devices

Improvements in efficiency, cost and emission coefficient of a device occur over time due to several factors including 'learning-by-doing' and 'economies of scale' effects. Following are some examples of approaches that can be used for these projections:

3.6.1 Obtaining improvement targets from authoritative sources

Sometimes near-term improvement targets can be found in publications of sectoral plans by governments/government-agencies.

3.6.2 Top-runner methodology

An approach popularly known as 'top-runner' methodology in Japan, assumes that the 'best' performing technology in a particular year will become 'average' performing after a certain period of time. This period of time can vary from 2-5 years in case of a rapidly growing industry in a growing economy to 10-20 years in case of a matured industry in a slow economy. Judgment of user team is crucial in making these assumptions.

3.6.3 Assuming fixed rates of improvement

Fixed rates of improvement can be assumed for each technology based on recent trends and considering efficient technologies from developed countries as benchmarks. Appendix F provides improvement potential of selected advanced technologies. Appendix M provides characteristics of technologies in AIM-Japan including advanced technologies in Japan.

3.7 Projection of Maximum Share of Devices

Ideally, maximum share of a device should be 100% from the year of its introduction. This implies that service demand in a given year will be met by most economic devices (with least annualized capital cost and running cost for newly recruited devices, and least running cost for old devices). However this phenomenon is not fully observed in reality especially in developing economies.

In the past, several socio-economic-institutional barriers existing in developing countries have prevented early penetration of efficient devices. Although this situation is changing with ongoing economic reforms, several barriers continue to exist. Therefore, upper bounds on penetration of new and future technologies need to be introduced in different sectors. User teams should use their understanding of the domestic socio-economic-institutional context and opinions of experts to decide upper bounds on future share on devices.

Several factors that are not incorporated in costs, play crucial role in determining competitiveness of technologies in certain markets even in developed countries. For example, reliability of performance during use, service quality including perceived quality, effectiveness of delivery and maintenance services, and constraints of production or delivery capacities, are among such factors. User teams are advised to consider all such factors while applying upper bounds on shares of technologies.

In addition to projecting maximum shares under business-as-usual scenario, user teams should build additional scenarios based on different forecasts about progress of economic and institutional reforms in future.

4. Illustration of AIM-India

This chapter explains the case of setting up AIM-India. Process of data estimation, scenario design, and selected tentative results are described in following sections.

4.1 Start Year, End Year, and Discount Rate

Start year was chosen as 1995, mainly because it was the latest year for which reasonable data for services and technologies were available for India. This will be updated to 2000 once the data for 2000 is available from various national and international publications. End year was chosen as 2035.

Discount rate was set at 6% per annum. This is close to the interest rate set by the Reserve Bank of India. However, average cost of capital in the Indian industry is much higher – in the range of 8-12% (net of inflation). Thus, additional scenarios were analyzed with higher discount rates, as explained in section 4.7.2 of this chapter.

4.2 Choice of Sectors and Services

Following sectors were chosen for India:

- *Energy demand sectors*: Agriculture, Transport (further divided into Road, Rail, Air, and Water), Residential, Commercial, and Industry (further divided into Iron & steel, Aluminium, Cement, Brick, Nitrogenous fertilizer, Pulp & paper, Caustic soda, Soda ash, Cotton textiles, Sugar, and Other industries).

- *Energy conversion and supply sectors*: Electricity generation and supply, Petroleum refining & supply, Natural gas production & supply.

Choice of services in each sector was based on i) importance of each service in contribution to sector's energy use, and ii) availability of data for service demand and technologies. Table 4.1 shows this classification for AIM-India:

Table 4.1. Classification of sectors and services in AIM-India

Sector	Service	Unit of service demand
Agriculture	Farm land preparation	1000 no. of tractors
	Irrigation	toe useful energy service
	Harvesting	toe useful energy service
Transport – road	Road passenger	Million person-km
	Road freight	Million ton-km
Transport – rail	Rail passenger	Million person-km
	Rail freight	Million ton-km
Transport – air	Air passenger	Million person-km
	Air freight	Million ton-km
Transport – water	Water freight+	Million ton-km

Table 4.1. Classification of sectors and services in AIM-India (continued)

Sector	Service	Unit of service demand
Residential	Cooking – urban	toe useful energy service
	Lighting – urban	Billion lumen-hours
	Cooking – rural	toe useful energy service
	Lighting – rural	Billion lumen-hours
	Electric fan[*]	1000 no.
	Air-conditioner[*]	1000 no.
	Refrigerator[*]	1000 no.
	TV[*]	1000 no.
	Washing machine[*]	1000 no.
	Other electric appliances[*]	toe energy use
Commercial	Commercial[++]	Million US$ Service value added (1995 price)
Industry – iron & steel	Finished steel products	Tons
	Crude steel[**]	Tons
	Sinter[**]	Tons
	Pig iron[**]	Tons
	Blast furnace gas[**]	kgoe
	Coke gas[**]	kgoe
	Coke[**]	Tons
Industry – aluminium	Finished aluminium products	Tons
	Molten aluminium[**]	Tons
	Alumina[**]	Tons
Industry – cement	Cement	Tons
	Clinker[**]	Tons
	Preheated powder[**]	Tons
	Powder grinded from limestone, clay and sand[**]	Tons
	Slurry[**]	Tons
Industry – brick	Brick	Million no.
Industry – nitrogenous fertilizer	Urea	Tons
	Ammonia[**]	m^3
	Synthesis gas[**]	m^3
	Condensate[**]	Tons
Industry – pulp and paper	Finished paper	Tons
	Stock[**]	Tons
	Bleached pulp[**]	Tons
	Waste pulp[**]	Tons
	Semi chemical pulp[**]	Tons
	Mechanical pulp[**]	Tons
	Kraft pulp[**]	Tons
Industry – caustic soda	Caustic soda	Tons
	Brine[**]	Tons
	Cell liquor[**]	Tons
Industry – soda ash	Soda ash	Tons
	Sodium bicarbonate[**]	Tons
	Mother liquor[**]	Tons
	Ammoniated brine[**]	Tons
	Ammonia[**]	m^3

Table 4.1. Classification of sectors and services in AIM-India (continued)

Sector	Service	Unit of service demand
Industry – textiles	Finished cotton cloth	1000 m^2
	Weaved yarn**	1000 m^2
	Spinned yarn**	1000 m^2
Industry – sugar	Sugar	Tons
	Sugar crystals**	Tons
	Sugar slurry**	Tons
	Juice**	Tons
	Steam**	kgoe
	Electricity**	kgoe
Industry – others	Other industries – commercial fuels	Index (= 1000 for 1995)
	Other industries – biomass	toe energy use
Electricity generation	Electricity**	kgoe
	Heat**	kgoe
Oil refining & supply	Gasoline**	kgoe
	Diesel**	kgoe
	Heavy oil / Fuel oil**	kgoe
	Kerosene**	kgoe
	LPG**	kgoe
	Other petroleum products	kgoe
Natural gas production & supply	Natural gas (uncompressed)**	kgoe
	CNG**	kgoe
	Hydrogen**	kgoe

[+] Service for water passenger transport, being insignificant, was ignored.

[*] These services were defined for the entire residential sector since urban:rural break-up of data on appliances was not easily available.

[++] Value added was chosen as aggregate measure of all services in commercial sector since disaggregated data was not easily available.

[**] These are internal services (certain internal services like captive production of electricity and heat in enduse industries are not shown in this table)

4.3 Classification and Definition of Technologies

Appendix J shows the technology systems for various services considered in AIM-India. The extent of detail or disaggregation while identifying the technologies depended on the extent of technology level data available from published sources. Technology representation for industrial services was more complex than that for other sectors' services as it involved detailed mapping of process flows in order to enable evaluation of specific technological improvement options for GHG mitigation. For example, aluminium production technology was represented as a sequence of three operations (or sub-technologies) – Bayer's process (producing alumina), Smelting (producing molten aluminium), and Casting & Rolling (producing finished aluminium products) (see Fig. J.7 in Appendix J). Energy inputs and service/energy outputs were identified in each operation.

4.4 Estimation of Data for Start Year

We illustrate this methodology for two services – road passenger (transport) and cooking-rural (residential) – and their technologies.

Step 1.   Estimation of service demands

Data for service demands in start year were obtained from various published sources. Definition of services and unit of service demands were driven mainly by the availability of data. Table 4.2 shows the service demands and sources for their estimation.

Table 4.2. Example of service demand data for start year

Service	Service demand in 1995	Source
Road passenger transport	1550 billion person-km	GOI (1997a)
Cooking – rural residential	16110 ktoe	GOI (1997b)

Step 2.   Estimation of technology level data for AIM-India

Final estimation of technology level data was carried out by simultaneously i) considering the data obtained from published sources, ii) making reasonable assumptions about unknown parameters, and iii) ensuring that fundamental relationship between different parameters is not violated. Tables 4.3 and 4.4 illustrate this approach for road passenger and cooking-rural services respectively. Appendix K shows the estimates for various technologies.

Table 4.3. Estimation of technology data for road passenger service in India in 1995

Device	Bill.km. pers / year	Avg. km/l	Avg. calorific value of fuel (kgoe/l)	Approx. stock (1000 no. of vehicles)	Avg. no. of pers / vehicle	Avg. km travel / vehicle / day	Specific energy input (kgoe / vehicle / year)	Specific service output (km.pers / vehicle / year)	Stock (1000 no. of vehicles)
	A	B	C		D	E	= $360*C*E/B$	F = $360*D*E$	= A / $(F*10^6)$
2-wheeler	263	60	0.78	32,000	2.5	9	42	8,100	32,531
Diesel 3-wheeler	31	15	0.78		3	50	936	54,000	574
Kerosene 3-wheeler	46.5	7	0.76		3	50	1,954	54,000	861
Pvt. 4-wheeler	108.5	12	0.78		4	10	234	14,400	7,535
Diesel bus	1,100.5	5	0.86		40	180	11,146	2,592,000	425
				Total = 40,000					Total = 41,925

Note: A, B, C, and Approx. Stock data are obtained from published sources GOI (1997a), MOST (2000), and IRF (2001); D and E are assumptions; and last three columns are final estimations for AIM-India database.

Table 4.4. Estimation of technology data for cooking-rural service in India in 1995

Service demand (ktoe delivered)	Device	Share in service (%)	Heat energy delivered (ktoe / year)	Stove energy efficiency (%)	Hourly fuel use by one stove (kg / stove-hr)	Approx stock (Mill. no. of stoves)	Avg. daily hours of cooking by a household	Hourly energy use by one stove (kgoe / stove-hr)$^{+}$	Spec. energy input (kgoe / stove / year)	Spec. service output (kgoe / stove / year)	Stock (Mill. no. of stoves)
A		B	C = A*B	D	E		F	G	= 360*G*F	H = 360*D* G*F	= C / (H*10^6)
	LPG stove	0.2	32.2	44	0.19		4	0.23	331	145	0.22
	Kerosene stove	8.3	1,353	36			4	0.28	403	145	9.32
6,110	Coal stove	0.5	80.6	25			5	0.32	576	145	0.56
	Bio-mass stove	91.0	14,660	11			5	0.73	1314	145	101.4
							Total = 100				Total = 111.5

$^{+}$ First, hourly energy use for LPG stove was estimated as follows: $G = E * 1.13$, where heat value of LPG = 1.13 kgoe/kg. Then, specific service output for LPG stove (H) was estimated as shown. Specific service output of other stoves was set equal to that of LPG stove, assuming that delivered cooking energy requirement of a typical household is independent of the efficiency of stove. Finally, hourly energy use by each of the other stove-types (G) was derived as follows: $G = (H / (D * F * 360))$

Note: (i) Last three columns are final estimates used in AIM-India database; remaining columns are either assumptions or obtained from published sources NSSO (1997a, 1997b), TERI (2001), and FAO (1997). (ii) Aggregate consistency check was performed by comparing value of aggregate parameter whose rough estimate was available from published sources (e.g. total energy use in rural residential sector). (iii) In case of new and improved versions of each technology, a certain level of improvement in vehicle performance was assumed.

4.5 Projection of Service Demands

A top-down disaggregation method, starting from projection of India's GDP, followed by projection of drivers, and finally arriving at projection of each service demand, was adopted for AIM-India. This approach was adopted because i) adequate historical data for several service demands were not available to allow their direct projection, and ii) it provided a framework to ensure top-down consistency among the projections. Following steps outline this approach. It must be noted that this particular approach is by no means most desirable methodology for long-term demand projection. It was adopted in case of India based on several factors including convenience of data availability and expert opinion.

Step 1. Projection of GDP

First, India's GDP was projected over long run using logistic regression. Logistic function is sown in equation (4.1)

$$GDP_t = GDP_0 \cdot \left\{ \frac{\exp(a + b \cdot t)}{1 + \exp(a + b \cdot t)} \right\} \qquad (4.1)$$

Where,
GDP_t = GDP in time period (year) t
GDP_0 = Expected asymptotic limit for GDP_t

Parameters a and b were estimated by linear regression of the log-log form of equation (4.1), given in equation (4.2), based on time series data. Data points from historical data as well as from projections available from authoritative sources were used for regression.

$$\ln \left\{ \frac{GDP_t/GDP_0}{1 - GDP_t/GDP_0} \right\} = a + b \cdot t \qquad (4.2)$$

Step 2. Projection of drivers for service demands

Following drivers were selected for projecting services in different sectors of India: Agriculture value added (for services in agriculture); Commercial value added (for services in commercial); Industry value added (for services in industry); Transport value added (for services in transport); and Private final consumption expenditure (for services in residential). These drivers were projected from GDP projections by assuming their intensity with GDP.

$$DRV_t = m_t \cdot GDP_t \qquad (4.3)$$

Where,
DRV_t = Level of driver in time period (year) t
m_t = Coefficient representing intensity of DRV_t with respect to GDP_t

Coefficients m_t were projected by linear regression with historical data.

Step 3. Projection of service demands

Finally, each service demand was projected by assuming its intensity with the driver.

$$Y_t = C_t \cdot DRV_t \qquad (4.4)$$

Where,
Y_t = Level of service demand in time period (year) t
DRV_t = Value of driver for service demand Y_t in time period t
C_t = Coefficient representing intensity of Y_t with respect to DRV_t

C_t was either assumed constant and equal to C_{t0}, the intensity of Y with respect to DRV in the reference year t_0, or (SER_{t0} / DRV_{t0}), or projected assuming a rate of change.

4.6 Improvement of Technological Performance and Service Efficiency

Autonomous improvements in performance of technologies were assumed as follows:

- Energy efficiency of devices: Firstly, predictions of improvement in efficiency of selected devices were obtained from IPCC (2001). For old and existing devices, 5% improvement in energy efficiency in every 10 years was assumed. For new devices, 10% improvement in energy efficiency in first 10 years after introduction, and 5% improvement in every subsequent decade, was assumed.
- Fixed cost of devices: Reductions over time in fixed cost of new and future devices were assumed based on information provided in IPCC (2001).
- Operating rate of power plants: Improvements in operating rate of power plant technologies were assumed based on existing performance in more industrialized countries and targets set by the central and state governments in India as part of commitments under ongoing power reforms.
- Service efficiency: Improvements in transmission and distribution efficiency of electricity supply were assumed based on existing performance in more industrialized countries and targets set by the central and state governments in India as part of commitments under ongoing power reforms.

4.7 Design of Scenarios

4.7.1 Business-as-Usual (BaU) scenario

Assumptions used in BaU scenario were as follows:

- No policy intervention or countermeasure for mitigating emissions.
- GDP growth: Compounded annual growth rate of 5% in 1995-2030, decreasing from 5.7% in 1995-2010 to 4.0% in 2020-2030. This assumption was used to project GDP. Drivers and service demands were in turn projected using the methodology described in the Sec. 4.5.
- Discount rate: 6% per annum.

4.7.2 Policy scenarios

Several scenarios related to policy intervention for GHG mitigation in India were built and evaluated using AIM/Enduse model. Under each scenario, impacts on carbon emissions, local emissions (SO_2 and NO_x), system cost, average abatement cost, marginal abatement cost, energy-mix and technology-mix were analyzed. The scenarios included:

- *Carbon-tax scenarios*: Different levels of carbon tax were analyzed. In some scenarios it was assumed that India will join the mitigation commitment after 2010.
- *Carbon constraint scenarios*: Regulatory constraints on quantity of carbon emissions were analyzed. Marginal cost of carbon constraint gives an idea of the level carbon tax required to achieve corresponding mitigation target. Sector-wise carbon constraints were also analyzed. Studying technology-mix results under sector-wide constraint gives an idea of new technologies that need to be promoted over BaU through policy interventions like CDM.
- *SO_2 constraint scenarios*: High quantity of sulfur emissions is a major local pollution concern in India. Hence regulatory constraint on SO_2 emissions is likely to be a more acceptable domestic policy intervention than constraint on carbon emissions. Synergy of such domestic countermeasures for local pollution control with carbon emissions was analyzed.
- *CDM incentives for transfer of clean technologies*: Effect of penetration of certain technologies that are likely candidates under CDM was analyzed. This was done in two ways – by providing subsidies to fixed cost of such technologies and increasing upper bounds on their share.
- *Subsidies for promotion of efficient and clean technologies*: Effect of different levels of subsidy on fixed cost of certain clean technologies was analyzed. Effectiveness of 'subsidy on capital cost of clean technologies' and 'subsidy on price of clean fuels' were compared.
- *Demand-side management scenarios*: Specific demand-side management scenarios including 'efficiency improvement in certain devices at end-use stage', and 'reduced energy use in certain sectors' (through constraint on energy use) were analyzed.

In addition to policy scenarios, certain external scenarios like 'natural gas price scenarios', 'economic growth' scenarios and 'discount rate' scenarios were also analyzed. These external scenarios correspond to different levels of certain parameters whose forecasts are highly uncertain in Indian context but they have an important bearing on model results. Thus, robustness of mitigation countermeasures with respect to fluctuations in such parameters was studied.

4.8 Tentative Results from AIM-India

For purpose of illustration, selected results from AIM-India are described in this section. Let us consider following three scenarios:

1. BaU
2. Carbon-tax of US$100 per ton-carbon applied from 2010 onwards (C-Tax)
3. Sulfur-constraint corresponding to sulfur emissions pathway of C-Tax scenario (S-Constraint)

Figures 4.1 and 4.2 show fuel-wise primary energy use and sector-wise carbon emissions under BaU. They indicate dominance of coal and oil in the Indian energy system over next 30 years, leading to similar levels of rise in both primary energy use and carbon emissions. Power sector, road transport sector, and certain core industries like iron and steel are likely to be major contributors to carbon emissions from India. Carbon emissions from India are expected to be over 700 Mt-C in 2032 – a 3.5 fold rise over 1995 level.

Figures 4.3 and 4.4 compare carbon emissions and carbon intensity of primary energy under the three scenarios. Carbon emissions under C-Tax scenario decline by almost 30 percent in 2032 compared to BaU. This scenario is compared to S-Constraint scenario where a domestic regulation of constraint on sulfur emissions corresponding to their pathway in C-Tax scenario is introduced. Limiting sulfur to the same levels as in C-Tax scenario also results in a decline in carbon emissions, but to a little lesser extent than in C-Tax scenario. This is clearly because carbon intensity of primary energy declines to a slightly greater extent in C-Tax scenario as compared to S-Constraint scenario. Study of fuel-mix and technology-mix results reveals that this minor difference is due to greater use of natural gas (in power and industry sectors) and sulfur removal processes (in power sector) in S-Constraint scenario as compared to C-Tax scenario.

Figure 4.5 shows the primary energy switch under application of C-Tax. A major decline in use of coal is compensated by rise in use of natural gas, renewables and nuclear energy after 2010. Use of renewables and nuclear energy increases significantly towards later years.

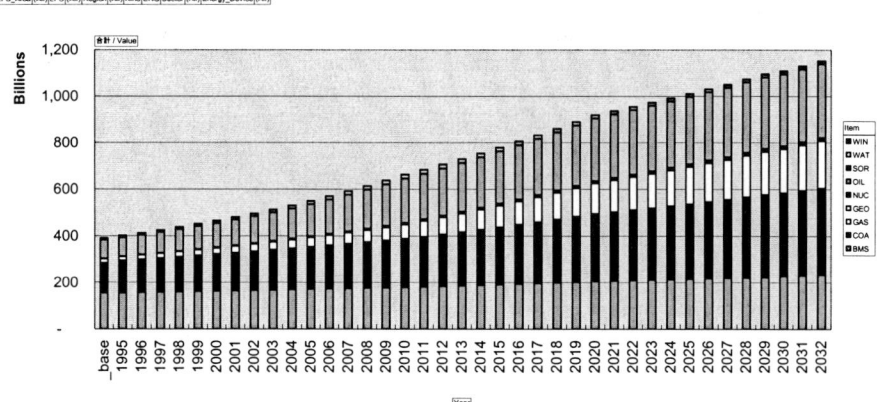

Fig. 4.1. Primary energy use in India under BaU (in Billion kgoe or Million toe)

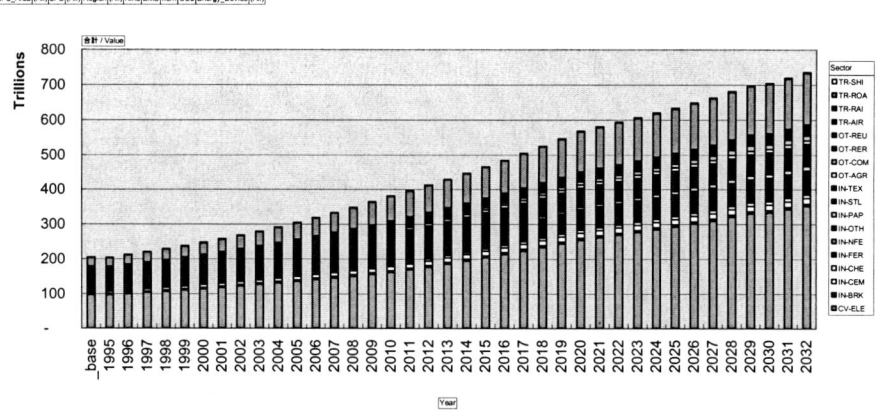

Fig. 4.2. Carbon emissions in India under BaU (in Trillion g-C or Mt-C)

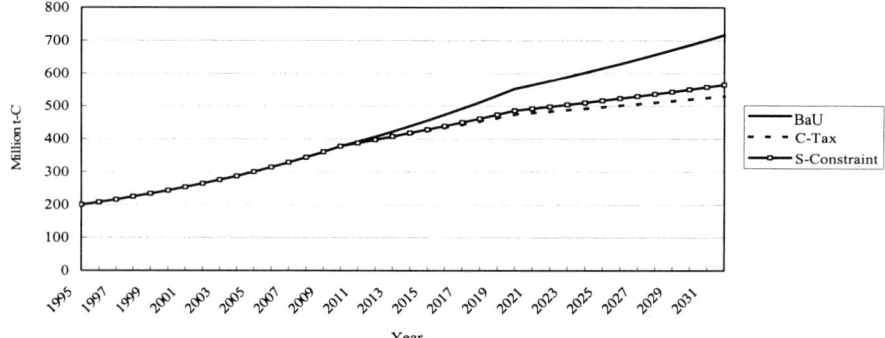

Fig. 4.3. Carbon emissions under three scenarios

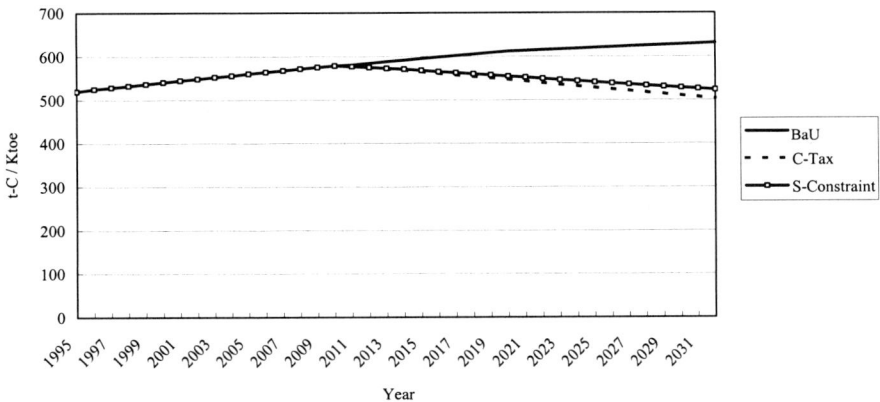

Fig. 4.4. Carbon intensity of primary energy under three scenarios

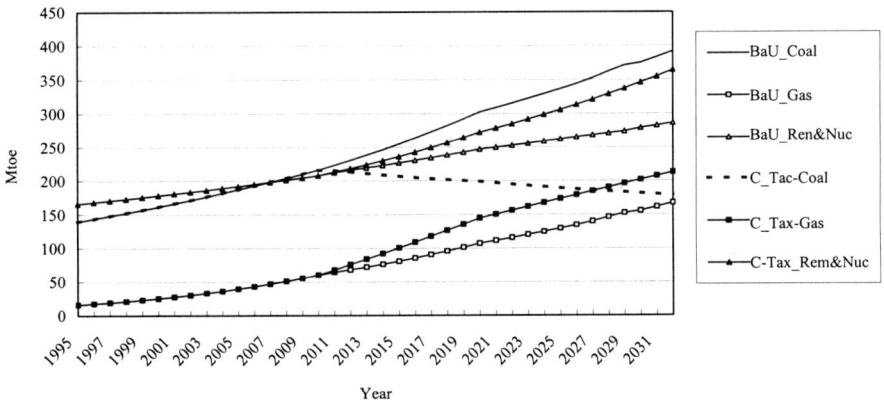

Fig. 4.5. Primary energy use under BaU and C-Tax scenarios
(Note: Ren&Nuc includes renewables, nuclear and biomass energy)

5. Illustration of AIM-Japan

This chapter explains the case of setting up AIM-Japan. Process of data estimation, scenario design, and selected tentative results are described in following sections.

5.1 Start year, End Year and Discount Rate

Start year is chosen as 1998. Calibration for energy consumption and CO_2 emission is done for 2000. End year is chosen as 2020.

The discount rate is assumed as 30%. The questionnaire-survey we executed for companies and households in Japan showed that the payback period of energy saving technology was about three years. We set the discount rate so as to fit the rate on payback period.

5.2 Choice of Sectors and Services

5.2.1 Sectors

Following sectors are chosen for Japan:

Energy demand sectors. Industry (further divided into Steel, Cement, Petrochemicals, Paper, Agriculture, Mining, Construction, Food, Textile, Other chemicals, Other ceramics, Non Ferrous, Metal and Machinery, Other manufacture), Residential, Commercial and Transportation.

Energy production, conversion and supply sectors. Electricity generation, Oil refinery and Town gas.

5.2.2 Services in industrial sector

In steel, cement, petrochemicals and paper, the outputs of main products denoted by weight are used as final service. In other industries, value added denoted by monetary account is used as the final service.

5.2.3 Services in residential sector

25 types of services are chosen in residential sector. As for cooling, warming, hot water and lighting, effective (or useful) energy demand is used as the final service demand. The statistics of effective energy demand is not available. Therefore it is estimated by the method of dividing energy consumption by service efficiency. As for other types of services, mainly electric appliances, total number of holding is used as the index indicating final service demand. In case a service is denoted by total appliance holding comprising different sizes of appliances, unit of service demand is defined as number of appliances equivalent to a particular size. This

particular size is normally the average size in the start year. If average size of appliances changes in future, the future demand is still expressed as number of appliances weighted with change of average size.

5.2.4 Services in commercial sector

15 types of services are chosen in commercial sector. As for cooling, warming and hot water, effective energy demand is used as the final service demand. Effective energy demand is estimated using same method as in the case of residential sector. As for other types of services, floor space area weighted with service intensity change is used as the measure of final service demand.

5.2.5 Services in transportation sector

10 and 9 types of services are chosen in passenger and freight transportation sectors respectively. Volume of transportation denoted by person-km and ton-km are used as the service demand in passenger and freight transportation sectors, respectively.

5.2.6. Services in energy conversion sector

Demand for electricity, oil products and town gas are used as the final service demand, respectively.

5.3 Estimation of Energy Data

5.3.1 Energy price

Fuel import price. Crude oil import CIF (cost including insurance and freight) price in 2010 is quoted from the estimate by METI (2001). As for the price after 2010, it is assumed that the trend from 2000 to 2010 would continue until 2020. LNG price would be linked with crude oil price. The rate of increase in coal price would be half of that in oil price.

Secondary energy price. The elasticity between crude oil price and secondary energy price without tax is estimated from EDMC (2001) data for the past 15 years. Then the future price of oil products is estimated from the elasticity and the future crude oil price. Finally nominal price is converted to real one with 1% deflator.

5.3.2 Emission factor

CO_2 emission factors estimated by Ministry of Environment for Japan's National Communication are used in AIM-Japan. Factors for coal and oil products are estimated based on weighted averages.

SO$_2$ emission factors estimated in Japan's National Communication are used in AIM-Japan. N-content of fuel is assumed to be zero. The NO$_x$ emission, which originates due to interaction of air with combustion, is listed by energy device.

Table 5.1. Classification of sectors and services in AIM-Japan

Sector and Final Service	Service Unit	Sector and Service	Service Unit
Industry - Steel		Commercial	
Hot rolled products	Tons	Cooling	10^5 kcal
Cold rolled products	Tons	Warming	10^5 kcal
Industry - Cement		Hot water (town gas area)	10^5 kcal
Cement	Tons	Hot water (LPG area)	10^5 kcal
Portland clinker for export	Tons	Cooking	100m^2
Industry - Petrochemistry		Lighting(fluorescent)	100m^2
Ethylene	Tons	Lighting(incandescent)	100m^2
Low density polyethylene	Tons	Fire exit light	100m^2
High density polyethylene	Tons	Mainframe	100m^2
Polypropylene	Tons	Duplicator	100m^2
Polystylen	Tons	Elevator	100m^2
Industry - Paper		FAX	100m^2
Paper and Board	Tons	Personal Computer	100m^2
Industry - Others		Pumping power for AC	100m^2
Value added	Million yen	Others	100m^2
Residential		Transportation - Passengers	
Cooling	10^5 kcal	Mini-sized vehicle	100 prs.-km
Warming	10^5 kcal	Small-sized vehicle	100 prs.-km
Hot water (town gas area)	10^5 kcal	Regular-sized vehicle	100 prs.-km
Hot water (LPG area)	10^5 kcal	Commercial car	100 prs.-km
Lighting (fluorescent)	10^5 kcal	Private bus	100 prs.-km
Lighting (incandescent)	10^5 kcal	Commercial bus	100 prs.-km
Conventional refrigerator	Unit number	Private truck	100 prs.-km
Kotatsu	Unit number	Railroad	100 prs.-km
Fan	Unit number	Ship	100 prs.-km
Electric blanket	Unit number	Air	100 prs.-km
Electric fan heater	Unit number	Transportation – Freight	
Washing machine	Unit number	Mini vehicle (private)	100 ton-km
Vacuum cleaner	Unit number	Small vehicle (private)	100 ton-km
Microwave oven	Unit number	Regular vehicle (private)	100 ton-km
Clothing drier	Unit number	Mini vehicle (commercial)	100 ton-km
Electric carpet	Unit number	Small vehicle (commercial)	100 ton-km
TV	Unit number	Regular vehicle (commercial)	100 ton-km
VTR	Unit number	Railroad	100 ton-km
Stereo	Unit number	Ship	100 ton-km
Tape recorder-cum-radio	Unit number	Air	100 ton-km
Desktop personal computer	Unit number	Electricity generation	
Note personal computer	Unit number	Electricity	10^5 kcal
Word processor	Unit number	Oil refinery	
Toilet bow with warm water cleaner	Unit number	Oil products	10^5 kcal
		Town gas	
Other electricity use	Household	Town gas	10^5 kcal

Table 5.2. Energy price in AIM-Japan (Nominal)

Energy Type	Unit	1990	1995	2000	2010	2020
Crude oil	$/Barrel	23	18	27	30	35
Coal	$/t	51	50	35	36	40
LNG	$/t	23	18	27	30	35
Gasoline	Yen/l	122	107	119	125	129
Diesel	Yen/l	72	71	85	89	93
Kerosene	Yen/l	45	40	51	55	59
Heavy oil (A)	Yen/l	35	24	36	40	44
Heavy oil (C)	Yen/l	26	16	27	32	36
LPG	Yen/kg	218	238	228	230	231
Electricity (Lighting)	Yen/kWh	25	25	26	26	27
Electricity (Power)	Yen/kWh	17	17	18	19	20
Electricity (Industrial Use)	Yen/kWh	13	13	14	15	15
Town Gas	Yen/Mcal	10	9	11	12	12

Table 5.3. Emission factor in AIM-Japan

	Emission Factor		
	CO_2 (g-C/10^2Mcal)	SO_2 (g-SO_2/10^2Mcal)	NO_x (g-NO_2/10^2Mcal)
Coal	10,120	139	0
Coal Products	12,300	172	0
Crude Oil	7,811	15	0
Oil products	7,743	83	0
Gasoline	7,658	2	0
Naphtha	7,605	1	0
Jet fuel	7,665	22	0
Kerosene	7,748	1	0
Diesel	7,839	25	0
Heavy oil(A)	7,911	87	0
Heavy oil(C)	8,180	277	0
LPG	6,833	0	0
Natural gas	5,639	1	0
Town Gas	5,500	0	0

Note: NO_x factors are specified by device (see Appendix M).

5.4 Classification and Definition of Technology Systems

Appendix L shows the technology systems considered in AIM-Japan. The extent of detail or disaggregation while identifying the devices depends on the extent of device level data available from published sources. We briefly describe some characteristics of technology systems below.

5.4.1 Technology systems in industrial sector

As for steel, cement, petrochemicals and paper, we assume the typical manufacturing process based on material flows with internal products to combine energy devices. In case of other industrial sectors, it is difficult to assume typical manufacturing processes due to variety of processes and products. Therefore we illustrate the systems based on energy use in the other industrial sectors. Energy use is disaggregated into steam, direct heat, power and other uses, and energy devices are selected to fit the system.

5.4.2 Technology systems in residential and commercial sectors

Technology systems in residential and commercial sectors are rather simpler than industrial sector. Internal energy or service does not exist except electricity. Electricity obtained from purchase, co-generation or solar power is defined as endogenous electricity and energy devices, e.g. air conditioner, refrigerator and personal computer, use the electricity as input.

5.4.3 Technology system of transportation sector

Technology systems in transportation sector are also simple. Internal energy or service does not exist. Energy devices in this sector need only one type of energy and supply only one type of service.

5.5 Estimation of Energy Device Data

5.5.1 Energy devices in industrial sector

Table 5.4 shows examples of estimation of specific service output, specific energy consumption, cost and other data for DC electric furnace and high performance pulp washing device. They are substitutional energy saving devices for AC electric furnace in steel sector and conventional pulp washing device in paper sector, respectively. In case of DC electric furnace, energy saving quantity can be obtained from published sources, but specific energy consumption data does not exist. We estimated energy consumption from the data on service share and energy consumption of related process. Estimation of data for several energy devices in industrial sector requires this kind of method.

Table 5.4. Estimation examples of energy device data in industrial sector of AIM-Japan

		ex.1) DC electric furnace	ex.2) High performance pulp dashing device
Data from published sources and assumption method			
A	Energy saving quantity per products	85 Mcal/t-steel	-
B	Energy consumption in related process	373 Mcal/t-steel ('90)	-
C	Service share of device ('90)	1 % ('90)	12% ('90)
D	Energy consumption per products	341 Mcal/t-steel	5 kWh/t-pulp
Final estimates used in AIM-Japan			
E	Service share of device ('98)	4 % ('98)	38% ('98)
F	Energy consumption per device unit	341 Mcal/t-steel	4 Mcal/t-steel
G	Service supply per device unit	Crude Steel 1 t	Pulp 1 t
H	Price per device unit	5250 yen/t-steel	3143 yen/t-pulp

ex.1) A : MITI (1997); B : MITI (1991); C, E : MOE (2001) ; D : Estimate from A,B,C; F : =E;
H : Japan Iron and Steel Federation (1997)
ex.2) C, E, H : MOE (2001), EA; D : Quesionaire to Japan Paper Association; F : E*0.86

5.5.2 Energy devices in residential and commercial sectors

Table 5.5 shows estimated specific service output, specific energy input, cost and other data for air conditioner with highest efficiency in 1999 and conventional refrigerator. Values for air conditioner are estimated from product brochure. Values for stove, fan heater and water heater are estimated similarly. On the other hand, in case of refrigerator, average energy consumption value from published statistics is used. Data for all the electrical appliances are estimated by the same method.

Table 5.5. Estimation examples of energy device data in residential sector of AIM-Japan

			ex.1) Air conditioner (Highest Eff. in 2000)	ex.2) Conventional refrigerator
Data from published sources and assumption method				
A	Capacity	cool	2.8kW	-
		warm	4.2kW	
B	Operating hours per year	cool	551 hours	-
		warm	1,073 hours	
C	Efficiency	cool	5.42	-
		warm	5.99	
D	Demand of service which a device supplies	cool	37.8 Tcal	56.0 million
Final estimates used in AIM-Japan				
E	Service share of device		0%	100%
F	Service supply per device unit	cool	1,376 Mcal/yr.	1 household
		warm	3,876 Mcal/yr.	
G	Energy consumption per device unit		341 Mcal/yr.	721 Mcal/yr.
H	Price per device unit		290,000 yen	180,000 yen
I	Stock quantity of devices		0	56.0 million

ex.1) A,B,C,H: Product brochure; D: Estimate (cf. Chapter 5.6 of Part IV); F: A*B; G: F*C; I: D*E/F
ex.2) D: Estimate (cf. Chapter 6.6 of Part IV), G: MITI(1998); H: Product brochure, I: D*E/F

5.5.3 Energy devices in transportation sector

Table 5.6 shows estimation of specific service output, specific energy consumption, cost and other data for small-sized passenger vehicle and regular-sized freight vehicle. As shown in the table, the Japan's average value from published statistics is used for specific energy consumption and specific service output of vehicle. In the case of advanced vehicles (ex. gasoline hybrid, CNG, fuel cell), the value is estimated from the product brochure.

Table 5.6. Examples of estimating energy device data in transport sector of AIM-Japan

		ex.1) Passenger vehicle (Small-sized, Gasoline)		ex.2) Freight vehicle (Regular-sized, Diesel)	
Data from published sources and assumption method					
A	Transportation volume by service	631.5	10^9 Person-km	201.3	10^9 ton-km
		410.9	10^9 km	49.9	10^9 km
B	Stock by device type	26.0	Million	0.88	million
C	Stock share of device	62.7	%	100.0	%
D	Transportation volume by device type	396.0	10^9 Person-km	201.3	10^9 ton-km
		257.6	10^9 km	49.9	10^9 km
E	Energy consumption by device type	217.9	10^{12} kcal	129.0	10^{12} kcal
F	NO_x emission factor per energy use	0.17	g/km	3.61	g/km
Final estimates used in AIM-Japan					
G	Price per device	1,527	10^3 yen	4,600	10^3 yen
H	Service supply per device unit	15.2	10^3 person-km/yr.	227.6	10^3 ton-km/yr.
		9.9	10^3 km/yr.	56.5	10^3 km/yr.
I	Energy consumption per device unit	8.38	10^6 kcal/yr.	145.9	10^6 kcal/yr.
J	NO_x emission per device unit	1.68	kg/yr.	203.8	kg/yr.

ex.1,2)   A, B, C, E: Estimate based on MOT(2000);   D : A*C;   F: NRI(1998);   G: Product brochure;
H: A/B;   I: E/B;   J: F*H

5.5.4 Energy devices in energy conversion sector

Table 5.7 shows estimation of specific service output, specific energy consumption, cost and other data for coal power plant and nuclear power plant. In case of coal power plant, Japan's average data from published statistics are used for specific energy consumption and specific service output of plant. Data for oil and gas plant are estimated by the same method.

Table 5.7. Examples of estimating energy device data in conversion sector of AIM-Japan

		Unit	Coal power	Nuclear
Data from published sources and assumption method				
A	Electricity power generation	GWh	135,997	316,818
B	Installed capacity of generation	MW	24,511	44,917
C	Energy consumption	PJ	1,234	-
D	NO_x emission factor	kg/10^8kcal	71.0	-
E	Own use	-	5%	4.4%
F	Thermal efficiency		39.7%	

Table 5.7. Examples of estimating energy device data in conversion sector of AIM-Japan (continued)

		Unit	Coal power	Nuclear
	Final estimates used in AIM-Japan			
G	Fixed cost	1000yen/kW	289	430
H	Service supply per device unit	100Mcal/yr.	71.57	72.03
I	Energy consumption per device unit	100Mcal/yr.	180.4	
J	Stock quantity	kW	24,511	44,917
K	Operating rate	-	63.3%	84.2%

A, B, C: MITI(1998); D: Environmental Research and Control Center (2000); E: (Coal) Estimate based on the data from MITI(1998), (Nuclear) Estimate based the data from Federation of Electric Power Companies of Japan (2001); F: C/A; G: Estimate based on data from Nikkan Denki Tsushinsha, (2000); H: 365*24*860*(1-E); I: (Coal) 365*24*860/F , (Nuclear) 365*24*2250; J: B, K: A/(B*365*24)

5.6 Projection of Service Demands

5.6.1 GDP

GDP does not directly correspond to the service demand for energy device in AIM-Japan. But it is used to estimate various service demands. Economic Council estimated in 1999 that Japan's economic average growth rate would be about 2% per annum by 2010. This estimate is used in AIM-Japan. After 2010, we assume that the growth rate per capita will continue at same level.

Table 5.8. Economic growth rate in Japan

	'91-'95	'96-'00	'01-'05	'06-'10	'11-'15	'16-'20
Economic growth rate	1.4%	1.0%	2.0%	2.0%	1.8%	1.6%

5.6.2 Population, Household

The projections of population and household are taken from National Institute of Population and Social Security Research (1997, 1998).

Table 5.9. Population and household in Japan

		1990	1995	2000	2010	2020
Population	Million	123,611	125,570	126,892	127,623	124,133
Household	Million	40,670	43,900	46,407	49,142	48,853

5.6.3 Service demands in industrial sector 1 – Steel, cement, petrochemicals, paper

The output projection of crude steel and ethylene are taken from Meeting for the Study of Material Industrial Structure (1999). The output of hot / cold rolled products and other petrochemicals are assumed to increase at the same rate as crude steel and ethylene, respectively. As for cement, the rate of increase is assumed to be same as for crude steel. The total output of paper and board are assumed to increase at the same rate as in 1990's. After 2010, decrease of population is assumed to restrain the increase rate of paper and board demand.

Table 5.10. Industrial Production in AIM-Japan

Industrial Products	Unit	1990	1995	2000	2010	2020
Crude steel	Million ton	111.71	100.02	106.44	96.51	93.87
Hot rolled products	Million ton	80.57	71.34	75.69	68.63	66.75
Cold rolled products	Million ton	28.35	27.89	27.33	24.78	24.10
Cement	Million ton	86.85	91.50	81.06	80.53	78.33
Ethylene	Million ton	5.81	6.94	7.61	6.66	6.48
Low density polyethylene	Million ton	1.78	1.75	1.89	1.61	1.56
High density polyethylene	Million ton	1.10	1.24	1.25	1.13	1.10
Polypropylene	Million ton	1.94	2.50	2.72	2.28	2.21
Polystylene	Million ton	1.21	1.28	1.16	1.02	1.00
Paper and board	Million ton	28.09	29.66	31.83	33.74	35.57

5.6.4 Service demands in industrial sector 2 – Other industries

The GDP elasticity of value added by type of industry, "β" in the following equation, is estimated from the statistics, EPA(2000), in the 1990s. Then the future value added is estimated from the elasticity and the future GDP.

$$Y = \alpha X^{\beta} \tag{5.1}$$

Where,
Y: Value added by type of industry
X: GDP
β: Elasticity

Table 5.11. Value added by type of industry in AIM-Japan

Type of industry	Unit	1990	1995	2000	2010	2020
Agriculture	Trillion yen	10.9	9.7	9.5	8.1	7.1
Mining	Trillion yen	1.1	0.9	0.9	0.8	0.7
Food	Trillion yen	12.3	12.8	12.7	13.4	14.0
Textile	Trillion yen	2.5	2.1	1.9	1.3	0.9
Paper	Trillion yen	3.4	3.1	3.0	2.7	2.3
Chemicals	Trillion yen	9.4	11.0	13.0	18.3	24.4
Oil and coal products	Trillion yen	4.1	3.7	4.4	4.7	5.0
Ceramics	Trillion yen	4.4	4.3	4.3	4.4	4.5
Steel	Trillion yen	7.1	6.4	5.8	5.3	4.5
Non ferrous	Trillion yen	2.4	2.3	2.3	2.7	3.1
Metal and machinery	Trillion yen	56.5	62.9	65.7	93.5	125.2
Other manufacture	Trillion yen	19.2	16.7	17.0	14.8	13.2
Construction	Trillion yen	43.4	44.8	44.2	45.0	45.6

5.6.5 Service demands in residential sector

As for cooling, warming, hot water and lighting in residential sector, the service demand is projected using following equation.

$$SRV_t = HH_t * (SRV_0 / HH_0) * SI_t \tag{5.2}$$

$$SRV_0 = \sum_l (ENE_{0,l} / EF_{0,l} * SH_{0,l}) \tag{5.3}$$

Where,
SRV_t = Service demand in time period t
HH_t = Household number in time period t
SI_t = Change of service demand intensity in time period t (1990's value = 1.0)
$ENE_{0,l}$ = Energy consumption of all the stock of device l in 1990.
$EF_{0,l}$ = Average service supply per energy input of device l in 1990.
$SH_{0,l}$ = Service share of device l in 1990.

As for other services in residential sector, the service demand is projected using following equation.

$$SRV_t = HH_t * HR_t * IS_t \tag{5.4}$$

Where,
SRV_t = Service demand in time period t

HH_t = Household number in time period t
HR_t = Holding rate in time period t
IS_t = Index of size in time period t (1990's value = 1.0)

5.6.6 Service demands in commercial sector

As for cooling, warming, and hot water in commercial sector, the service demand is projected using following equation. The projection of floor space is estimated with the GDP elasticity of 0.7.

$$SRV_t = FS_t * (SRV_0 / FS_0) \tag{5.5}$$

$$SRV_0 = \sum_l (ENE_{0,l} / EF_{0,l} * SH_{0,l}) \tag{5.6}$$

Where,
SRV_t = Service demand in time period t
FS_t = Floor space in time period t
$ENE_{0,l}$ = Energy consumption of all the stock of device l in 1995.
$EF_{0,l}$ = Average service supply per energy input of device l in 1995.
$SH_{0,l}$ = Service share of device l in 1995.

As for other services in commercial sector, the service demand is projected using following equation.

$$SRV_t = FS_t * SI_t \tag{5.7}$$

Where,
SRV_t = Service demand in time period t
FS_t = Floor space in time period t
SI_t = Change of service demand intensity in time period t (1990's value = 1.0)

5.6.7 Service demands in transport sector – Passenger and freight

The projection of transportation volume is based on the estimate made by Council for Transport Policy (2000). Although the service classification of automobile is set by size in AIM-Japan, the projections are given by rough classification. Therefore projection for each service demand is derived by assuming the shares in 1998.

5.6.8 Service demands in energy conversion sector – Electricity generation, oil refinery and town gas

The service demands in energy conversion sector are the total demands for electricity, oil products and town gas in all the end-use sectors.

5.7 Case Study of AIM-Japan

5.7.1 Design of scenarios

Simulations are performed for the following three scenarios (cases):

- *Fixed case*: Current technologies continue to be selected because of a lack of understanding and/or for social reasons, even though there are economic benefits in changing the technologies. No countermeasures such as carbon / energy taxes or subsidies are assumed.
- *Market case*: Technology selection is based solely on a reasonable policy of economic efficiency.
- *Tax case*: A carbon tax is introduced beginning in 2001. 30,000 yen per ton of carbon is imposed on secondary fuels. As for electricity, energy tax is imposed at the same level as carbon tax.
- *Tax + Lifestyle change case*: Change of lifestyle for energy consumption is assumed in addition to a carbon tax. Concretely, the countermeasures shown in table 5.12 are considered. The existing implementation rate is based on the questionnaire-survey of the general public in Aichi prefecture. We assume that all the countermeasures will be completely implemented in 2010 due to lifestyle change.

Table 5.12. Countermeasure at used stage through lifestyle change in AIM-Japan

Countermeasures	Reduction rate of energy consumption	Existing implementation rate
1 degree reduction in temperature setting of air conditioner (heat)	3.00%	24%
1 degree increase in temperature setting of air conditioner (cool)	5.00%	24%
Cut down turning on lights	10.00%	40%
Reduction in hot water	5.00%	15%
Use refrigerator efficiently	5.00%	17%
Setting temperature to a more appropriate one	10.00%	42%
Turn off lights during lunch	5.00%	40%
Removal of any unnecessary luggage prior to driving	0.40%	27%
Driving with appropriate pressure	3.80%	40%
Stop idling	3.00%	33%

5.7.2 Results of the simulations

Figure 5.1 shows CO_2 emissions under four cases. CO_2 emission of fixed case would increase by 17% in 2010 and 21% in 2020 compared to 1990. Driving force in residential and transportation sectors will not expand after 2010 due to decrease of population in Japan. That is why increase of CO_2 emission is not significant after 2010. Under the condition of market case, increase in rate of CO_2 emission is 4% in 2010 and 0% in 2020 compared to 1990. Guideline for Measures to Prevent

Global Warming was concluded in 2002 so as to achieve the target of Kyoto Protocol in Japan. The guideline provided the target of CO_2 emission originated from fuel combustion. The target is that the emission in 2010 should be at the same level as in 1990's. The result of market case shows that this target would be achieved without countermeasures. CO_2 emission in tax case would decrease by 1% in 2010 and 3% in 2020 compared to 1990. The result shows that carbon tax is a useful policy option to achieve the target of Kyoto Protocol. Moreover, the last case shows that lifestyle change would cause an additional 1% of CO_2 reduction.

Figure 5.2 shows SO_2 and NO_x emission. SO_2 emission, like CO_2 emission, would also decrease sufficiently in market case and tax case. On the other hand, NO_x emission would not decrease sufficiently. Figures 5.3 and 5.4 show sector-wise CO_2 and NO_x emissions, respectively. The structures of emission differ vastly. NO_x emission in transportation sector shares large part of total NO_x emission. Hybrid and fuel cell vehicles which reduce both NO_x and CO_2 emissions would not be adopted even with carbon tax due to high installation costs. Lean-burn engine which would be economically efficient option to reduce CO_2 emission under the tax case is always not useful to reduce NO_x emission. That is why NO_x emission would not decrease in the tax case.

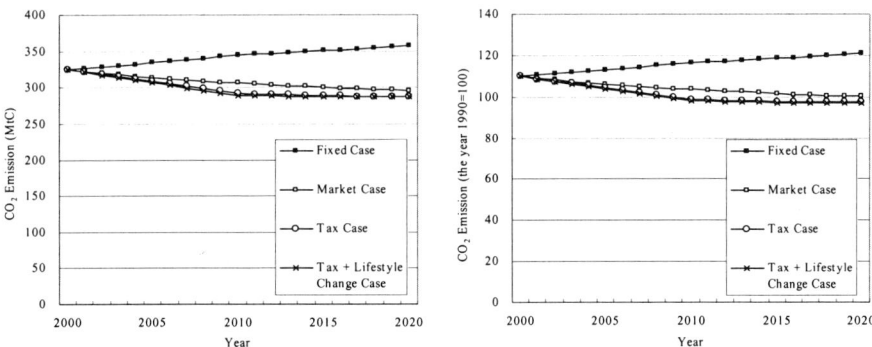

Fig. 5.1. CO_2 emission in AIM-Japan

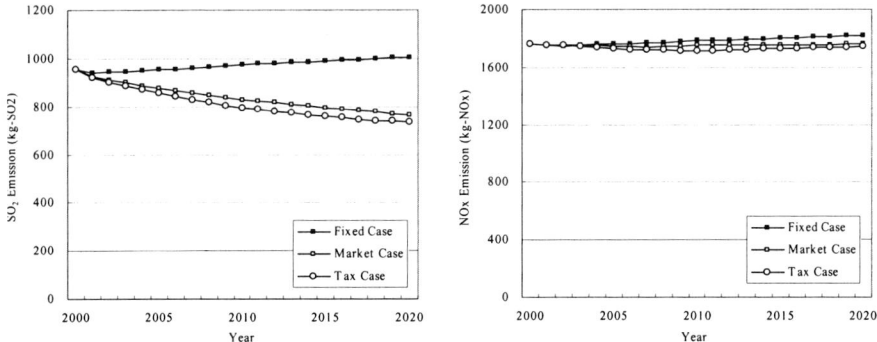

Fig. 5.2. SO_2 and NO_x emission in AIM-Japan

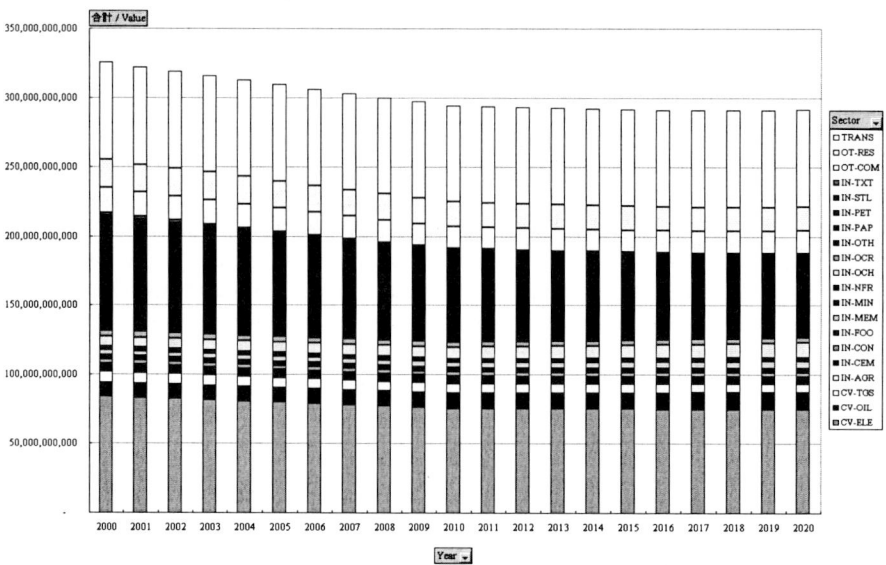

Fig. 5.3. Sector-wise CO_2 emission in AIM-Japan

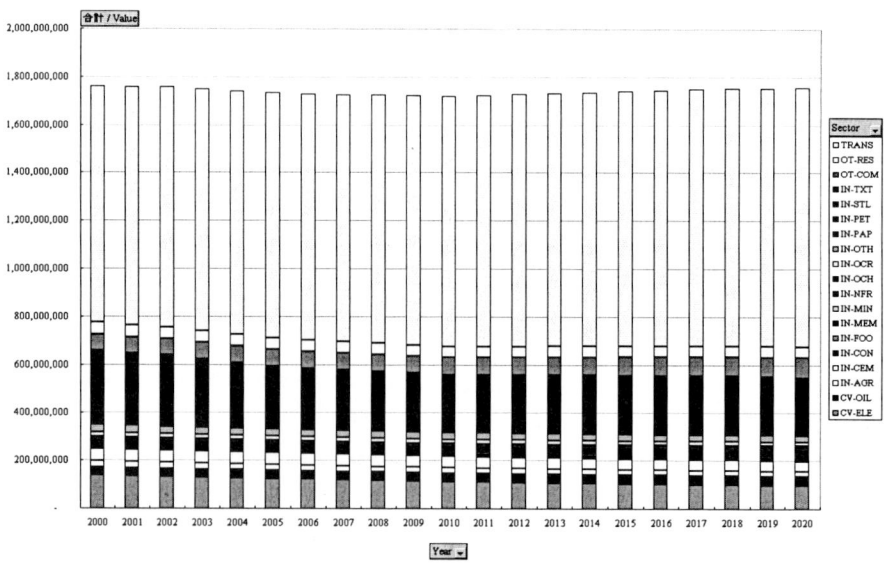

Fig. 5.4. Sector-wise NO_x emission in AIM-Japan

References

Blok K, Turkenburg WC, Eichhammer W, Farinelli U, Johansson TB (eds) (1995) Overview of energy R&D options for a sustainable future. European Commission, Directorate-General XII, Brussels, Belgium

Brown MA, Levine MD, Romm JP, Rosenfeld AH, Koomey JG (1998) Engineering-economic studies of energy technologies to reduce greenhouse gas emissions: Opportunities and challenges. Annual Review of Energy and Environment 23:287-385

Central Environment Council (2001) Interim report of sub-committee for developing scenario to achieve the target. http://www.env.go.jp/council/06earth/r062-01/index.html

Cofala J, Syri S (1998a) Sulfur emissions, abatement technologies and related costs for Europe in the RAINS model database. Interim Report, IR-98-035, IIASA

Cofala J, Syri S (1998b) Nitrogen oxides emissions, abatement technologies and related costs for Europe in the RAINS model database, Interim Report. IR-98-035, IIASA

Council for Transport Policy (2000) Long term projection of transportation demand. http://www.mlit.go.jp/kisha/oldmot/kisha00/21koutu/chouki-c_.htm

ECC (2002). Energy conservation performance catalog. Energy Conservation Center, Japan

EDMC (eds) (2001) Handbook of energy & economic statistics in Japan. Energy Conservation Center, Japan

EDMC (eds) (2002) Handbook of energy & economic statistics in Japan. Energy Conversation Center, Japan

Environmental Research and Control Center (2000) Regulation of total NO_x emission manual. Environmental Research and Control Center, Japan

EPA (eds) (2000) Annual report on national accounts. Printing Bureau, Ministry of Finance, Japan (CD-ROM)

Euromonitor (2002) World marketing data and statistics 2002. Euromonitor, London, UK (http://www.statistischedaten.de/products/wmd.htm)

FAO (1997) The role of wood energy in Asia. FAO working paper FOPW/97/2. Food and Agriculture Organization, UN, Italy

FAO (2001) FAOSTAT 2001. Food and Agriculture Organization, UN, Italy (http://apps.fao.org/)

Federation of Electric Power Companies of Japan (eds) (2001) Handbook of electric power industry. Japan Electric Association, Japan (in Japanese)

Garg A, Shukla PR (2002) Emissions inventory of India. Tata McGraw-Hill Publishing Company Limited, New Delhi

GOI (1997a) Motor transport statistics of India. Transport Research Wing, Ministry of Surface Transport, Govt. of India, New Delhi

GOI (1997b) Energy use by Indian households. National Sample Survey Organization, Department of Statistics, Govt. of India, New Delhi

IEA (1999) Engineering-economic analyses of conserved energy and carbon. International Workshop on Technologies to Reduce Greenhouse Gas Emissions, Washington D.C., 5-7 May 1999, IEA, Paris

IEA (2001) Energy balances of Non-OECD countries, 2000-2001. International Energy Agency, France

IISI (2001) Steel statistical yearbook 2001. International Iron and Steel Institute, Brussels

IMF (2001) International financial statistics yearbook 2001. International Monetary Fund

IPCC (1996a) Climate change 1995: Impacts, adaptation and mitigation of climate change: Scientific-technical analyses. Contribution of Working Group II to the second assessment report of the Intergovernmental Panel on Climate Change. Cambridge University Press, Cambridge, UK

IPCC (1996b) IPCC guidelines for national greenhouse gas inventories: Workbook and manual. Intergovernmental Panel on Climate Change

IPCC (2001) Climate change 2001: Mitigation. Contribution of Working Group III to the third assessment report of the Intergovernmental Panel on Climate Change. Cambridge University Press, Cambridge, UK

IRF (2001) World road statistics 2001. International Road Federation, Washington, D.C. (http://www.irfnet.org/wrs.asp)

Ishitani H, Johansson TB (1996) Energy supply mitigation options. In Climate change 1995: Impacts, adaptation and mitigation of climate change: Scientific-technical analyses. Contribution of Working Group II to the second assessment report of the Intergovernmental Panel on Climate Change. Cambridge University Press, Cambridge, UK

JAMA (2001) Automotive guidebook of Japan. Japan Automobile Manufactures Association, Inc., Japan

Japan Iron and Steel Federation (1997) Study on environmental technology in steel industry. Nikkiren 8 Sentan-31 (in Japanese)

JAPAN TAPPI (1990) Summarization of responses to the 6th questionnaire sent to member companies on the energy conservation. JAPAN TAPPI Journal 44: 327-341, 447-549 (in Japanese)

JEA (1992) Estimation of CO_2 emission in Japan. Japan Environment Agency, Japan

JEA (1998) Investigation of air pollutants emission amounts. Japan Environment Agency, Japan

Kaplan N (1994) Integrated air pollution control system version 5.0, vol.2, technical documentation manual. PB96-157391, Radian Corporation, Research Triangle Park, North Carolia. M.Gundappa, L.Gideon, and E.Soderberg, EPA Norman Kaplan, EOP 68-D1-0031

Meeting for the Study of Material Industrial Structure (1999) Interim report of meeting for the study of material industrial structure, Japan (in Japanese)

MITI (eds) (1997) Report of global environment committee, Industrial structure council. Research Institute of Economy, Trade & Industry, Japan (in Japanese)

METI (2001) Long-term energy supply-demand outlook of Japan. Research Institute of Economy, Trade & Industry, Japan (in Japanese)

MITI (1991) Yearbook of iron and steel statistics. Research Institute of Economy, Trade & Industry, Japan (in Japanese)

MITI (eds) (1998) Outlook of electricity supply and demand. Chuwa Insantsu, Japan (in Japanese)

MOST (2000) Handbook of transport statistics 1998/99. Transport Research Wing, Ministry of Surface Transport, Govt. of India, New Delhi

MOT (2000) Annual statistics report for car transportation. Ministry of Transport, Japan (in Japanese)

MLIT (2001) Domestic transportation statistics handbook. Institute for Transport Policy Studies, Japan (in Japanese)

MOE (2001) Report of the study group on scenarios of technologies to reduce GHG emissions. Ministry of Environment, Japan (in Japanese)

National Institute of Population and Social Security Research (1997) Population projections for Japan: 1996-2100. Health and Wealth Statistics Association, Japan (in Japanese)

National Institute of Population and Social Security Research (1998) Household projections for Japan by prefecture: 1995-2020. Health and Wealth Statistics Association, Japan (in Japanese)

Nikkan Denki Tsushinsha (2000) Directory of new electric power plants. Nikkan Denki Tsushinnsha, Japan (in Japanese)

NRI (1998) Report for emission factor and total emission from automobile. Nomura Research Institute Corporation, Japan (in Japanese)

NSSO (1997a) Energy use by Indian households. National Sample Survey Organization, Department of Statistics, Govt. of India, New Delhi

NSSO (1997b) Sarvekshana. National Sample Survey Organization, Department of Statistics, Ministry of Planning and Programme Implementation, Govt. of India, New Delhi

Penner JE, Lister DH, Griggs DJ, Dokken DJ, McFarland M (1999) Aviation and global atmosphere. Intergovernmental Panel on Climate Change. Cambridge University Press, Cambridge, UK

Perkins JE (1998) Focus on transport emissions needed if Kyoto's CO_2 targets are to be met. Oil and Gas Journal 96(3): 36-39

Phylipsen GJM, Blok K, Worrell E (1998) Handbook on international comparisons of energy efficiency in the manufacturing industry. Department of Science, Technology and Society, Utrecht University, The Netherlands

Smil V (1999) Energies: An illustrated guide to the biosphere and civilization. The MIT Press, Massachusetts Institute of Technology, Cambridge, USA

TERI (2001) Tata energy data directory & yearbook 2000/2001. Tata Energy Research Institute, New Delhi

UN (2000) Long-range world population projections: 1998. World population prospects: 2000. United Nations Population Division, UN, New York, USA (http://www.un.org/esa/population/publications/longrange/longrangeOrder.pdf)

UN (2001a) Industrial commodity statistics yearbook 1999. United Nations Statistics Division, UN, New York, USA (http://unstats.un.org/unsd/industry/)

UN (2001b) Statistical yearbook. United Nations Statistics Division, UN, New York, USA

WB (2002) World development indicators 2002. The World Bank, (http://www.worldbank.org/data/wdi2002/index.htm)

WEC (1995) Efficient use of energy using high technology – An assessment of energy use in industry and buildings. M D Levine, N Martin, L Price, and E Worrell (eds), World Energy Council, London, UK

Appendix A: Terms of Reference

Combination of device and removal process: It is defined as the combination of a device and a combination of removal process.

Combination of removal process: Combination of different removal processes at the stages of pre-combustion, in-site combustion and post-combustion, can be defined in AIM/Enduse. For each such combination, data for Removal rates for SO_2 and NO_x, Fixed cost, Operational cost, Energy consumption and Retrofit factor are specified by the user. A combination of removal process can be attached to a main device.

Countermeasure: A countermeasure is a type of intervention aimed at mitigating emission. Four types of countermeasures are defined in AIM/Enduse – Countermeasure at use stage, Tax, Regulation, and Subsidy.

Device or Energy device: A 'device' refers to a set of procedures/operations and/or machines/equipments. Unlike a technology that supplies final service, a service output from a device can be either final or intermediate. It must be noted that in AIM/Enduse a device is defined as an indivisible part of a technology, whereas in reality a device may itself comprise multiple parts (e.g. multiple machines). Thus the term 'device' is an imaginary concept used only for convenience of definition in AIM/Enduse. In AIM/Enduse, a device is represented by its unit, life, capital cost, operational cost, operating rate, specific inputs and outputs and their quantities used/produced by a unit of device. Examples of device include sintering machine (a single equipment to produce an intermediate service as part of a steel making technology), alumina production process (a set of equipments and operations to produce an intermediate service as a part of an aluminium production technology), incandescent lamp (a single equipment that entirely constitutes a technology to deliver a final service), wet process for making cement (a set of equipments and operations that entirely constitutes a technology to produce a final service).

Discount rate: 'Discount rate' is a way to measure how much we value future benefits today. For the purpose of evaluating competing technologies with different lives, in a particular year, AIM/Enduse model uses discount rate to convert investments required in capital, operational and energy costs in multiple years to the same base year. It can be defined as an average estimate of the real market interest rate or the discount rate on goods that the user team expects to prevail in a country over the time horizon of the model (typically next few decades). This definition is a reasonable representative of the cost of capital to the economy, and the returns to direct investments.

Efficiency of device: Efficiency of a device can be defined in multiple ways. Although 'efficiency' is not defined as a parameter in AIM/Enduse, it is captured in the combined definitions of specific service output and specific energy input. Efficiency can be defined as the quantity of service output produced or delivered by one unit of energy input. This definition is useful for data preparation since for most devices, figures for specific service output and specific energy input cannot be obtained directly but will need to be estimated from efficiency figures. Unit: Unit of service output per unit of energy input.

Emission coefficient of device: AIM/Enduse permits defining emission coefficients for a device for SO_2 and NO_x emissions that arise due to the process of operation of device. Unit: Grams (or Kilograms or Tons) of gas per unit of device.

Emission coefficient of energy: For each of the emissions – CO_2, SO_2 and NO_x – emission coefficient of energy is the quantity of emissions per unit of energy (fuel or material) used. Unit: Grams (or Kilograms or Tons) of gas emitted per unit of energy burnt (in case of fuels); or Grams (or Kilograms or Tons) of gas emitted per unit of material used (in case of materials like limestone).

End year of calculation: It is the last year of calculation in the model.

Energy or Energy kind or Energy type: In AIM/Enduse, 'energy' is an extended concept that refers to an input to a device. Such an input can be a primary fuel, a secondary fuel, electricity, heat, or a material. Examples include wood, coal, natural gas, crude oil (primary fuels); gasoline, kerosene, diesel (secondary fuels); and iron ore, crude steel, clinker (materials). An energy can be represented as either internal energy or external energy in AIM/Enduse. Possible units: Joules (or KJ, MJ, or GJ) or Kgoe (or Toe, Ktoe, or Mtoe) or Kgce (or Tce, Ktce, or Mtce) for energy inputs (fuels, electricity, heat); Units of mass, volume, area, length, or physical numbers for material inputs.

Exchange of combinations of removal process: In AIM/Enduse, a particular combination of removal process attached to a device in a given year can be replaced by another combination of removal process in the next year. If a device is not attached to any combination of removal process in a given year, then it is said to be attached to 'Non' (dummy combination signifying no removal process) in that year. Thus, in order for a combination of removal process to be introduced with a particular device in a given year, it has to either replace an already attached combination of removal process to that device, or replace 'Non'.

External service and External energy: An external service is a service that is produced in the model but not used as input in any device. Thus, an external service is a final service whose demand is provided by the user external to the model. Similarly, an external energy is an energy that is not produced from any device but is supplied from outside the model.

Fixed cost or Capital cost of device: It is the initial investment cost required to recruit one unit of a device. Unit: US$ (or 1000 US$) or domestic currency per unit of device in 1995 price (or 2000 price).

Fixed cost of removal process: It is the installation cost of a removal process per unit of energy input (to the main device). Unit: US$ (or 1000 US$) or domestic currency per energy unit.

GAMS: General Algebraic Modeling System (GAMS) is a software designed by GAMS Development Corporation (www.gams.com) for modeling linear, non-linear and mixed-integer optimization problems. Linear optimization formulation of AIM/Enduse model is written and solved in GAMS.

IDRISI32: IDRISI32 is a Geographical Information System (GIS) and image processing software designed by Clark Labs, Clark University (www.clarklabs.org) for supporting special analysis and environmental decision making.

Improvement of device: It is defined as the improvement in a device's performance over time on any of the following parameters – Fixed cost, Operational cost, Specific service output, Specific energy input, Emission coefficient.

Internal service and Internal energy: An intermediate service, i.e. a service output from a device that goes as an energy input to another device, is classified as both internal service and internal energy in the model. The quantities of flow of such internal service and internal energy have to be balanced inside the model. Examples include electricity (output from power plants, input to enduse devices like refrigerator in households or smelting process in aluminium plant), alumina (output from Bayer's process, input to smelting process, both in aluminium plant), and gasoline (output from refineries, input to enduse devices like two-wheelers and cars).

Life of device: 'Life' of a device is the number of years over which the device remains operationally useful. Unit: Number of years.

Material: 'Material' is not used as a term in AIM/Enduse model. In this manual we sometimes use 'material' to refer to non-energy inputs or outputs. In AIM/Enduse, all material inputs are identified as 'energy' and all material outputs are identified as 'services'.

Maximum share of device: It is the upper bound on share of a device.

Operating rate of device: It is defined as the operating capacity of a device divided by its rated capacity. Normally it is less than 100% because of loss of capacity due to scheduled and unscheduled maintenance or outage of a device. Therefore, total operating capacity of a device available in a given year is estimated by multiplying its stock with operating rate.

Operation of device: It is defined as the process of running a device to produce a service.

Operational cost of device: It is the annual cost incurred in operating one unit of a device. This includes fixed operational and maintenance cost including wages, variable operational and maintenance cost, overhead cost, logistics costs, and other costs that are not included in 'Fixed cost' and 'Price of energy'. Unit: US$ (or 1000 US$) or domestic currency per unit of device per year, in 1995 price (or 2000 price).

Operational cost of removal process: It is the running cost of a removal process per year. Unit: US$ (or 1000 US$) or domestic currency.

Price of energy: It is the price of purchase (or market price) of one unit of a fuel. Preferable unit: US$ (or 1000 US$) or domestic currency per unit of energy, in 1995 price (or 2000 price).

Recruitment of device: It is defined as the process of purchase and installation of a device.

Reference year: It is the year for which the input data in the model is calibrated with actual figures. Normally, it is same as the start year of calculation in the model.

Region: It refers to a geographical area defined for the purpose of carrying out emission mitigation analysis. In AIM/Enduse, regions are classified by country (region classification 1).

Region classification 1: It refers to aggregate regional classification defined by the user in AIM/Enduse database system. A country level classification is an example of Region classification 1. Region classification 2, a disaggregate regional classification (for example, district level), is not used in the current version of AIM/Enduse. Although Region classification 2 is not used in this manual, it will be introduced in future versions of the model.

Regulation: A regulation is a constraint or upper limit imposed on quantity of use of energy or quantity of emission.

Removal process: In AIM/Enduse, an equipment or set of equipments and/or procedures, attached to a main device, for sole purpose of reducing SO_2 or NO_x from the main device, is called a 'removal process'. Three kinds of removal processes – introduced at pre-combustion, in-situ combustion, and post-combustion stages – can be defined in AIM/Enduse. For example, coal washing equipment attached before the boiler (pre-combustion stage) in a coal-fired power plant is a removal process that reduces sulphur content of coal, and hence, SO_2 emissions. Limestone addition during combustion of coal is an example of in-situ stage removal process, while Flue gas desulfurization is an example of post-combustion removal process.

Removal rate of removal process: Removal rate of a removal process is defined as the percentage of SO_2 or NO_x emission that it reduces from any device to which it is attached.

Retrofit factor of combination of device and removal process: Retrofit factor is defined as the proportion of a device's stock that is combined with a combination of removal process.

Sector: A 'sector' refers to a portion of an economy. For convenience, an economy is classified into a finite number of sectors. In bottom-up energy models like AIM/Enduse, this classification is based on the energy dynamics. Therefore, each sector is unique with respect to following characteristics: sources of service demands, patterns of energy flows, and technologies or devices used to satisfy service demands. Refer to Appendix D for examples of classification of sectors.

Service or Service kind or Service type: A 'service' refers to a measurable need within a sector that has to be satisfied. A service can be satisfied by supplying an output from a device. Thus, in AIM/Enduse, a service is identified by such an output. A service output from a device can be either a final output (external service) that satisfies a need of an enduse consumer, or an intermediate output (internal service) that goes as an input to another device. A service can be defined in either tangible or abstract terms. Examples include finished steel products (tangible, final output of steel making technologies), person-km traveled by road (abstract, final output of road transport vehicles), crude steel (tangible, intermediate output from blast furnace and converter), and heat energy for raising superheated steam (abstract, intermediate output from heat exchanger in combined cycle power plant). It must be noted here that concepts of 'final service' and 'intermediate service' are defined by the user team for convenience of energy system modeling, and may not necessarily imply real-life interpretations of these terms. For example, finished steel products like bars and beams may be defined as final service in AIM/Enduse model, whereas in reality they are intermediate services in an economy. Refer to Appendix D for examples of classification of services.

Service demand: It refers to the quantified demand created by a service. Service outputs from devices satisfy service demands. Possible units: Refer to Appendix D for a list of possible units of service demands.

Service efficiency: This concept is defined as opposite to service loss. It is used in this manual to signify the efficiency of supply or delivery of a service. Please refer to 'Service loss' for further description.

Service loss: It is a term used for representing transmission & distribution loss (in percentage) of electricity or loss during supply of any other service in AIM/Enduse.

Share of device: It is the share of a device's contribution in meeting a service demand. In case a device can produce multiple service outputs, its share can be defined separately for each service.

Specific energy input or Specific energy consumption of device: It is the quantity of energy input used by one unit of device in one year. It can be defined in AIM/Enduse as either the maximum potential energy input, or the net energy input, used by one unit of device in a year. It is separately defined for each fuel or material that a device uses. Unit: Unit of energy per unit of device.

Specific service output of device: It is the quantity of service output produced or delivered by one unit of device in one year. It can be defined in AIM/Enduse as either the maximum potential service output, or the net utilized service output, produced by one unit of device in a year. It is separately defined for each service that a device produces. Unit: Unit of service output per unit of device.

Start year of calculation: It is the starting year of calculation in the model.

Stock of device: It is the quantity of a device (expressed in units of device) available in a given year. It is equivalent to the rated capacity of a device available in a country in a given year. In AIM/Enduse, stock of each device needs to be specified at the beginning of reference year. Stock of a device in future years is calculated by the model. Unit: Same as unit of device.

Subsidy rate: A subsidy is concession (in percentage) on cost offered to promote a technology. In AIM/Enduse, subsidy can be applied to fixed cost or operational cost of a device or a combination of removal process.

Tax rate: A tax rate can be applied to an energy or an emission. Unit: US$ (or 1000 US$) or domestic currency per unit of energy or emission.

Technology or Technology system: A 'technology' or 'technology system' refers to a set of devices, coupled with a combination of removal processes, which can be employed to convert material and energy inputs to produce a service output for meeting a final service demand. A device implies a set of procedures and/or machines. A technology can be represented by material and energy flows through a sequence of devices, where each device may or may not be attached to a combination of removal processes (see Figure 2.2 in Chapter 2). Examples of technology with multiple devices include steel making by basic oxygen process, and aluminium production by bayer's process. In certain cases a technology may refer to a single device. For example, electric pump for pumping water for irrigation, and incandescent lamp for lighting.

Unit of device: 'Unit' of a device is defined to measure its stock or quantity in a given year. Possible units: Physical unit of device (e.g. number of devices); Unit of rated operational capacity of device (e.g. MW for power plant); or Unit of service output of device (i.e. same as the unit of service demand). Examples include: Number of incandescent lamps; Mega-Watts of coal-fired power plant; Tons of capacity (equivalent to tons of crude steel production) of electric arc furnace used for steel making.

Useful energy service: It refers to the service output of a device expressed in equivalent energy units. Although this term is not used in AIM/Enduse model, it is used in this document to distinguish energy services like heat or electricity from non-energy services like steel or road passenger transport.

Utility: It is a general term for fuels for boiler, industrial-owned power generation and industrial furnace.

Appendix B: Description of GAMS Files and Code

B.1 Input Parameter

B.1.1 'Set' file

M	Emission gas type
I	Sector
J	Service type
L	Energy device
P	Removal process
K	Energy type
INT	Internal service and energy pairs
ME	Energy constraint group
MQ	Emission constraint group
MR	Region / sector group for combination of internal energy and internal service
JE_KE(INT,J,K)	Mapping among internal service J and internal energy K (INT J K)
M_MQ(I,MQ)	Mapping of I to emission constraint group
M_ME(I,ME)	Mapping of I to energy constraint group
M_MR(I,MR)	Mapping of I to region or sector group for combination of internal energy and internal service

B.1.2 '_1.Gms' file

EN_T(K,L,P,YV,Y)	Energy use by recruited technology per unit operation (K L P YV Y)
F0_T(L,M,YV,Y)	Gas generated per unit operation, independent with energy
AN_T(L,J,YV,Y)	Service quantity per unit operation of recruited energy device (L J YV Y)
BN_T(L,P,YV,Y)	Initial cost of the recruited technology (L)
GAS_T(I,K,M,YV,Y)	CO_2, SO_2, NO_x Emission factor (I K M YV Y)
SERV_T(I,J,YV,Y)	Service amount required (I J YV Y)

B.1.3 '_2.Gms' file

S(I,L,P)	Stock quantity (I L P)
E(I,K,L,P)	Energy use by stocked technology per unit operation (I K L P)
DN(L,P,M)	Pollutant removal ratio(L P M)
D(I,L,P,M)	Pollutant removal ratio of stocked technology (I L P M)

A(I,L,J)	Service quantity per unit operation of stock technology (I L J)
TN(L)	Life span of the recruited energy device (L)
T(I,L)	Life span of the stocked energy device (I L)
ALPHA	Interest rate
BX_T(L,P,P1)	Initial cost of exchanging
G0_T(I,L,P,YV,Y)	Operation cost per unit operating excluding energy cost (I L P YV Y)
GE_T(I,K,YV,Y)	Energy price (I K YV Y)
PHIN_T(I,L,YV,Y)	Service improvement (I L YV Y)
GAM_T(I,L,YV,Y)	Operating efficiency Loss (I L YV Y)
XI_T(I,L,YV,Y)	Countermeasure for energy efficiency change by maintenance (I L YV Y)
TH_T(I,L,J,YV,Y)	Maximum allowable service share (I L J YV Y)
TAX_T(I,M,YV,Y)	Emission tax prescribed (I M YV Y)
TAXE_T(I,K,YV,Y)	Energy tax prescribed (I K YV Y)
QMAX_T(MQ,M,YV,Y)	Maximum allowable emission of gas (MQ M YV Y)
EMAX_T(ME,K,YV,Y)	Maximum allowable energy supply (ME K YV Y)
SCN_T(L,P,YV,Y)	Subsidy rate for recruited technology (L P YV Y)
SCO_T(L,P,YV,Y)	Subsidy rate for operation cost (L P YV Y)
SCX_T(L,P,P1,YV,Y)	Subsidy rate for exchange (L P P1 YV Y)
UB(K,L)	Ratio of material use to total input (K L)

B.2 Output Files

FUEL.CSV	Characteristics of fuels
DEVICE.CSV	Characteristics of devices
ENGBAL.CSV	Energy balance table
DETAIL.CSV	Detailed output table
SERVICE.CSV	Device shares of each services
AGGREG.CSV	Aggregated output table
ERROR.TXT	This is output from _ errorout.gms

After these files are output to the execution directory once they are stored applying %INPUT%.
For instance, %INPUT% is defined by $SETGLOBAL INPUT ..\data\aim_local_china .

B.3 Source of program

```
$title AIM_CMB complex combination of devices version 20020109
$eolcom !
$SETGLOBAL DEBUG
```

```
*$SETGLOBAL ADJUST
$if not setglobal debug $offlisting offinclude
$if not setglobal debug $goto debug_end
   OPTION PROFILE=2;
   OPTION PROFILETOL=0.01;
$label debug_end
* set constant values
PARAMETER SMALL "small number";
   SMALL=1.E-6;
$INCLUDE ..\data\AIM_Local.inp
*$SETGLOBAL INPUT ..\data\AIM-Local_TEST
*
* Calibration of service and stock compatibility
* Calibration of emission amount
*
* Included libraries are
* _interp  : interpolation of input data, linearly and geometrically
* _printout: write files of calculation
* _errout.txt: parameter error output
*        FUEL.CSV    characteristics of fuels
*        DEVICE.CSV  characteristics of devices
*        ENGBAL.CSV  Energy balance table
*        DETAIL.CSV  Detailed output table
*        SERVICE.CSV Device shares of each services
*        AGGREG.CSV  Aggregated output table
* ---
* Following sets and parameters are determined in %INPUT%.set
*SET M "Emission gas type" /CO2, SO2/;
*SET I "Region or LPS identifier" /R1,R2/;
*SET J "Service type" /PWR/;
*SET L "Technology" /OILBLR,COLBLR,GASTBN/;
*SET P "Removal technology" /NON,FGD,WSH/;
*SET K "Energy type" /COL,GAS,OIL/;
*SET  YEAR "Simulation years" /1995*2020/;
*Parameter JE_KE(INT,J,K) "Mapping among internal service J and internal energy K(INT J K)"
SET YV "value and year" /V,Y/;
SET Y  "interpolated year" /Y1*Y8/;
*SET M_MQ(I,MQ)  "Mapping of i to emission constraint group;
*SET M_ME(I,ME)  "Mapping of i to Energy constraint group;
*SET M_MR(I,MR)  "Mapping of i to Region group;
$INCLUDE %INPUT%.set

* In case of no internal energy and service
$if not setglobal INT set INT "Internal service and energy pairs" /DUMMY/;
$if not setglobal INT Parameter JE_KE(INT,J,K) "Mapping among internal service J and internal en-
ergy K(INT J K)";
$if not setglobal INT JE_KE(INT,J,K)=0;

ALIAS (P,P1);
ALIAS (L,L1);

* data read in include file
PARAMETER EN_T(K,L,P,YV,Y) "Energy consumption of recruited technology per unit operation
(K L P YV Y)"
        F0_T(L,M,YV,Y) "Gas generated per unit operation, independent with energy (L M YV Y)"
        GAS_T(I,K,M,YV,Y) "Gas material contained in unit energy(I K M YV Y)"
        AN_T(L,J,YV,Y)  "Service quantity per unit operation of recruited technology(L J)"
```

BN_T(L,P,YV,Y) "Initial cost of the recruited technology(L YV Y)"
BX_T(L,P,P1,YV,Y) "Initial cost of exchanging (L P)->(L P1) (L P P1 YV Y)"
G0_T(I,L,P,YV,Y) "Operating cost per unit operating excluding energy cost(I L P)"
GE_T(I,K,YV,Y)   "Energy price(I K YV Y)"
PHI_T(I,J,YV,Y) "Social service efficiency improvement(I J YV Y)"
GAM_T(I,L,YV,Y) "Operating efficiency improvement(I L YV Y)"
QMAX_T(MQ,M,YV,Y) "Maximum allowable emission of Gas(MQ M YV Y)"
TH_T(I,L,J,YV,Y) "Maximum allowable survice share(I L J YV Y)"
EMAX_T(ME,K,YV,Y) "Maximum allowable energy supply(ME,K YV Y)"
XI_T(I,L,YV,Y) "Energy coefficient change by maintenance(I K L YV Y)"
TAX_T(I,M,YV,Y) "Emission tax prescribed(I M YV Y)"
TAXE_T(I,K,YV,Y) "Energy tax prescribed(I K YV Y)"
SERV_T(I,J,YV,Y) "Service amount required(I J YV Y)"
SCN_T(L,P,YV,Y) "Subsidy ratio of the recruited technology(L P YV Y)"
SCO_T(L,P,YV,Y) "Subsidy ratio of operation cost(L P YV Y)"
SCX_T(L,P,P1,YV,Y) "Subsidy ratio of the recruited technology(L P P1 YV Y)"
DN(L,P,M) "Pollutant removal ratio of recruited technology(L P M)"
TN(L) "Life span of the recruited technology(L)"
E(I,K,L,P) "Energy consumption of stocked technology per unit operating(I K L P)"
D(I,L,P,M) "Pollutant removal ratio of stocked technology(I L P M)"
S(I,L,P)   "Stock quantity(I L P)"
A(I,L,J) "Service quantity per unit operation of stock technology(I L J)"
T(I,L) "Life span of the stocked technology(I L)"
ALPHA   "Interest rate"
V_YEAR(YEAR) "Numeric of year"
Q(I,M) "Observed emission (I M)"
ENG(I,K) "Observed energy consumption(I K)"
;

Q(I,M)=0;
ENG(I,K)=0;
PARAMETER SERV_C(I,J) "Calibration parameter of service amount (I J)"
Q_C(I,M)   "Calibration parameter of emission (I M)"
ENG_C(I,K) "Calibration parameter of energy (I K)"
;
PARAMETER WORK_IL(I,L) "Work dimension"
WORK_IJ(I,J) "Work dimension"
WORK_F   "Work dimension"
WORK_LP(L,P) "Work dimension"
WORK_ILP(I,L,P) "Work dimension"
;

* parameter updated in this program
PARAMETER EN(K,L,P)   "Energy consumption of recruited technology per unit operating(K L P)"
F0(L,M)   "Gas generated per unit operation, independent with energy (L M)"
GAS(I,K,M)   "Gas material contained in unit energy(I K M)"
AN(L,J)   "Service quantity per unit operation of recruited technology(L J)"
BN(L,P)   "Initial cost of the recruited technology(L)"
CN(L,P)   "Annualized initial cost(L P)"
BX(L,P,P1)   "Initial cost of exchanging (L P)->(L P1) (L P P1)"
CX(L,P,P1)   "Annualized initial cost of exchanging (L P)->(L P1) (L P P1)"
G0(I,L,P)   "Operating cost per unit operating excluding energy cost(I L P)"
GE(I,K)   "Energy price(I K)"
PHI(I,J)   "Social service efficiency improvement(I J)"
GAM(I,L)   "Operating efficiency improvement(I L)"
QMAX(MQ,M)   "Maximum allowable emission of Gas(MQ M)"
TH(I,L,J)   "Maximum allowable survice share(I L J)"
EMAX(ME,K)   "Maximum allowable energy supply(K)"

```
        XI(I,L)      "Energy coefficient change by maintenance(I K L)"
        TAX(I,M)     "Emission tax prescribed(I M)"
        TAXE(I,K)    "Energy tax prescribed(K M)"
        SERV(I,J)    "Service amount required(I J)"
        SCN(L,P)     "Subsidy ratio of the recruited technology(L P)"
        SCO(L,P)     "Subsidy ratio of operation cost(L P)"
        SCX(L,P,P1)  "Subsidy ratio of the recruited technology(L P P1)"
*       derived variables
        DE(K,L,P,P1) "Energy consumption change multiplier from (L P) to (L P1)"
        EM(I,L,P,M)  "Emission of per unit operation(I,L,P,M)"
        X(I,L,P)     "Operating amount(I,L,P)"
        R(I,L,P)     "Recruited amount(I,L,P)"
        R_MAX(I,L,P) "Maximum recruit quantity(I,L,P)"
        YEAR_S       "Start year of calculation"
        SS(I,L,P)    "Serviced stock after one year"
*       check parameters
        DLT_IJ(I,J)     Allowable conbination of I J
        DLT_JL(J,L)     Allowable conbination of J L
        DLT_LP(L,P)     Allowable conbination of L P
        DLT_IL(I,L)     Allowable conbination of I L
        DLT_IK(I,K)     Allowable conbination of I K
        DLT_KLP(K,L,P)  Allowable conbination of K L P
        DLT_ILP(I,L,P)  Allowable conbination of I L P
        DLT_IJL(I,J,L)  Allowable conbination of I J L
        DLT_IJLP(I,J,L,P) Allowable conbination of I J L P
        DLT_IKLP(I,K,L,P) Allowable conbination of I K L P
        DLT_IKM(I,K,M)  Allowable conbination of I K M
        DLT_IKL(I,K,L)  Allowable conbination of I K L
        DLT_LPP(L,P,P1) Allowable conbination of L P P1
*       Active flags
        ACT_I(I)        Activity of I
;
* parameter updated in this program
PARAMETER EN(K,L,P)  "Energy consumption of recruited technology per unit operating(K L P)"
        F0(L,M)      "Gas generated per unit operation, independent with energy (L M)"
;
FILE F_CHECK /check.txt/;F_CHECK.PW=2000;
*Default values of activities
    ACT_I(I)=1;
*   ---
*   ACT_I(I)=0;
*   ACT_I('IN-STL_JPN')=1;
*   ACT_I('IN-CEM_JPN')=1;
*   ACT_I('IN-PET_JPN')=1;
*   ACT_I('IN-PAP_JPN')=1;
*   ACT_I('IN-AGR_JPN')=1;
*   ACT_I('IN-MIN_JPN')=1;
*   ACT_I('IN-CON_JPN')=1;
*   ACT_I('IN-FOO_JPN')=1;
*   ACT_I('IN-TXT_JPN')=1;
*   ACT_I('IN-OCR_JPN')=1;
*   ACT_I('IN-OCH_JPN')=1;
*   ACT_I('IN-NFR_JPN')=1;
*   ACT_I('IN-MEM_JPN')=1;
*   ACT_I('IN-OTH_JPN')=1;
*   ACT_I('OT-RES_JPN')=1;
*   ACT_I('OT-COM_JPN')=1;
```

```
*    ACT_I('TRANS_JPN')=1;
*    ACT_I('CV-OIL_JPN')=1;
*    ACT_I('CV-ELE_JPN')=1;
*    ACT_I('CV-COL_JPN')=1;
*    ---
* Endogenous flag parameters
PARAMETER END_K(K)    Endogenous flag of energy K
      END_J(J)    Endogenous flag of service J
;
      END_K(K)=SUM((INT,J),JE_KE(INT,J,K));
      END_J(J)=SUM((INT,K),JE_KE(INT,J,K));
*
$INCLUDE %INPUT%_1.gms
*----
$  batinclude _errorout "START"
*Input parameter check
    DLT_IJ(I,J)=1$(SUM(Y,SERV_T(I,J,'V',Y)) and ACT_I(I));
    DLT_JL(J,L)=1$SUM(Y,AN_T(L,J,'V',Y));
    DLT_LP(L,P)=1$sum(Y,BN_T(L,P,'V',Y));
    EN_T(K,L,P,'V',Y)$(not DLT_LP(L,P))=0;
    DLT_KLP(K,L,P)=1$sum(Y,EN_T(K,L,P,'V',Y));
    DLT_IJL(I,J,L)=1$(DLT_IJ(I,J) and DLT_JL(J,L));
    DLT_IL(I,L)=1$SUM(J,DLT_IJL(I,J,L));
    DLT_ILP(I,L,P)$DLT_IL(I,L)=1$(DLT_IL(I,L) and DLT_LP(L,P));
    DLT_IJLP(I,J,L,P)$DLT_IJL(I,J,L)=1$(DLT_IJL(I,J,L) and DLT_LP(L,P));
    DLT_IKLP(I,K,L,P)$DLT_ILP(I,L,P)=1$(DLT_ILP(I,L,P) and DLT_KLP(K,L,P));
    DLT_IKL(I,K,L)=1$SUM(P,DLT_IKLP(I,K,L,P));
    DLT_IK(I,K)   =1$SUM(L,DLT_IKL(I,K,L));
    GAS_T(I,K,M,'V',Y)$(not DLT_IK(I,K))=0;

DLT_IKM(I,K,M)=1$SUM(Y,GAS_T(I,K,M,'V',Y)+SUM(L$DLT_IKL(I,K,L),F0_T(L,M,'Y',Y)));

*    Default values
    TH_T(I,L,J,'V',Y)$DLT_IJL(I,J,L)=1;
    TAX_T(I,M,'V',Y)=0;
    TAXE_T(I,K,'V',Y)=0;
$INCLUDE %INPUT%_2.gms
*    Material use of energy
$if not setglobal MATERIAL PARAMETER UB(K,L) ratio of material use to total input;
$if not setglobal MATERIAL UB(K,L)$SUM(I,DLT_IKL(I,K,L))=0;
*
    DLT_LPP(L,P,P1)=1$(SUM(Y,BX_T(L,P,P1,'V',Y)) and DLT_LP(L,P) and DLT_LP(L,P1));
*    ----
    G0_T(I,L,P,'V',Y)$(not DLT_ILP(I,L,P))=0;
    DN(L,P,M)       $(not DLT_LP(L,P))=0;
    GE_T(I,K,'V',Y)  $(not DLT_IK(I,K))=0;
    PHI_T(I,J,'V',Y) $(not DLT_IJ(I,J))=0;
    GAM_T(I,L,'V',Y) $(not DLT_IL(I,L))=0;
    XI_T(I,L,'V',Y)  $(not DLT_IL(I,L))=0;
    TH_T(I,L,J,'V',Y)$(not DLT_IJL(I,J,L))=0;
    SCN_T(L,P,'V',Y) $(not DLT_LP(L,P))=0;
    SCO_T(L,P,'V',Y) $(not DLT_LP(L,P))=0;
    SCX_T(L,P,P1,'V',Y)$(not DLT_LPP(L,P,P1))=0;
*    ----
*    Check errors of stock and technical parameters
$  batinclude _errorout "COMMENT" '"Check errors of stock and technical parameters "'
    LOOP(L,
```

```
      IF(not TN(L),PUT 'ERROR_03I1 TN(L)=0 L:',L.TL/;
      );
   );
*=============================================================
POSITIVE VARIABLES
   VX(I,L,P)    "Operating quantity(I,L,P)"
   VR(I,L,P)    "Recruit quantity(I,L,P)"
   VS(I,L,P)    "Stock quantity(I,L,P)"
   VM(I,L,P,P1) "Exchanged stock of device (L P) to (L P1) (I L P P1)"
   VQ(I, M)     "Emission(I,M)"
   VE(I,K)      "Energy consumed"
   VSERV(I,J)   "Service supplied"
   RES_GEC(MQ,M)  Slack of gas mission constraints
   RES_SRC(I,J,L) Slack of share ratio constraints
   RES_ESC(ME,K)  Slack of energy supply constraints
   RES_SDC(I,J)   Slack of service demand conditions
   RES_OCC(I,L,P) Slack of operating capacity conditions
   RES_SEC(I,L,P) Slack of stock exchange constraints
   RES_SCS(I,L,P) Slack of stock quantity replacement
   RES_EMISS(I,M) Slack of emission
   RES_SVC(I,J)   Slack of supplied service amount
   RES_END(MR,INT) Slack of endogenous energy demand and service supply constraints
   RES_ENG(I,K)   Slack of energy consumption
;
VARIABLES
   VTTLCST     "Total cost"
;
EQUATIONS
   EQ_GEC(MQ,M)      Gas mission constraints
   EQ_SRC(I,J,L)     Share ratio constraints
   EQ_ESC(ME,K)      Energy supply constraints
   EQ_SDC(I,J)       Service demand conditions
   EQ_OCC(I,L,P)     Operating capacity conditions
   EQ_SEC(I,L,P)     Stock exchange constraints
   EQ_SCS(I,L,P)     Stock quantity replacement
   EQ_EMISS(I,M)     Emission
   EQ_SVC(I,J)       Supplied service amount
   EQ_END(MR,INT)    Endogenous energy demand and endogenous service supply constraints
   EQ_ENG(I,K)       Energy consumption
   EQ_TCE         Total cost
;
EQ_GEC(MQ,M)$(QMAX(MQ,M) NE INF) ..
   QMAX(MQ,M) =G= SUM(I$M_MQ(I,MQ),VQ(I,M)) ;
EQ_SRC(I,J,L)$((TH(I,L,J) LT 1) and DLT_IJL(I,J,L))..
   TH(I,L,J)*SUM((L1,P)$DLT_IJLP(I,J,L1,P),A(I,L1,J)*VX(I,L1,P))
      =G= A(I,L,J)*SUM(P$DLT_IJLP(I,J,L,P),VX(I,L,P));
EQ_SVC(I,J)$DLT_IJ(I,J) ..
   RES_SVC(I,J) =E= (1+PHI(I,J))*SUM((L,P)$DLT_IJLP(I,J,L,P),A(I,L,J)*VX(I,L,P))-
VSERV(I,J)*(1+SERV_C(I,J));
EQ_SDC(I,J)$(DLT_IJ(I,J) AND NOT END_J(J)) ..
   RES_SDC(I,J) =E= VSERV(I,J)-SERV(I,J);
EQ_END(MR,INT)$SUM(I,M_MR(I,MR))..
   RES_END(MR,INT) =E= SUM(I$M_MR(I,MR),SUM(J$(SUM(K,JE_KE(INT,J,K)) and
DLT_IJ(I,J)),VSERV(I,J))-SUM(K$SUM(J,JE_KE(INT,J,K)),VE(I,K)));
   RES_END.L(MR,INT) =0;
EQ_ESC(ME,K)$(EMAX(ME,K) NE INF) ..
   RES_ESC(ME,K) =E= EMAX(ME,K) - SUM(I$M_ME(I,ME), VE(I,K));
```

```
EQ_OCC(I,L,P)$DLT_ILP(I,L,P)..
    RES_OCC(I,L,P) =E= (1+GAM(I,L))*VS(I,L,P) - VX(I,L,P) ;
EQ_SEC(I,L,P)$DLT_ILP(I,L,P)..
    RES_SEC(I,L,P) =E= SS(I,L,P)-SUM(P1$CX(L,P,P1),VM(I,L,P,P1));
EQ_SCS(I,L,P)$DLT_ILP(I,L,P)..
    RES_SCS(I,L,P) =E=SS(I,L,P)+VR(I,L,P)+SUM(P1,VM(I,L,P1,P)$CX(L,P1,P)-
VM(I,L,P,P1)$CX(L,P,P1))-VS(I,L,P) ;
EQ_EMISS(I,M)$ACT_I(I)..
    RES_EMISS(I,M) =E= VQ(I,M)*(1+Q_C(I,M)) -
SUM((L,P)$EM(I,L,P,M),VX(I,L,P)*EM(I,L,P,M));
    RES_EMISS.FX(I,M)=0;
EQ_ENG(I,K)$DLT_IK(I,K)..
    VE(I,K) =E= SUM((L,P),(1-XI(I,L))*E(I,K,L,P)*VX(I,L,P))/(1+ENG_C(I,K));
EQ_TCE ..
    VTTLCST =E=SUM(I,
            SUM((L,P)$DLT_ILP(I,L,P),
                CN(L,P)*VR(I,L,P)+SUM(P1$CX(L,P,P1),CX(L,P,P1)*VM(I,L,P,P1))
                +(G0(I,L,P)+(1-
XI(I,L))*SUM(K$E(I,K,L,P),GE(I,K)*E(I,K,L,P)/(1+ENG_C(I,K))))*VX(I,L,P)
                *(1-SCO(L,P))
            )
        ) +SUM((I,M)$(TAX(I,M) and ACT_I(I)),VQ(I,M)*TAX(I,M))
        +SUM((I,K)$(TAXE(I,K) and ACT_I(I)),VE(I,K)*TAXE(I,K))
        +SUM((I,L,P)$DLT_ILP(I,L,P),VX(I,L,P))*SMALL;


MODEL AIM_LOCAL / EQ_GEC, EQ_SRC, EQ_SVC, EQ_SDC, EQ_END, EQ_ENG, EQ_ESC,
EQ_OCC, EQ_SEC, EQ_SCS, EQ_EMISS, EQ_TCE /;
* ===================== Solve equations year by year ========================
YEAR_S=SUM(YEAR$(ORD(YEAR) = 1),V_YEAR(YEAR));

$batinclude  _printout '"_base"' 0

$IF NOT SETGLOBAL ADJUST LOOP(YEAR,
$IF    SETGLOBAL ADJUST LOOP(YEAR$(ORD(YEAR)<2),

$  batinclude  _interp  EN  K,L,P   0  DLT_KLP(K,L,P)
$  batinclude  _interp  F0  L,M    0  1
$  batinclude  _interp  GAS I,K,M  0  DLT_IKM(I,K,M)
$  batinclude  _interp  AN  L,J    0  DLT_JL(J,L)
$  batinclude  _interp  BN  L,P    0  DLT_LP(L,P)
$  batinclude  _interp  BX  L,P,P1  0  DLT_LPP(L,P,P1)
$  batinclude  _interp  G0  I,L,P   0  DLT_ILP(I,L,P)
$  batinclude  _interp  GE  I,K    0  DLT_IK(I,K)
$  batinclude  _interp  PHI I,J    0  DLT_IJ(I,J)
$  batinclude  _interp  GAM I,L    0  DLT_IL(I,L)
$  batinclude  _interp  QMAX MQ,M   0  1
$  batinclude  _interp  TH  I,L,J   0  DLT_IJL(I,J,L)
$  batinclude  _interp  EMAX ME,K   0  1
$  batinclude  _interp  XI  I,L    0  DLT_IL(I,L)
$  batinclude  _interp  TAX  I,M   0  1
$  batinclude  _interp  TAXE I,K   0  1
$  batinclude  _interp  SERV I,J   1  DLT_IJ(I,J)
$  batinclude  _interp  SCN  L,P   0  DLT_LP(L,P)
$  batinclude  _interp  SCO  L,P   0  DLT_LP(L,P)
$  batinclude  _interp  SCX  L,P,P1  0  DLT_LPP(L,P,P1)
*  ----
$  batinclude _errorout "COMMENT" '"Check errors of running parameters "'
```

```
    LOOP((I,J)$DLT_IJ(I,J),                           !Error check 03T1
      IF( not SUM(L$DLT_IJL(I,J,L),(1+PHI(I,J))*(1+GAM(I,L))*TH(I,L,J)),
         PUT 'ERROR_03T1 sum(L,(1+PHI)*(1+GAM)*TH)=0 Y:',YEAR.TL,' I:',I.TL,' J:',J.TL/;
      );
    );
    LOOP((L,P,P1)$BX(L,P,P1),                         !Error check 03T2
      IF( not BN(L,P) or not BN(L,P1),
         PUT 'ERROR_03T2 BX(L,P,P1) incompatible with BN Y:',YEAR.TL,' L:',L.TL,'
P:',P.TL,P1.TL/;
      );
    );
    LOOP((I,L)$(sum(P,S(I,L,P)) and ACT_I(I)),
      IF(not sum((J,P)$DLT_IJLP(I,J,L,P),A(I,L,J)*S(I,L,P)),        !Error check 03T3
         PUT 'ERROR_03T3 sum((J,P),A(I,L,P)*S(I,L,P))=0 Y:',YEAR.TL,' I:',I.TL,' L:',L.TL/;
      );
      IF(not T(I,L),                       !Error check 03T4
         PUT 'ERROR_03T4 T(I,L)=0 Y:',YEAR.TL,' I:',I.TL,' L:',L.TL/;
      );
    );
    LOOP((I,L,P)$(S(I,L,P) and ACT_I(I)),
      IF(not sum(K$DLT_IKLP(I,K,L,P),E(I,K,L,P)),              !Error check 03T5
         PUT 'ERROR_03T5 sum(K,E(I,K,L,P))=0 Y:',YEAR.TL,' I:',I.TL,' L:',L.TL,' P:',P.TL/;
      );
    );
    LOOP((I,K)$DLT_IK(I,K),
      IF(not GE(I,K),                          !Error check 03T6
         PUT 'ERROR_03T6 GE(I,K)=0 Y:',YEAR.TL,' I:',I.TL,' K:',K.TL/;
      );
    );
*    --- Depreciation of the year ---
    SS(I,L,P)=0;
    SS(I,L,P)$(S(I,L,P) and DLT_ILP(I,L,P))=S(I,L,P)*max((1.-1/T(I,L)),0);
    CN(L,P)$BN(L,P) = (BN(L,P)*(1-
SCN(L,P)))*ALPHA*EXP(TN(L)*LOG(1+ALPHA))/(EXP(TN(L)*LOG(1+ALPHA))-1);
    CX(L,P,P1)$BX(L,P,P1)=(BX(L,P,P1)*(1-
SCX(L,P,P1)))*ALPHA*EXP(TN(L)*LOG(1+ALPHA))/(EXP(TN(L)*LOG(1+ALPHA))-1);
    DE(K,L,P1,P)$(EN(K,L,P) and EN(K,L,P1))=EN(K,L,P)-EN(K,L,P1);
*    --- Default parameters of recruited devices ---
    A(I,L,J)      $(DLT_IJL(I,J,L) and not sum(P$DLT_ILP(I,L,P),SS(I,L,P)))=AN(L,J);
    D(I,L,P,M)      $(DLT_ILP(I,L,P) and not SS(I,L,P))=DN(L,P,M);
    T(I,L)        $(DLT_IL(I,L) and not SUM(P,SS(I,L,P)))=TN(L);
    E(I,K,L,P)      $(DLT_IKLP(I,K,L,P) and not SS(I,L,P)) =EN(K,L,P);
    EM(I,L,P,M)$DLT_ILP(I,L,P)
       =(F0(L,M)+SUM(K$E(I,K,L,P),GAS(I,K,M)*(1-XI(I,L))*E(I,K,L,P)*(1-UB(K,L))))*(1-
D(I,L,P,M));
*    ---
    IF(ORD(YEAR)=1,
*    Check parameter
$    batinclude _errorout "COMMENT" '"Calibration parameter for initial year"'
    LOOP((I,K)$(ENG(I,K) and ACT_I(I)),
      IF(not DLT_IK(I,K),                         !Error check 03I7
         PUT 'ERROR_03I7 ENG(I,K)>0 and DLT_IK>0 I:',I.TL,' K:',K.TL/;
      );
    );
    LOOP((I,M)$(Q(I,M) and ACT_I(I)),
      IF(not SUM(K,DLT_IKM(I,K,M)),                    !Error check 03I8
         PUT 'ERROR_03I8 Q(I,M)>0 and DLT_IM>0 I:',I.TL,' M:',M.TL/;
```

```
    );
  );
  LOOP(J,
    IF(END_J(J) and SUM(INT$SUM(K,JE_KE(INT,J,K)),1) > 1,            !Error check 03I9
      PUT 'ERROR_03I9 J appears more than one time in JE_KE J:',J.TL/;
    );
  );
  LOOP(K,
    IF(END_K(K) and SUM(INT$SUM(J,JE_KE(INT,J,K)),1) > 1,            !Error check 03I9
      PUT 'ERROR_03I9 K appears more than one time in JE_KE K:',K.TL/;
    );
  );
$ batinclude _errorout "COMMENT" '"Service balance check for initial year"'
  LOOP((I,J)$(SERV(I,J) and (not END_J(J)) and ACT_I(I)),
    WORK_IJ(I,J)=(1+PHI(I,J))*SUM((L,P)$DLT_IJLP(I,J,L,P),
A(I,L,J)*(1+GAM(I,L))*S(I,L,P));
    IF(WORK_IJ(I,J) > SERV(I,J)*1.1 or SERV(I,J)*.9 > WORK_IJ(I,J)*1.1,            !
ERROR_04I1
      PUT 'ERROR_04I1 I:',I.TL,' J:',J.TL,' SERV:',SERV(I,J):10:5,' SUPLY:',WORK_IJ(I,J):10:5;
      PUT ' # of TH(ILJ):',SUM(L$TH(I,L,J),1):5:0,' # of A(ILJ):',SUM(L$A(I,L,J),1):5:0,
        ' # of non(1+GAM(IL)):',SUM(L$(DLT_IJL(I,J,L)*(1+GAM(I,L))),1):5:0,
        ' # of non S(ILP):',SUM((L,P)$(DLT_IJL(I,J,L) and S(I,L,P)),1):5:0/;
    );
    SERV_C(I,J)=WORK_IJ(I,J)/SERV(I,J)-1;
  );
* ----
$ batinclude _errorout "COMMENT" '"Service share check for initial year"'
  LOOP((I,J)$(SERV(I,J) and (not END_J(J)) and ACT_I(I)),
    LOOP(L$(not TH(I,L,J)=1 and DLT_IJL(I,J,L)),            !ERROR_05I1:SERV(IJ)
    WORK_F=(1+PHI(I,J))*A(I,L,J)*(1+GAM(I,L))*SUM(P$DLT_IJLP(I,J,L,P),S(I,L,P));
    IF( WORK_F > TH(I,L,J)*WORK_IJ(I,J),
      PUT 'ERROR_05I1 SERV(IJ) I:',I.TL,' J:',J.TL,' L:',L.TL,' SUP(IJL):',WORK_F:10:5,
                    ' SUP*TH(IJL):',(TH(I,L,J)*WORK_IJ(I,J)):10:5/;
    );
    );
  );
* ----
$ batinclude _errorout "COMMENT" '"Emission check for initial year"'
  LOOP((I,M)$(Q(I,M) and ACT_I(I)),
    WORK_F=SUM((L,P)$EM(I,L,P,M),(1+GAM(I,L))*S(I,L,P)*EM(I,L,P,M));
    IF(WORK_F > Q(I,M)*1.1 or Q(I,M)*0.9 > WORK_F,            !ERROR_06I1:Q(IM)
      PUT 'ERROR_06I1 Q(IM) I:',I.TL,' M:',M.TL,' Q:',Q(I,M):10:5,' Q_calc:',WORK_F:10:5/;
    );
    Q_C(I,M) = WORK_F/Q(I,M)-1;
  );
* ----
$ batinclude _errorout "COMMENT" '"Energy check for initial year"'
  PUT Z2_DEVICE;
  PUT 'GH Check' /;
  LOOP((I,K)$(ENG(I,K) and ACT_I(I)),
    WORK_F=SUM((L,P)$E(I,K,L,P),(1-XI(I,L))*E(I,K,L,P)*(1+GAM(I,L))*S(I,L,P));
    IF(WORK_F > ENG(I,K)*1.2 or ENG(I,K)*0.8 >
WORK_F,            !ERROR_07I1:ENG(IM)
      PUT 'ERROR_07I1 ENG(IK) I:',I.TL,' K:',K.TL,' ENG:',ENG(I,K):10:5,'
ENG_calc:',WORK_F:10:5/;
    );
    ENG_C(I,K) = WORK_F/ENG(I,K)-1;
```

```
    PUT Z2_DEVICE;
    PUT 'GH Check', ENG_C(I,K) /;
  );
*   ----
    VX.L(I,L,P)$DLT_ILP(I,L,P)=(1+GAM(I,L))*S(I,L,P);
    VR.L(I,L,P)$DLT_ILP(I,L,P)=0;
    R(I,L,P)$DLT_ILP(I,L,P)=0;
    VM.L(I,L,P,P1)$DLT_ILP(I,L,P)=0;
    VQ.L(I,M) = SUM((L,P)$DLT_ILP(I,L,P),VX.L(I,L,P)*EM(I,L,P,M))/(1+Q_C(I,M));
    DISPLAY 'Calibration in base year';
    DISPLAY SERV_C,Q_C,ENG_C;
*   ----
$    batinclude _printout '"_base'" 1
*   ----
$IF SETGLOBAL ADJUST );
$IF SETGLOBAL ADJUST $INCLUDE ADJUST.GMS
$IF SETGLOBAL ADJUST LOOP(YEAR,
*   ----
    SERV(I,J)$END_J(J)=SMALL;

    SOLVE AIM_LOCAL USING LP MINIMIZING VTTLCST;
*   ----------
    PUT Z2_DEVICE;
    LOOP(MR,LOOP(INT,
      PUT 'Yr=',YEAR.TL,
        ' Res=' ,RES_END.L(MR,INT),
        ' Vserv=',SUM(I$M_MR(I,MR),SUM(J$SUM(K,JE_KE(INT,J,K)),VSERV.L(I,J))),
        ' Veng=', SUM(I$M_MR(I,MR),SUM(K$SUM(J,JE_KE(INT,J,K)),VE.L   (I,K)));
    ));
*   ----------
    IF (AIM_LOCAL.modelstat>2,
      DISPLAY 'Solver error : AIM_LOCAL ****';
      ABORT 'abort';
    );
* --- Recruit and revise stock characteristics ---
    WORK_IL(I,L)$DLT_IL(I,L)=SUM(P,VS.L(I,L,P));
    LOOP((I,L)$WORK_IL(I,L),
     S(I,L,P)$DLT_ILP(I,L,P)=VS.L(I,L,P);
     R(I,L,P)$DLT_ILP(I,L,P)=VR.L(I,L,P);
    );
$   batinclude _printout YEAR.TL 1
    LOOP((I,L)$WORK_IL(I,L),
     WORK_ILP(I,L,P)$VS.L(I,L,P)=SS(I,L,P)-SUM(P1$CX(L,P,P1),VM.L(I,L,P,P1));
     D(I,L,P,M)$VS.L(I,L,P)=
          ( D(I,L,P,M)*WORK_ILP(I,L,P)
          +DN(L,P,M) *( R(I,L,P) +SUM(P1$CX(L,P1,P),VM.L(I,L,P1,P)) )
          )/VS.L(I,L,P);
     E(I,K,L,P)$(VS.L(I,L,P) and DLT_IKLP(I,K,L,P))=
          ( E(I,K,L,P)*WORK_ILP(I,L,P)
          +SUM(P1$VM.L(I,L,P1,P),(E(I,K,L,P1)+DE(K,L,P1,P))*VM.L(I,L,P1,P))
          +EN(K,L,P) *R(I,L,P)
          )/VS.L(I,L,P);

     A(I,L,J)$DLT_IJL(I,J,L)=
         ( A(I,L,J)*SUM(P$DLT_IJLP(I,J,L,P),SS(I,L,P))
         +AN(L,J) *SUM(P$DLT_IJLP(I,J,L,P),R(I,L,P)))
         /WORK_IL(I,L);
```

```
    T(I,L)=(T(I,L)*SUM(P,SS(I,L,P))+TN(L)*SUM(P,R(I,L,P)))
        /WORK_IL(I,L);
  );
*----
);
* --- Solve equations year by year END ---
$batinclude _errorout "END"
$batinclude _printout "_base" 2
```

Appendix C: AIM/Enduse Database System and Implementation

C.1 Description of Input Data

The window as shown in Fig. C.1 appears on clicking at file AIM_Enduse.mdb on Windows explorer. On clicking at "Main" in the list of forms, the form 'Main' as shown in Fig. C.2 can be seen. Input data is entered by clicking on the buttons in the main form.

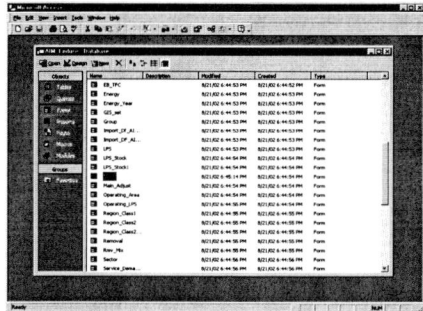

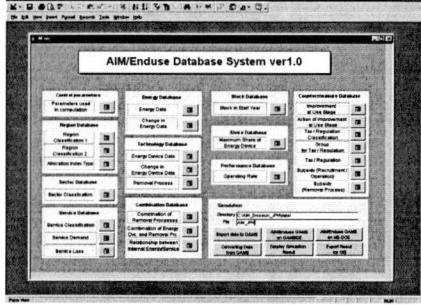

Fig. C.1. After clicking AIM_Enduse.mdb   **Fig. C.2.** After clicking "Main"

C.1.1 Control parameters

C.1.1.1 Parameters used in computation

The list of parameters used in computation is shown in the following table.

Table C.1. List of items in "Parameters used in Computation"

Items	Format	Contents
Start year of calculation *	Integer	Year from which AIM/Enduse calculates CO_2, SO_2, NO_x emissions. It corresponds to the base year.
End year of calculation *	Integer	Year to which AIM/Enduse calculates CO_2, SO_2, NO_x emissions.
Discount rate *	Percent	Rate is used for economic criteria of technology selection based life cycle cost.
Unit of price	Character	Unit is shown in the footer of other forms. Change of unit
Unit of energy	Character	could not cause the exchange of values.
Unit of CO_2	Character	
Unit of SO_2	Character	
Unit of NO_x	Character	

Item with * : Code or value of the item is indispensable for database system and calculation.

C.1.2 Region database

C.1.2.1 Region classification 1

This table specifies the aggregate region classification. Service demand, technologies' share and countermeasure etc. should be listed by this classification.

Table C.2. List of items in "Region Classification 1"

Items	Format	Contents
No.	Integer (Max 32768)	Number of the coarse region classification. It is an independent number and is not used in the calculation.
Region 1 code *	Character (Max 6)	Code of the coarse region classification. Every code must be unique in the list. - "ALL" could not be permitted to use as the code. - "_" could not be permitted to use as a part of the code.
Region 1 name	Character (Max 40)	Name of the coarse region classification.

Item with * : Code or value of the item is indispensable for database system and calculation.

C.1.3 Sector database

C.1.3.1 Sector classification

This table specifies the sector classification. AIM/Enduse shows emission by each sector based on this classification. Also, countermeasure scenarios can be set for each sector.

Table C.3. List of items in "Sector Classification"

Items	Format	Contents
No.	Integer (Max 32768)	Number of the sector classification. It is an independent number and is not used in the calculation.
Sector code *	Character (Max 6)	Code of the sector classification. Every code must be unique in the list. - "ALL" could not be permitted to use as the code. - "_" could not be permitted to use as a part of the code.
Sector name	Character (Max 40)	Name of the sector classification.

Item with * : Code or value of the item is indispensable for database system and calculation.

C.1.4 Service database

C.1.4.1 Service classification

This table specifies the service classification.

Table C.4. List of items in "Service Classification"

Items	Format	Contents
No.	Integer (Max 32768)	Number of the energy classification. It is an independent number and is not used in the calculation.

Table C.4. List of items in "Service Classification" (continued)

Items	Format	Contents
Service code *	Character (Max 6)	Code of the service classification. Every code must be unique in the list.
Service name	Character (Max 40)	Name of the service classification.
Service unit	Character (Max 10)	Unit of the service classification.
Sector name *	-	Select the sector that the service belongs in the list.
Allocation index	-	In case of area source, firstly, its emission is calculated by each region classification 1. Then emission is allocated to the region classification 2 in proportion to allocation index. Select the number and the name of the allocation index by which the emission is divided to the classification 2.

Item with * : Code or value of the item is indispensable for database system and calculation.

C.1.4.2 Service demand

The data for service demand in reference year as well as its projections in future years are entered in this table.

Table C.5. List of items in "Service Demand"

Items	Format	Contents
No.	Integer (Max 32768)	Number of the data sets. It is an independent number and is not used in the calculation.
Region 1 *	-	Select the coarse region classification in the list.
Service *	-	Select the service classification in the list. Every pair of 'Region 1' and 'Service' must be unique.
Service demand (year) *	Integer	Enter data sets given by pairs of year and quantity. The total sets are less than four. Temporary value may be entered as the demand of internal service so that the one is decided endogenously in the model.
Service demand (value) *	Single (>0)	

Item with * : Code or value of the item is indispensable for database system and calculation.

C.1.4.3 Service loss

Data for transmission and distribution loss between the producer and the receiver of energy service is entered in this table. If no data in entered in this table for a service, its transmission and distribution loss is assumed as 0%.

Table C.6. List of items in "Service Loss"

Items	Format	Contents
No.	Integer (Max 32768)	Number of the data sets. It is an independent number and is not used in the calculation.
Region 1 *	-	Select the region classification 1 in the list.
Service *	-	Select the service type in the list.
Service loss rate (year) *	Integer	Enter data sets given by pairs of year and quantity.
Service loss rate (value) *	Single	The total sets are less than four.

Item with * : Code or value of the item is indispensable for database system and calculation.

C.1.5 Energy database

C.1.5.1 Energy data

Energy data is listed in this table.

Table C.7. List of items in "Energy Data"

Items	Format	Contents
No.	Integer (Max 32768)	Number of the energy classification. It is an independent number and is not used in the calculation.
Energy code *	Character (Max 6)	Code of the energy classification. Every combination of energy code and region 1 be unique in the list.
Energy name	Character (Max 40)	Name of the energy classification.
Energy price *,+	Single	Energy price by energy type with the unit shown in the heading row.
CO_2 emission factor +	Single	CO_2 emission factor by energy type with the unit shown in the heading row.
SO_2 emission factor +	Single	SO_2 emission factor by energy type with the unit shown in the heading row.
NO_x emission factor +	Single	NO_x emission factor by energy type with the unit shown in the heading row.

Item with * : Code or value of the item is indispensable for database system and calculation.
Item with + : If the factor is not constant year by year, check the box and enter the value in "Energy Data by year" table.

Explanation of the command button, "Update energy factor with checking"
If the check box is not checked, the corresponding data is assumed constant for all the years. If this box is checked, the corresponding data in "Energy Data by year" table is referred. After clicking the command button, the starting year's value is calculated with using the data in "Energy Data by year" table and the data in "Energy Classification" table is updated.

C.1.5.2 Change in energy data

This table shows energy price, CO_2 emission factor, SO_2 emission factor and NO_x emission factor by year. If this data is constant by year, this table is not needed. Data in this table is considered only when corresponding row in table "Energy Classification" is checked.

Table C.8. List of items in "Change in Energy Data"

Items	Format	Contents
No.	Integer (Max 32768)	Number of the energy classification. It is an independent number and is not used in the calculation.
Energy *	-	Select the energy type of which energy price or emission factor changes year by year in the list.
Price or gas *	-	Select energy price or type of emission gas in the list.
Energy factor (year) *	Integer	A set of data is given by a pair of year (year) and
Energy factor (year) *	Single	price/emission factor (value). The total sets are less than four. The value is entered with the units shown in the footer of the form.

Item with * : Code or value of the item is indispensable for database system and calculation.

C.1.6 Technology database

"Technology" includes two types of devices, one is energy devices and another is removal processes. Energy device refers to the technology which consumes energy and supply service in order to satisfy service demand. Removal process refers to the technology which removes air pollutants emitted by a energy device.

C.1.6.1 Energy device data

Energy device data is listed in this table.

Table C.9. List of items in "Energy Device Data"

Items	Format	Contents
No.	Integer (Max 32768)	Number of the energy device. It is an independent number and is not used in the calculation.
Energy device code *	Character (Max 6)	Code of the energy device. Every code must be unique in the list.
Energy device name	Character (Max 40)	Input the name of the technology. Maximum number of characters is 40.
Life time *	Single (>0)	Life time of the energy device.
Device unit	Character (Max 40)	Unit which service supply and energy consumption of each technology based on.
Specific service output (name) *	-	Select the service the technology supplies in the list.
Specific service output (value) *,+	Single (>0)	Service supply of an energy device per year and per device unit is listed.
Fixed cost *,+	Single ($\geqq 0$)	Fixed cost of the energy device per device unit. The value is entered with the units shown in the footer of the form.
Operation cost +	Single ($\geqq 0$)	Operation cost of the energy device. The value is entered with the units shown in the footer of the form.
Specific energy consumption (name) *	-	Select the type of energy or material that the energy device consumes in the list.
Specific energy consumption (value) *,+	Single	Energy or material consumption of the energy device per unit is listed. Energy consumption is entered with the unit shown in the footer of the form.
Specific energy (non-energy use)	Single	Ratio of energy or material except combustion use. If the rate is equal to 100%, each gas does not be emitted
SO_2/NO_x ex. fuel content +	Single	SO_2 (upper) and NO_x (lower) emission other than fuel content.

Item with * : Code or value of the item is indispensable for database system and calculation.
Item with + : If the value is not constant, click the box and enter time-series value in "Improvement of Energy Device" table.

Explanation of the command button, "Update energy factor with checking"
If the check box is not checked, the corresponding data is assumed constant for all the years. If this box is checked, the corresponding data in "Improvement of Energy Device" table is referred. After clicking the command button, the starting year's value is calculated with using the data in "Improvement of Energy Device" table and the data in "Energy Device Classification" table is updated.

C.1.6.2 Change in energy device data

Change in energy device data is listed in this table. Data in this table is considered only when corresponding row in "Energy Device Classification" table is checked.

Table C.10. List of items in "Change in Energy Device"

Items	Format	Contents
No.	Integer (Max 32768)	Number of the data sets. It is an independent number and is not used in the calculation.
Energy device *	-	Select the energy device in the list.
Improved item *	-	Select the item whose quantity is improved. Every pair of 'Energy Device' and 'Improved Item' must be unique.
Improvement (year) *	Integer	Input data sets given by pairs of year and quantity. The
Improvement (value) *	Single	total sets are less than four. The value is entered with the units shown in the footer of the form.

Item with * : Code or value of the item is indispensable for database system and calculation.

C.1.6.3 Removal process

This table specifies the classification of air pollution removal processes.

Table C.11. List of items in "Removal Process"

Items	Format	Contents
No.	Integer (Max 32768)	Number of the removal process classification. It is an independent number and is not used in calculation.
Removal process code *	Character (Max 6)	Code of the removal process classification. Every code must be unique in the list.
Removal process name *	Character (Max 40)	Name of the removal process.
Stage of control *	-	Select at which stage the control is done. Pre-combustion: coal screening, coal washing etc. In Situ Combustion: lime stone injection into furnace etc. Post-combustion: flue gas desulfurization, selective catalytic reduction etc.
SO_2 / NO_x *	-	Select which SO_2 or NO_x removal process mitigates.
Fixed cost	Single	Fixed cost of the removal process per energy consumption of energy device with the unit shown in the heading row.
Operation cost	Single	Operation cost of the removal process per year and per energy consumption of energy device with the unit shown in the heading row.
Energy consumption	Single	Energy consumption of the removal process per energy consumption of energy device with the unit shown in the heading row.
Removal rate	Percentage	Removal rate to mitigate for air pollution emission by control.

Item with * : Code or value of the item is indispensable for database system and calculation.

C.1.7 Combination database

In combination database, users must set two types of combination (see Fig. 1.2 in Part IV). One is the combination concerning removal process. Firstly, users set the combination of removal processes by selecting each process at pre-combustion, in-site and post-combustion in "Combination of Removal Process" table. Next, users set the combination of energy devices and removal processes in "Combination of Energy Devices and Removal Processes".

Another is the combination concerning energy devices. In case that a device consumes output from others as input (see Fig. 1.3 in Part IV), users set the linkage between input and output in "Combination of Input and Output of Energy Devices".

C.1.7.1 Combination of removal processes

Combination of removal processes is assembled in this table. In case that removal processes are not introduced, this table can be ignored.

Table C.12. List of items in "Combination of Removal Processes"

Items	Format	Contents
No.	Integer (Max 32768)	Number of the combination of removal process. It is an independent number and is not used in the calculation.
Combination of removal processes Code *	Character (Max 6)	Code of the combination of removal process. Every code must be unique. "NON" cannot be used as a code.
Name	Character (Max 40)	Name of the combination of the removal process.
Removal process *	-	Select the removal process at each stage.
Removal rate	-	Removal rate of the combination is calculated automatically after clicking "UPDATE".
Fixed cost	-	Fixed cost of the combination is calculated automatically after clicking "UPDATE".
Operation cost	-	Operation cost of the combination is calculated automatically after clicking "UPDATE".
Energy consumption	-	Energy consumption of the combination is calculated automatically after clicking "UPDATE".

Item with * : Code or value of the item is indispensable for database system and calculation.

C.1.7.2 Combination of energy device and removal processes

Combination of energy device and removal processes is assembled in this table.

Table C.13. List of items in "Combination of Energy Device and Removal Processes"

Items	Format	Contents
Energy device *	-	Select the energy device from the upper box.
No.	Integer (Max 32768)	Number of the combination of energy device and removal process. It is an independent number and is not used in the calculation.
Code	-	Code of the combination of energy device and removal process is determined automatically after selecting removal process.
Combination of energy device and removal processes	Character (Max 40)	Name of the combination of energy device and removal process.
Combination of removal processes *	-	Select the combination of the removal process.

Item with * : Code or value of the item is indispensable for database system and calculation.

C.1.7.3 Relationship between internal energy/service

Relationship between internal energy/service is assembled in this table.

Table C.14. List of items in "Relationship between Internal Energy/Service"

Items	Format	Contents
No.	Integer (Max 32768)	Number of the combination of input and output of energy devices. It is an independent number and is not used in the calculation.
Internal energy *	-	Internal energy is selected to combine with internal service.
Internal service*	-	Internal service is selected to combine with internal service.

Item with * : Code or value of the item is indispensable for database system and calculation.

C.1.8 Stock database

C.1.8.1 Stock in start year

This table specifies the stock of each combination of device and removal process in start year of calculation.

Table C.15. List of items in "Area Stock"

Items	Format	Contents
No.	Integer (Max 32768)	Number of the dataset. It is an independent number and is not used in the calculation.
Region 1 *	-	Select the coarse region classification.
Energy device *	-	Select the energy device.
Combination of removal process *	-	Select the combination of removal process.
Stock	Single	Stock quantity in the start year of calculation.
Specific service output (name)	-	Service name is shown automatically after selecting an energy device.
Specific service output (value)$^+$	Single	Average service supply of the stocked devices in starting year.
Fixed cost / operation cost	Single	Average cost of the stock devices in the starting year.
Specific energy consumption (name)	Single	Energy name is shown automatically after selecting an energy device.
Specific energy consumption (value) $^+$	Single	Average energy consumption of the stocked devices in starting year.
Specific energy consumption (non ene.) $^+$	Single	Value is shown as same as the one in energy device table after clicking the Update button.
Removal rate $^+$ (DeSlr/DeNtr)	Percent	"DeSlr" and "DeNtr" refer to average desulfurization rate and average denitration rate of the stocked removal process respectively.
Emission except fuel content (SO$_2$/NO$_x$ ex. fl.) $^+$	Single	"SO$_2$ ex. fl." and "NO$_x$ ex. fl." refer to the average SO$_2$ and NO$_x$ emission except fuel content from the stocked devices in starting year, respectively.

Item with * : Code or value of the item is indispensable for database system and calculation.
Item with + : If the option button is false, the value is calculated automatically with using the value of energy device table after clicking Update button.

C.1.9 Share database

C.1.9.1 Maximum share of energy device

Maximum share of energy device in satisfying a service is entered in this table. This table is necessary only for the devices whose share is bound by upper limit. No such limit is assumed for devices whose data is not entered in this table.

Table C.16. List of items in "Maximum Share of Energy Device"

Items	Format	Contents
No.	Integer (Max 32768)	Number of the data sets. It is an independent number and is not used in the calculation.
Region 1 *	-	Select the coarse region classification.
Service *	-	Select the service classification in the list.
Energy device *	-	Select the energy device in the list.
Maximum share * (year)	Integer	Enter data sets given by pairs of year and quantity. The
Maximum share * (value)	Single (%)	total sets are less than four.

Item with * : Code or value of the item is indispensable for database system and calculation.

C.1.10 Performance database

C.1.10.1 Operating rate

Operating rate of a device is entered in this table. If this value is not entered by the user, default value of operating rate is assumed as 100% in the model.

Table C.17. List of items in "Operating Rate"

Items	Format	Contents
No.	Integer (Max 32768)	Number of the data sets. It is an independent number and is not used in the calculation.
Region 1 *	-	Select the region classification 1 in the list.
Energy device *	-	Select the energy device in the list.
Operation rate * (year)	Integer	Enter data sets given by pairs of year and quantity. The
Operation rate * (value)	Single	total sets are less than four.

Item with * : Code or value of the item is indispensable for database system and calculation.

C.1.11 Countermeasure database

AIM/Enduse can estimate CO_2, SO_2, NO_x emissions with countermeasures. Following countermeasures can be set in the model: (i) Countermeasure at use stage; (ii) Tax for energy consumption and CO_2, SO_2, NO_x emission; (iii) Regulation for energy consumption and CO_2, SO_2, NO_x emission; (iv) Subsidy for recruited technology and exchange etc.

C.1.11.1 Improvement at use stage

Change in life style and method of use and maintenance of devices can result in conservation of energy at use stage. This table shows improvement at use stage of energy device.

Table C.18. List of items in "Improvement at Use Stage"

Items	Format	Contents
No.	Integer (Max 32768)	Number of the data sets. It is an independent number and is not used in the calculation.
Code *	-	Code of the countermeasure at used stage.
Content of countermeasure	Character (Max 60)	Content of the countermeasure at used stage.
Energy device *	-	Select the energy device in the list.
Reduction rate *	Single	Rate of reduction of service supply or energy use.

Item with * : Code or value of the item is indispensable for database system and calculation.

C.1.11.2 Action of improvement at use stage

Action rate of improvement at use stage in the region is listed in the table.

Table C.19. List of items in "Action of Improvement at Use Stage"

Items	Format	Contents
No.	Integer (Max 32768)	Number of the data sets. It is an independent number and is not used in the calculation.
Region 1 *	-	Select the coarse region classification in the list.
Improvement at use stage *	-	Select the countermeasure menu in the list.
action rate (year) *	Integer	Enter data sets given by pairs of year and quantity.
action rate (value) *	Percent	The total sets are less than four.

Item with * : Code or value of the item is indispensable for database system and calculation.

C.1.11.3 Tax/regulation classification

User can set the countermeasure as tax or restriction (constraint) by emission type and/or energy.

Table C.20. List of items in "Tax/Regulation Classification"

Items	Format	Contents
No.	Integer (Max 32768)	Number of the data sets. It is an independent number and is not used in the calculation.
Group code *	Character (Max 6)	Code of the group.
Countermeasure type *	-	Select the countermeasure type in the following choices. - Energy Tax: Tax is imposed on energy use - CO_2 Tax: Tax is imposed on CO_2 emission - SO_2 Tax: Tax is imposed on SO_2 emission - NO_x Tax: Tax is imposed on NO_x emission - Energy constraint: Energy use is restricted - CO_2 constraint: CO_2 emission is restricted. - SO_2 constraint: SO_2 emission is restricted. - NO_x constraint: NO_x emission is restricted.
Group name *	Character (Max 40)	Name of group.

Item with * : Code or value of the item is indispensable for database system and calculation.

C.1.11.4 Group for tax/regulation

The tax and constraints defined in previous section can be applied to selected sectors in this table.

Explanation of the command button, "Same group"
If a group is selected in a cell of a column and this button is clicked, all the cells of that column are changed to the same group

Table C.21. List of items in "Group for Tax/Regulation"

Items	Format	Contents
No.	-	These data are shown automatically.
Region 1 *	-	
Sector *	-	
Group for energy	-	Select group on measure.
Group for CO_2	-	
Group for SO_2	-	
Group for NO_x	-	

Item with * : Code or value of the item is indispensable for database system and calculation.

C.1.11.5 Tax / regulation

Tax rate or regulation for each group is listed in this table. If the tax rate or regulation is not entered for a group, tax rate is assumed zero or regulation is assumed infinite.

Table C.22. List of items in "Tax / Regulation"

Items	Format	Contents
(check box)	-	If you do not check, the rate is ignored.
Group *	-	Select the group in the list quoted from "Group on Measure Classification" table.
Type	-	Countermeasure type is shown after selecting group.
Energy *	-	If the countermeasure is energy tax or energy constraint, select the energy classification.
Tax rate/regulation (year) *	Integer	Input data sets given by pairs of year and quantity. The to-
Tax rate/regulation (value)*	Percent	tal number of sets is less than four. The unit is shown in the type field.

Item with * : Code or value of the item is indispensable for database system and calculation.

C.1.11.6 Subsidy (recruitment & operation)

The subsidy rate for recruitment (fixed cost) or operation (operational cost) can be entered in this table.

Table C.23. List of items in "Subsidy (Recruitment & Operation)"

Items	Format	Contents
(check box)	-	If you do not check, the rate is ignored.
No.	Integer (Max 32768)	The number of the data sets. It is an independent number and is not used in the calculation.
Energy device *	-	Select the energy device in the list.
Combination of removal processes *	-	Select the combination of removal processes.
For recruitment / operation	-	Select 'for recruited' for subsidy at recruited stage, and 'for operation' for subsidy at operation for technology.
Subsidy rate (year) *	Integer	Input data sets given by pairs of year and quantity. The
Subsidy rate (value) *	Percent	total sets are less than four.

Item with * : Code or value of the item is indispensable for database system and calculation.

C.1.11.7 Subsidy (removal process)

The subsidy rate at recruited or exchange stage for removal process is listed in this table.

Table C.24. List of items in "Subsidy (Removal Process)"

Items	Format	Contents
No.	Integer (Max 32768)	The number of the data sets. It is an independent number and is not used in the calculation.
Removal process*	-	Select the energy device in the list.
Combination of removal process 1*	-	Select the combination of removal processes that may be exchanged to new one.
Combination of removal process 2*	-	Select the combination of removal processes that may be exchanged from old one.
Subsidy rate (year) *	Integer	Input data sets given by pairs of year and quantity. The
Subsidy rate (value) *	Percent	total sets are less than four.

Item with * : Code or value of the item is indispensable for database system and calculation.

C.2 Implementation

C.2.1 How to export input files for AIM/Enduse

AIM/Enduse database system can supply the input data for the GAMS program.

1. The files in "data" directory and the ones in "src" directory are set in same directory as shown in Fig. C.3. User should confirm the following program sources in "src" directory.
   - AIM_CMB.gms
   - _interp.gms
   - _printout.gms
   - _errorout.gms

2. Input directory and file name of GAMS input file at the side of "Export data to GAMS" command button on main screen as shown in Fig. C.4. In this example, "AIM_Exe" files in "data" directory are used.

3. If user clicks on "Export data to GAMS", the interface makes input files for AIM-CMB.gms. After export of GAMS input files, the message as shown in Fig. C.5 appears. The new files, AIM_Exe_1.gms, AIM_Exe_2.gms, AIM_Exe.set and AIM_Exe.err, are made in "data" directory as shown in Fig. C.6.

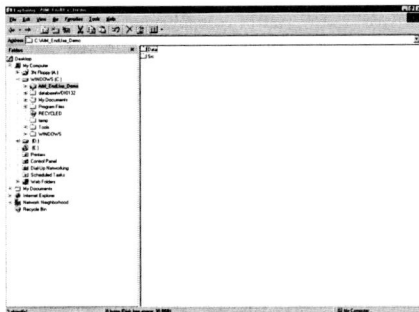

Fig. C.3. How to export input files
for AIM/Enduse GAMS version (1)

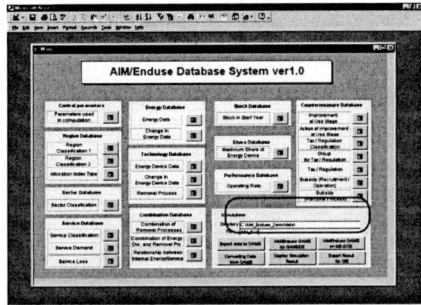

Fig. C.4. How to export input files
for AIM/Enduse GAMS version (2)

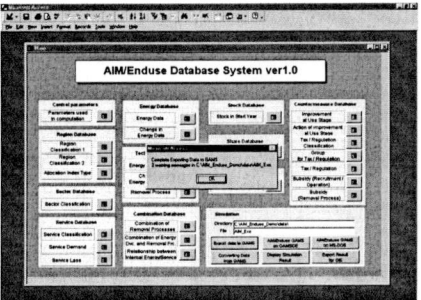

Fig. C.5. How to export input files
for AIM/Enduse GAMS version (3)

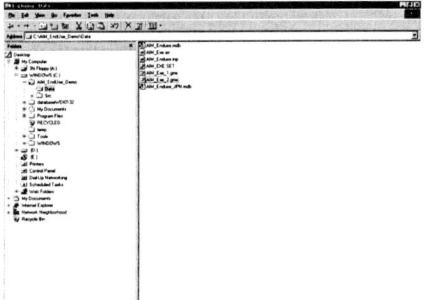

Fig. C.6. How to export input files
for AIM/Enduse GAMS version (4)

C.2.2 How to implement AIM/Enduse GAMS version

GAMS (http://www.gams.com/) must be installed on user's PC for implementation of AIM/Enduse
GAMS version.

C.2.2.1 GAMSIDE

GAMSIDE is a graphical interface to create, debug, edit and run GAMS files. If GAMSIDE is installed
on user's PC, user should click on "AIM/Enduse GAMS on GAMSIDE" command button on main
form. At first time user must create GAMS project file in 'src' directory as shown in Fig. C.7. After
creating the file in 'src', user must open 'AIM_CMB.gms' file and run it. If the model runs normally,
user can see the screen as shown in Fig. C.9. As for how to use GAMSIDE, see the following file:
GAMSIDE manual : http://www.gams.com/mccarl/useide.pdf

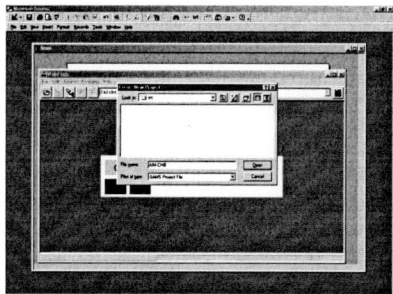

Fig. C.7. How to implement
AIM/Enduse GAMS version (1)

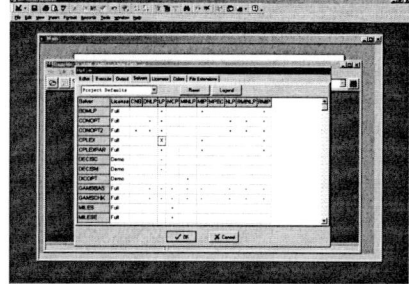

Fig. C.8. How to implement
AIM/Enduse GAMS version (2)

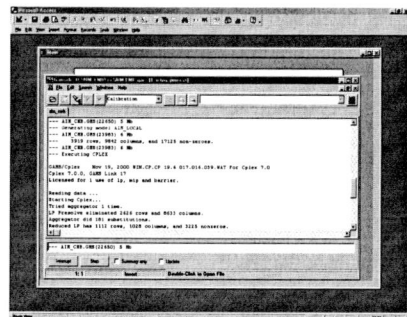

Fig. C.9. How to implement AIM/Enduse GAMS version (3)

C.2.2.2 MS-DOS

If user clicks on "AIM/Enduse GAMS on MS-DOS" command button, MS-DOS prompt appears as shown in Fig. C.10. If GAMS is installed on user's PC, user must input "gams AIM-CMB lp=cplex" in MS-DOS with the input data exported from the database system. GAMS would work as shown in Fig. C.11.

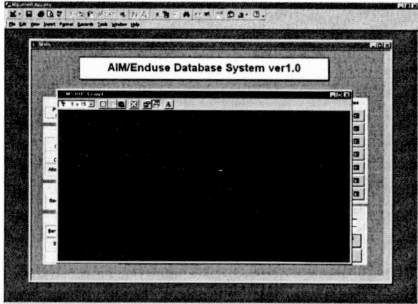

Fig. C.10. How to implement
AIM-Local GAMS version (4)

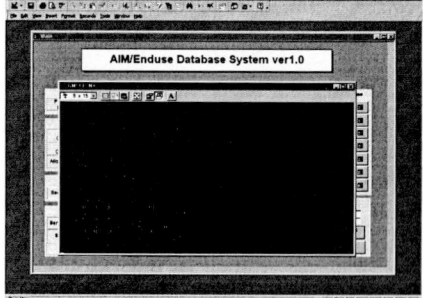

Fig. C.11. How to implement
AIM/Enduse GAMS version (5)

C.2.3 How to import output file of AIM/Enduse GAMS version

After implementation of simulation with AIM/Enduse GAMS version, the database system can import the output file of AIM/Enduse GAMS version. If user clicks on "Converting data from GAMS", the database system starts to import data from output file of AIM/Enduse GAMS version. After importing, the message as shown in Fig. C.12 appears.

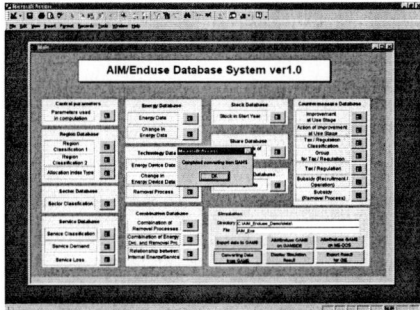

Fig. C.12. How to import output file of AIM/Enduse GAMS version

C.2.4 How to display simulation results

AIM/Enduse database system displays simulation results with pivot table after importing data from output file of AIM/Enduse GAMS version. If user clicks on "Display simulation result", the form as shown in Fig. C.13 appears. After simulating with AIM/Enduse GAMS version, user must refresh data. For this the user must select 'Table' worksheet and click right button on mouse. Then select "Refresh Data" on the list as shown in Fig. C.14.

Pivot table and chart allows the user to create dynamic summary data. For example, if user clicks the filter on the upper side of the form as shown in Fig. C.15, it shows the list. If user selects 'EMS', it shows emission results. Pivot table shows the result freely and easily with selecting the combination of items on the list of the filter (Table C.25).

Table C.25. Filter of pivot table

Filter		Choice	Content
Kind		EMS	Emission quantity (I,M,L,P,Y)
		ENG	Energy consumption (I,K,L,P,Y)
		SRV	Service supply (I,J,L,P,Y)
		STK	Stock quantity (I,L,P,Y)
		CST	Item = RCA, RCI, MDA, MDI
		RCT	Recruited amount (I,L,P,Y)
		DEV	Operating quantity (I,L,P,Y)
Item	Kind = EMS(M)	CO_2	CO_2 emission
		SO_2	SO_2 emission
		NO_x	NO_x emission
	Kind = ENG(K)	(Energy Type)	The energy code in "Energy Classification"
	Kind = SRV(J)	(Service Type)	The energy code in "Service Classification"
	Kind = STK	-	-
	Kind=CST	RCA	Total annualized investment cost(I,L,P,Y)
		RCI	Total initial investment cost(I,L,P,Y)
		MDA	Total annualized cost of exchanging removal process (I,L,P,Y)->(I,L,P1,Y)

Table C.25. Filter of pivot table (continued)

Filter		Choice	Content
Item	Kind = CST	MDI	Total initial cost of exchanging removal process (I,L,P,Y)->(I,L,P1,Y)
		MNT	Total operating cost including energy cost, material cost, maintenance cost etc.(I,L,P)
		TXE	Energy tax payment (I,L,P)
		TXM	Emission tax payment (I,L,P)
LPS_Area		LPS	Large point source
		Area	Area source
Region (I)		(Region 1)	The code you input in "Region Classification 1"
LPS (I)		(LPS)	The code you input in "LPS"
Sector (I)		(Sector type)	The code you input in "Sector Classification"
Energy_device (L)		(Energy device)	The code you input in "Energy Device Classification"
Removal (P)		(Removal process)	The code you input in "Removal Process"

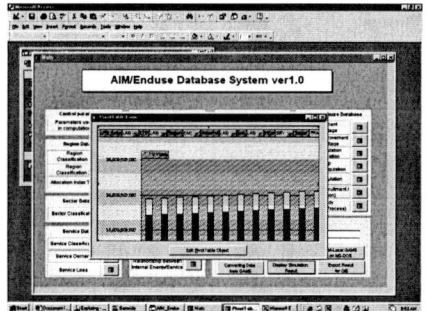

Fig. C.13. How to display simulation result (1)

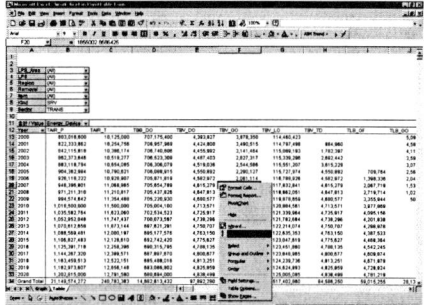

Fig. C.14. How to display simulation result (2)

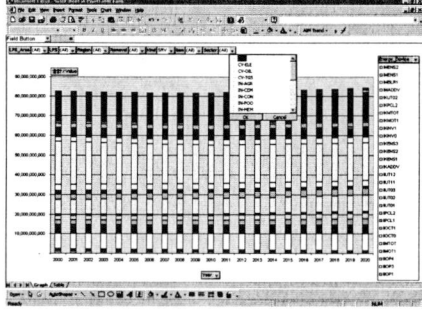

Fig. C.15. How to display simulation result (3)

Appendix D: Possible Classifications for Sectors and Services

Sector		Service		Possible measures/units of service demand	
Alternative 1	Alternative 2	Alternative 1	Alternative 2	Alternative 1	Alternative 2
Agriculture	Agriculture	Irrigation	Irrigation	Irrigated area; No. of pumps; Useful energy service for pumping	Irrigated area; No. of pumps; Useful energy service for pumping
		Other energy uses	Farm land preparation	Agriculture value added; Useful energy service for non-pumping agriculture activities	No. of tractors; Useful energy service for land preparation
			Harvesting		Area of crop harvested; Useful energy service for harvesting
Transport	Transport – road	Road passenger	Private passenger transport	Person-km; Vehicle-km; No.of vehicles	Person-km; No. of private vehicles
			Public passenger transport		Person-km; No. of public vehicles
		Road freight	Road freight	Ton-km; No.of freight vehicles	Ton-km; No. of freight vehicles
	Transport – rail	Rail passenger	Rail passenger	Person-km; No. of operational locomotives	Person-km; No. of operational locomotives
		Rail freight	Rail freight	Ton-km: No. of operational locomotives	Ton-km; No. of operational locomotives
	Transport – air	Air passenger	Air passenger – domestic travel	Person-km; Km of passenger flights	Person-km; Km of passenger flights
			Air passenger – international travel		Person-km; Km of passenger flights
		Air freight	Air freight – domestic travel	Ton-km; Km of cargo flights	Ton-km; Km of cargo flights
			Air freight – international travel		Ton-km; Km of cargo flights
	Transport – water	Water passenger	Water passenger	Person-km; Km of passenger movement by ships	Person-km; Km of passenger movement by ships
		Water freight	Water freight	Ton-km; Km of goods movement by ships	Ton-km; Km of goods movement by ships
Residential	Residential - rural	Cooking	Rural – cooking	Useful energy service for cooking; No. of households; Population	Useful energy service for rural cooking; No. of rural households; Rural population
		Lighting	Rural – lighting	Lumen-hr; No. of lamps, No. of households	Lumen-hr; No. of lamps in rural areas, No. of rural households
		Space cooling	Rural – fan	Useful energy service for cooling, No. of households	No. of fans; Fan hrs use in rural areas, No. of rural households
		Space heating	Rural – refrigerator	Useful energy service for heating, No. of households	No. of refrigerators in rural areas, No. of rural households
		Refrigerator	Rural – air conditioner	No. of refrigerators, No. of households	No. of ACs; AC hrs use in rural areas, No. of rural households
		Monochrome TV	Rural – monochrome TV	No. of TVs; TV-hrs use, No. of households	No. of TVs; TV-hrs use in rural areas

Sector		Service		Possible measures/units of service demand	
Alternative 1	Alternative 2	Alternative 1	Alternative 2	Alternative 1	Alternative 2
Residential		Color TV	Rural – color TV	No. of TVs; TV-hrs use, No. of households	No. of TVs; TV-hrs use in rural areas, No. of rural households
		Washing machine	Rural – washing machine	No. of washing machines; Washing machine-hrs use, No. of households	No. of washing machines; Washing machine-hrs use in rural areas, No. of rural households
		Other appliances	Rural – other appliances	Population; Energy use* in appliances, No. of households	Rural population; Energy use* in other appliances in rural areas, No. of rural households
	Residential - urban		Urban – cooking		Useful energy service for urban cooking; No. of urban households
			Urban – lighting		Lumen-hr; No. of lamps in urban areas, No. of urban households
			Urban – fan		No. of fans; Fan hrs use in urban areas, No. of urban households
			Urban – refrigerator		No. of refrigerators in urban areas, No. of urban households
			Urban – air conditioner		No. of ACs; AC hrs use in urban areas, No. of urban households
			Urban – monochrome TV		No. of TVs; TV-hrs use in urban areas, No. of urban households
			Urban – color TV		No. of TVs; TV-hrs use in urban areas, No. of urban households
			Urban – washing machine		No. of washing machines; Washing machine-hrs use in urban areas, No. of urban households
			Urban – other appliances		Urban population; Energy use* in other appliances in urban areas, No. of urban households
Commercial	Restaurants and hotels	Lighting	Cooking – restaurants	Lumen-hr; No. of lamps, Total floor area in commercial establishments	Useful energy service for cooking in restaurants; No. of restaurants
		Air conditioner	Lighting – restaurants	No. of Acs, Total floor area in commercial establishments	Lumen-hr; No. of lamps in restaurants, No. of restaurants
		Refrigerator	Air conditioner – restaurants	No. of refrigerators, Total floor area in commercial establishments	No. of ACs in restaurants, No. of restaurants
		Computer	Other appliances – restaurants	No. of computers, Total floor area in commercial establishments	No. of restaurants; Value added in restaurants; Energy use* in other appliances in restaurants

Sector		Service		Possible measures/units of service demand	
Alternative 1	Alternative 2	Alternative 1	Alternative 2	Alternative 1	Alternative 2
Commercial	Corporate and Government offices	Other appliances	Lighting – offices	Commercial value added; Energy use* in other appliances in commercial sector, Total floor area in commercial establishments	Lumen-hr; No. of lamps in offices, Total floor area in offices
			Air conditioner – offices		No. of ACs in offices, Total floor area in offices
			Computer – offices		No. of computers in offices, Total floor area in offices
			Other appliances – offices		Value added in offices; Energy use* in other appliances in offices, Total floor area in offices
	Hospitals and clinics		Lighting – hospitals		Lumen-hr; No. of lamps in hospitals, Total floor area in hospitals & clinics
			Air conditioner – hospitals		No. of ACs in hospitals, Total floor area in hospitals & clinics
			Other appliances – hospitals		Value added in hospitals; Energy use* in other appliances in hospitals, Total floor area in hospitals & clinics
	Other commercial establishments		Lighting – other commercial		Lumen-hr; No. of lamps in other commercial establishments, Total floor area in other establishments
			Air conditioner – other commercial		No. of ACs in other commercial establishments, Total floor area in other establishments
			Other appliances – other commercial		Value added in other commercial establishments; Energy use* in other appliances in other commercial establishments, Total floor area in other establishments
Industry	Industry – steel	Finished steel products	Hot rolled steel	Ton	Ton
			Cold rolled steel		Ton
			Crude steel**		Ton
			Sinter**		Ton
			Pig iron**		Ton
			Coke gas**		Cubic meters
			Coke**		Ton
			Heat**		Energy unit
			Electricity**		Energy unit
	Industry – aluminium	Finished aluminium products	Finished aluminium	Ton	Ton
			Molten aluminium**		Ton
			Alumina**		Ton
			Heat**		Energy unit
			Electricity**		Energy unit

Sector			Service		Possible measures/units of service demand	
Alternative 1	Alternative 2	Alternative 1		Alternative 2	Alternative 1	Alternative 2
Industry	Industry – cement	Cement	Cement		Ton	Ton
			Pre-grinded cement**			Ton
			Portland cement**			Ton
			Blast furnace ce-ment**			Ton
			Fly ash cement**			Ton
			Clinker**			Ton
			Heat**			Energy unit
			Electricity**			Energy unit
	Industry – paper	Paper	Finished paper		Ton	Ton
			Dried paper**			Ton
			Bleached pulp**			Ton
			Mechanical pulp**			Ton
			Waste pulp**			Ton
			Semi chemical pulp**			Ton
			Kraft pulp**			Ton
			Heat**			Energy unit
			Electricity**			Energy unit
	Industry – petrochemi-cals	Petrochemi-cal products	Ethylene		Ton	Ton
			Low density polyeth-ylene			Ton
			High density polyeth-ylene			Ton
			Polypropylene			Ton
			Polystylene			Ton
			Other petrochemical products			Ton
			Heat**			Energy unit
			Electricity**			Energy unit
	Industry – brick	Bricks	Bricks		No. of bricks	No. of bricks
			Heat**			Energy unit
	Industry – caustic soda	Caustic soda	Caustic soda		Ton	Ton
			Brine**			Ton
			Heat**			Energy unit
	Industry – soda ash	Soda ash	Soda ash		Ton	Ton
			Brine**			Ton
			Calcined sodium bi-carbonate**			Ton
			Heat**			Energy unit
			Electricity**			Energy unit
	Industry – fertilizer	Fertilizer products	Phosphatic fertilizer products		Ton	Ton
			Nitrogenous fertilizer products			Ton
			Ammonia**			Cubic meters
			Synthesis gas**			Cubic meters
			Heat**			Energy unit
			Electricity**			Energy unit
	Industry – textiles	Textile prod-ucts	Finished cotton cloth		Square metre; Ton	Square metre; Ton
			Weaved cloth**			Square metre; Ton
			Spinned cloth**			Square metre; Ton
			Electricity**			Energy unit
			Non-cotton textile products			Ton

Sector		Service		Possible measures/units of service demand	
Alternative 1	Alternative 2	Alternative 1	Alternative 2	Alternative 1	Alternative 2
Industry	Industry – sugar	Finished sugar	Finished sugar	Ton	Ton
			Crystallized sugar**		Ton
			Concentrated juice**		Cubic meters
			Clear juice**		Cubic meters
			Heat**		Energy unit
			Electricity**		Energy unit
	Industry – mining	Value added in mining	Value added in mining	Value added in US$	Value added in US$
			Heat		Energy unit
			Electricity		Energy unit
	Industry – construction	Value added in construction	Value added in construction	Value added in US$	Value added in US$
			Heat		Energy unit
			Electricity		Energy unit
	Other industries	Value added in other industries	Value added in other industries	Value added in US$	Value added in US$
			Heat demand in other industries		Energy unit
			Electricity demand in other industries		Energy unit
Electricity generation & supply	Electricity generation & supply	Electricity	Electricity	Energy unit	Energy unit
	Combined heat and electricity generation (CHP)	Heat	Electricity	Energy unit	Energy unit
			Heat		Energy unit
Oil refining & supply	Crude oil production	Petroleum products	Crude oil	Energy unit	Energy unit
	Oil refining & supply		Gasoline		Energy unit
			Diesel		Energy unit
			Heavy oil		Energy unit
			Kerosene		Energy unit
			LPG		Energy unit
			Other petroleum products		Energy unit
Natural gas production & supply	Natural gas production & supply	Natural gas	Natural gas (uncompressed)	Energy unit	Energy unit
			CNG		Energy unit
			Hydrogen		Energy unit
Coal production & supply	Coal production & supply	Coal products	Anthracite	Energy unit	Energy unit
			High grade bituminous coal		Energy unit
			Sub-bituminous coal		Energy unit
			Lignite		Energy unit
Biomass production & supply	Traditional biomass production & supply	Traditional biomass	Fuelwood	Energy unit	Energy unit
			Cattle-dung		Energy unit
			Crop residue		Energy unit
	Modern biomass production & supply	Modern biomass (commercial energy plantation)	Modern biomass products	Energy unit	Energy unit

Note:
i) Sectors, services, and measures/units under alternative 1 should be read together. Similarly, those under alternative 2 should be read together.
ii) The classifications and units given in this table are mere examples. Exact classification and choice

of measures and units should depend on the energy-economy context of a country, availability of data for services and technologies, and the user team's judgment.

iii) * In a bottom-up model like AIM/Enduse a service must be expressed in some measure of output of device, so that efficiency improvement options can be evaluated; Thus use of 'Energy use' as a measure of service must be avoided as far as possible.

iv) ** These items can be represented as internal services in AIM/Enduse. However, some of them can also be represented as external service or external energy if their supply is external to the model.

Appendix E: Partial List of Internationally Published Sources of Relevant Data

Publication	Possible source of following country level data	Remarks
Energy balance tables, International Energy Agency (IEA 2001)	* Sector-wise use of each fuel-category, electricity and heat[+]	* Available on CDROM * Aggregate estimates (e.g. use of a fuel in aggregate sectors like transport, or industry) are reliable, but disaggregated estimates (e.g. use of a fuel in sub-sectors like road transport or specific industry) may not be reliable for some countries
Steel statistical yearbook, International Iron and Steel Institute (IISI 2001)	* Production of crude steel * Process-wise production of crude steel * Production of pig iron and direct reduced iron	* Available on CDROM * A reliable source for iron and steel industry statistics
World development indicators, World Bank (WB 2002)	* Population, with urban:rural break-up[+] * Employment by economic activity[+] * Number of road vehicles per capita by vehicle category[+] * Road passenger traffic (in million vehicle-km)[+] * Road freight traffic (in million ton-km)[+] * Rail passenger traffic (in person-km per GDP in PPP)[+] * Rail freight traffic (ton-km per PPP $ million of GDP)[+] * Air passenger traffic (in 1000 passengers carried)[+] * Air freight traffic (in million ton-km)[+]	* Available on CDROM
World road statistics, International Road Federation (IRF 2001)	* Total road network (in km) * Road vehicles in use by vehicle category (passenger cars, buses & coaches, lorries & vans, road tractors, two-wheelers) * Road passenger traffic (in million vehicle-km) * Freight traffic for Road, Rail, and Water (in million ton-km) * Passenger traffic for Road and Rail (in million person-km)	* Available on CDROM * A reliable sources for road transport statistics
Industrial commodity statistics, United Nations (UN 2001a)	* Production of commodities by commodity categories[+]	* A reliable source for commodity-wise production statistics
Statistical yearbook, United Nations (UN 2001b)	* Population, with urban:rural break-up[+] * Agricultural production index[+] * Food production index[+] * Rates of discount of central banks[+] * Production of commodities by commodity category[+] * Industrial production index by industry categories[+] * Value added by industry categories[+] * GDP[+] * Road motor vehicles in use by vehicle category[+] * Rail passenger traffic (person-km)[+] * Rail freight traffic (ton-km)[+] * Merchant shipping fleet (tons)[+] * International maritime transport (tons)[+] * Air passenger traffic (person-km)[+] * Air freight traffic (ton-km)[+]	* Available on CDROM

Publication	Possible source of following country level data	Remarks
Long range world population projections, United Nations (UN 2000)	* Region-wise long term population projections under selected scenarios	* Available on CDROM
World marketing data and statistics, Euromonitor (Euromonitor 2002)	* Area of arable land (in hectares)[+] * Area of irrigated land (in hectares)[+] * Number of households[+] * Number of occupants per household[+] * Possession of appliances by appliance category (in no. per 100 households)[+] * Production of commodities by commodity category[+] * Total road network (in km)[+] * Density of road network (in km per km^2)[+] * Employment by sector[+] * GDP by sector[+] * Production indices by sector[+] * Rail passenger traffic (in person-km)[+] * Rail freight traffic (in ton-km)[+] * Air passenger traffic (in no. of persons carried; person-km)[+] * Air freight traffic (in tons; ton-km)[+]	* Available on CDROM
FAOSTAT, Food and Agricultural Organization (FAO 2001)	* Agricultural area (in hectares)[+] * Crop-wise production, yield, and area[+]	* Available on CDROM * A reliable source

[+] Historical time series data are available for these items.

Appendix F: Characteristics of Selected Technologies

Characteristics for selected technologies presented here are compiled from company brochures and other publications. Users are advised to use these numbers as guidelines or benchmarks while estimating technology specific numbers for their own countries. This list is partial. Refer to the accompanied CD for a more detailed list of technologies. Also refer to Appendices K and M for data on technologies in India and Japan.

Technology	Parameter	Typical value	Data source	Remarks
Residential sector				
Refrigerator – 141-200 liter capacity (Matsushita)	Electricity use (kWh/year)	390	ECC (2002)	These are high efficiency appliances; Valid for Japan
	Fixed cost (1000 JPY*)	40 to 65		
Refrigerator – 201-250 liter capacity (Toshiba, Hitachi)	Electricity use (kWh/year)	420 to 440	ECC (2002)	These are high efficiency appliances; Valid for Japan
	Fixed cost (1000 JPY*)	60 to 110	ECC (2002)	
Refrigerator – 251-300 liter capacity (Sanyo, National)	Electricity use (kWh/year)	390 to 400	ECC (2002)	These are high efficiency appliances; Valid for Japan
	Fixed cost (1000 JPY*)	64 to 120	ECC (2002)	
Refrigerator – 301-350 liter capacity (Hitachi, Matsushita)	Electricity use (kWh/year)	320 to 370	ECC (2002)	These are high efficiency appliances; Valid for Japan
	Fixed cost (1000 JPY*)	100 to 175	ECC (2002)	
Refrigerator – 351-400 liter capacity (Toshiba, Hitachi)	Electricity use (kWh/year)	300 to 310	ECC (2002)	These are high efficiency appliances; Valid for Japan
	Fixed cost (1000 JPY*)	118 to 220	ECC (2002)	
Refrigerator – 401-450 liter capacity (Toshiba, Matsushita)	Electricity use (kWh/year)	280	ECC (2002)	These are high efficiency appliances; Valid for Japan
	Fixed cost (1000 JPY*)	168 to 275	ECC (2002)	
Refrigerator – 451 and above liter capacity (Toshiba, Matsushita)	Electricity use (kWh/year)	280 to 290	ECC (2002)	These are high efficiency appliances; Valid for Japan
	Fixed cost (1000 JPY*)	188 to 300	ECC (2002)	
Color TV – standard, size: 14 to 20 (Aiwa, Sharp, Funai)	Rating (W)	49.1 to 66.1	ECC (2002)	These are high efficiency appliances; Valid for Japan
	Electricity use (kWh/year)	57 to 82	ECC (2002)	
	Fixed cost (1000 JPY*)	14 to 28	ECC (2002)	
Color TV – standard, size: 21 (Aiwa, Victor)	Rating (W)	72.6 to 76.6	ECC (2002)	These are high efficiency appliances; Valid for Japan
	Electricity use (kWh/year)	82 to 84	ECC (2002)	
	Fixed cost (1000 JPY*)	24 to 26	ECC (2002)	
Color TV – standard, size: 25 to 29 (Hitachi, Mitsubishi)	Rating (W)	110.1 to 12201	ECC (2002)	These are high efficiency appliances; Valid for Japan
	Electricity use (kWh/year)	121 to 132	ECC (2002)	
	Fixed cost (1000 JPY*)	36 to 70	ECC (2002)	
Color TV – wide with built-in BS tuner, size: 28 (Sanyo, Mitsubishi)	Rating (W)	120.1 to 135.4	ECC (2002)	These are high efficiency appliances; Valid for Japan
	Electricity use (kWh/year)	125 to 138	ECC (2002)	
	Fixed cost (1000 JPY*)	50 to 70	ECC (2002)	
Color TV – wide with built-in BS tuner, size: 32 to 36 (Toshiba)	Rating (W)	171.2 to 184.3	ECC (2002)	These are high efficiency appliances; Valid for Japan
	Electricity use (kWh/year)	186 to 201	ECC (2002)	
	Fixed cost (1000 JPY*)	144 to 277	ECC (2002)	
Air conditioner – 2.2 kW capacity (Toshiba, Hitachi, National)	Rating (W)	Cool: 365 to 370; Warm: 530 to 535	ECC (2002)	These are high efficiency appliances; Valid for Japan
	Electricity use (kWh/year)	727 to 760	ECC (2002)	
	Fixed cost (1000 JPY*)	10 to 20	ECC (2002)	

Technology	Parameter	Typical value	Data source	Remarks
Air conditioner – 2.5 to 2.8 kW capacity (Sanyo, Toshiba, Hitachi, National, Sharp)	Rating (W)	Cool: 430 to 470; Warm: 615 to 685	ECC (2002)	These are high efficiency appliances; Valid for Japan
	Electricity use (kWh/year)	827 to 941	ECC (2002)	
	Fixed cost (1000 JPY*)	10 to 25	ECC (2002)	
Air conditioner – 3.2 kW capacity (Mitsubishi)	Rating (W)	Cool: 1040; Warm: 1165	ECC (2002)	These are high efficiency appliances; Valid for Japan
	Electricity use (kWh/year)	1763	ECC (2002)	
	Fixed cost (1000 JPY*)	10 to 28	ECC (2002)	
Air conditioner – 3.6 4.0 kW capacity (Daikin)	Rating (W)	Cool: 780 to 895; Warm: 890 to 1215	ECC (2002)	These are high efficiency appliances; Valid for Japan
	Electricity use (kWh/year)	1309 to 1525	ECC (2002)	
	Fixed cost (1000 JPY*)	15 to 38	ECC (2002)	
Transport sector				
Gasoline car – curb wt.: 750-990 kg (Suzuki, Toyota, Nissan, Honda)	Fuel consumption at 10/15 mode (km/l)	19.0 to 23.0	JAMA (2001)	These are high efficiency vehicles; Valid for Japan
	Fixed cost (1000 JPY*)	914 to 1310	JAMA (2001)	
Gasoline car – curb wt.: 1000-1330 kg (Toyota, Honda, Volkswagen)	Fuel consumption at 10/15 mode (km/l)	12.4 to 15.0	JAMA (2001)	These are high efficiency vehicles; Valid for Japan
	Fixed cost (1000 JPY*)	1499 to 1998	JAMA (2001)	
Gasoline car – curb wt.: 1590-1840 kg (Toyota, Honda)	Fuel consumption at 10/15 mode (km/l)	9.2 to 9.5	JAMA (2001)	These are high efficiency vehicles; Valid for Japan
	Fixed cost (1000 JPY*)	2275 to 3760	JAMA (2001)	
Hybrid car (gasoline) – curb wt.: 820-1220 kg (Toyota, Honda)	Fuel consumption at 10/15 mode (km/l)	29.5 to 35.0	JAMA (2001)	These are high efficiency vehicles; Valid for Japan
	Fixed cost (1000 JPY*)	2090 to 2180	JAMA (2001)	
Hybrid car (gasoline) – curb wt.: 1670-1840 kg (Toyota)	Fuel consumption at 10/15 mode (km/l)	13.0 to 18.0	JAMA (2001)	These are high efficiency vehicles; Valid for Japan
	Fixed cost (1000 JPY*)	3630 to 4420	JAMA (2001)	
Electric car – curb wt.: 1270 kg (Suzuki)	Fuel consumption at 10/15 mode (km/l)	7.0	JAMA (2001)	These are high efficiency vehicles; Valid for Japan
	Fixed cost (1000 JPY*)	3000	JAMA (2001)	
CNG car – curb wt.: 770-810 kg (Daihatsu)	Fuel consumption at 10/15 mode (km/l)	18.8 to 22.6	JAMA (2001)	These are high efficiency vehicles; Valid for Japan
	Fixed cost (1000 JPY*)	1549 to 1550	JAMA (2001)	
CNG car – curb wt.: 1170-1190 kg (Honda, Nissan)	Fuel consumption at 10/15 mode (km/l)	17.2 to 18.8	JAMA (2001)	These are high efficiency vehicles; Valid for Japan
	Fixed cost (1000 JPY*)	2050 to 2660	JAMA (2001)	
Existing passenger cars	Average energy intensity (MJ/km-person)	1.73	IPCC (2001)	Europe, 1993
		2.59	IPCC (2001)	USA, 1994; High due to low passenger occupancy
		2.46	IPCC (2001)	Japan, 1994; High due to low passenger occupancy
Existing buses	Average energy intensity (MJ/km-person)	0.71	IPCC (2001)	Europe, 1993
		1.03	IPCC (2001)	USA, 1994; High due to low passenger occupancy
		0.73	IPCC (2001)	Japan, 1994

Technology	Parameter	Typical value	Data source	Remarks
Existing rail passenger loco-motives	Average energy inten-sity (MJ/km-person)	0.48	IPCC (2001)	Europe, 1993
		2.15	IPCC (2001)	USA, 1994; High due to low passenger occupancy
		0.19	IPCC (2001)	Japan, 1994; Low due to high passenger occupancy and high efficiency
Existing passenger aircrafts	Average energy inten-sity (MJ/km-person)	2.78	IPCC (2001)	Europe, 1993
		2.46	IPCC (2001)	USA, 1994
		2.13	IPCC (2001)	Japan, 1994
Average existing passenger cars with gasoline internal combustion engine	Average fuel consump-tion rate (l/100 km)	8.6	IPCC (2001)	This is average figure for passenger cars existing in ad-vanced markets in 1997
Most efficient models of exist-ing passenger cars with gaso-line internal combustion en-gine	Average fuel consump-tion rate (l/100 km)	7.0 to 7.5	Estimated based on IPCC (2001)	This is average figure for passenger cars introduced in Europe in the 1990s; It has remained essentially constant over past 10-15 years; EU has targeted to reduce average fuel consumption of new cars to 5.8 l/100 km by 2010
New car with aluminum-intensive material	Average fuel consump-tion rate (l/100 km)	3.0	IPCC (2001)	Few of such cars have been introduced in luxury-car segment in advanced coun-tries
Fuel cell passenger cars using hydrogen	Average fuel consump-tion rate (gasoline equivalent l/100 km)	About 2.5	IPCC (2001)	Likely to be introduced inter-nationally by 2005; Possible barriers include lack
Fuel cell passenger cars using methanol	Average fuel consump-tion rate (gasoline equivalent l/100 km)	About 3.2	IPCC (2001)	of hydrogen supply infra-structure and on-board stor-age for hydrogen fuel cells,
Fuel cell passenger cars using gasoline	Average fuel consump-tion rate (l/100 km)	About 4.0	IPCC (2001)	and on-board reforming for methanol and gasoline
Best models of existing light trucks	Average fuel consump-tion (l/100 km)	11.5	IPCC (2001)	Valid for advanced country markets in 1997
Best models of existing pas-senger aircrafts	Average fuel consump-tion (seat-l/100 km)	4.5	IPCC (2001)	Valid for 1997
Industry / Manufacturing sector				
Iron & steel industry				
	Average specific carbon emission (kg-C/GJ)	23.6	IPCC (2001)	Applicable for OECD coun-tries
	Best specific carbon emission (kg-C/GJ)	19.8	IPCC (2001)	Applicable for OECD coun-tries
Existing process of making pig iron from iron ore, coke, etc. (BOF route)	Average energy inten-sity (MJ/kg of iron)	20 to 25	Smil (1999),	
	Pelletized ore intensity (kg/kg of iron)	1.6	Smil (1999)	
	Coke intensity (kg/kg of iron)	0.4	Smil (1999)	
	Injected coal intensity (kg/kg of iron)	0.1	Smil (1999)	Alternatively, 0.06 kg of fuel oil is used
	Limestone intensity (kg/kg of iron)	0.2	Smil (1999)	
Existing process of steel-making from iron (BOF route)	Energy intensity (MJ/kg of steel)	20 to 50	Smil (1999)	
Best performing BOF process (from iron ore to rolled steel)	Energy intensity (MJ/kg of steel)	15 to 20	Phylipsen et al. (1998)	Based on a comparison of steel making for various countries in 1988
Best performing EAF process	Energy intensity (MJ/kg of steel)	4 to 9	Phylipsen et al. (1998)	Based on a comparison of steel making for various countries in 1988

Technology	Parameter	Typical value	Data source	Remarks
Aluminium industry				
Existing (Hall-Heroult) process for production of aluminum from bauxite	Energy intensity (MJ/kg of aluminum)	200 to 342	Smil (1999)	
	Dry bauxite intensity (kg/kg of aluminum)	4.47	Smil (1999)	
	Sodium hydroxide intensity (kg/kg of aluminum)	0.23	Smil (1999)	
	Steam intensity (kg/kg of aluminum)	6.60	Smil (1999)	
	Cryolite intensity (kg/kg of aluminum)	0.04	Smil (1999)	
	Carbon anode intensity (kg/kg of aluminum)	1.00	Smil (1999)	
	Alumina intensity (kg/kg of aluminum)	1.90	Smil (1999)	
Best performing aluminium production process in Germany	Fuel intensity of alumina production (MJ/kg Al)	5.6	Phylipsen et al. (1998)	Valid for Germany in 1989
	Electricity intensity of alumina (MJ/kg Al)	2.2	Phylipsen et al. (1998)	Valid for Germany in 1989
	Fuel intensity of anode production (MJ/kg Al)	3.7	Phylipsen et al. (1998)	Valid for Germany in 1989
	Fuel intensity of primary aluminium production (MJ/kg Al)	3.0	Phylipsen et al. (1998)	Valid for Germany in 1989
	Electricity intensity of primary aluminium production (MJ/kg Al)	55.1	Phylipsen et al. (1998)	Valid for Germany in 1989
Copper industry				
Existing process of copper production from sulfide ore	Energy intensity (MJ/kg of copper)	60 to 125	Smil (1999)	
Titanium industry				
Existing process of producing titanium from ore concentrate	Energy intensity (MJ/kg of titanium)	900 to 940	Smil (1999)	
Cement industry				
Existing cement production using limestone and clay as raw material	Energy intensity (MJ/kg of cement)	5 to 9	Smil (1999)	
Best performing cement production	Fuel intensity of process with very high clinker:cement ratio (MJ/kg of cement)	2.5 to 3.5	Phylipsen et al. (1998)	Based on a comparison of cement making in various countries in 1988/89
	Fuel intensity of process with very low clinker:cement ratio (MJ/kg of cement)	0.8 to 1.5	Phylipsen et al. (1998)	Based on a comparison of cement making in various countries in 1988/89
Existing cement production using limestone and clay as raw material, and coal as fuel	Average carbon emission intensity (t-C/ton cement)	0.34	Watson et al (1996)	This is average for existing process; 60% of this is from energy used in production, and 40% as process gas
Brick industry				
Existing brink-making from clay	Energy intensity (MJ/kg of brick)	2 to 5	Smil (1999)	
Glass industry				
Existing glass-making from sand, etc.	Energy intensity (MJ/kg of glass)	18 to 35	Smil (1999)	

Technology	Parameter	Typical value	Data source	Remarks
Limestone industry				
Existing process of limestone extraction from sedimentary rock	Energy intensity (MJ/kg of limestone)	0.07 to 0.10	Smil (1999)	
Sulfuric acid industry				
Existing process of producing sulfuric acid from sulfur	Energy intensity (MJ/kg of sulfuric acid)	2 to 3	Smil (1999)	
Ammonia industry				
Existing ammonia production process	Energy intensity (GJ/ton)	33.0 to 46.0	IPCC (2001)	Applicable for OECD countries
		39.5	IPCC (2001)	Applicable for South Asia
		44.0	IPCC (2001)	Applicable for Indonesia
Nitrogenous fertilizer industry				
Existing process of urea making (including ammonia making)	Energy intensity (MJ/kg of nitrogen)	70 to 110	Smil (1999)	Nitrogen comprises about 47% of urea
Pulp & paper industry				
Existing process of papermaking from standing timbre	Energy intensity (MJ/kg of paper)	25 to 50	Smil (1999)	
Best performing pulp and paper industry	Primary energy intensity (MJ/kg of paper)	12 to 20	Phylipsen et al. (1998)	Based on a comparison of pulp and paper making in various countries in 1988
Petroleum refining sector				
Best performing oil refining	Energy intensity for gasoline (MJ/kg of gas)	3.8	Phylipsen et al. (1998)	Based on a comparison of oil refining in various countries in 1988
	Energy intensity for kerosene (MJ/kg of gas)	1.6	Phylipsen et al. (1998)	Based on a comparison of oil refining in various countries in 1988
	Energy intensity for gasoil including naphtha (MJ/kg of gas)	3.2	Phylipsen et al. (1998)	Based on a comparison of oil refining in various countries in 1988
	Energy intensity for fuel oil (MJ/kg of gas)	1.8	Phylipsen et al. (1998)	Based on a comparison of oil refining in various countries in 1988
	Energy intensity for other petroleum products (MJ/kg of gas)	1.8	Phylipsen et al. (1998)	Based on a comparison of oil refining in various countries in 1988
Electricity generation sector				
Existing typical pulverized coal fired power plant	Efficiency (J/J)	0.30	IPCC (2001)	Existing worldwide average
	Fixed cost ($/kW)	1300	IPCC (2001)	This is typical cost of modern plant with SO2 and NO$_x$ controls; Can be 50% higher depending on location; Less efficient designs with fewer pollution controls are cheaper
Best existing pulverized coal fired power plant (with new materials allowing higher temperatures and pressures)	Efficiency (J/J)	0.45	IPCC (2001)	Efficiency of up to 0.55 is possible by 2020
	Fixed cost ($/kW)	1740	IPCC (2001)	
Best existing CCGT (combined cycle gas turbine) power plant	Efficiency (J/J)	0.60	IPCC (2001)	Efficiency has been improving at 0.01 per year in the past decade; Efficiency of up to 0.70 is possible in next 10-20 years
	Fixed cost ($/kW)	450 to 500	IPCC (2001)	This includes cost of selective catalytic reduction for NO$_x$, dry cooling, and switchyard; Cost can be higher in some regions if new gas supply infrastructure is required
New IGCC (Integrated Gasification Combined Cycle) power plant (using gasification of coal or biomass or other fuel, and then running a CCGT)	Efficiency (J/J)	0.51	IPCC (2001)	Efficiency of up to 0.60 is possible by 2020
	Fixed cost ($/kW)	About 2000	IPCC (2001)	Fixed cost is expected to decline to 1100 by 2030

Technology	Parameter	Typical value	Data source	Remarks
CHP (Combined Heat and Power) plants (using any fuel)	Total efficiency including both heat and electricity utilization (J/J)	0.90	IPCC (2001)	This high efficiency is possible only if there are sufficiently high heating/cooling load densities available; Various factors including regulation can facilitate or hinder viability of CHPs
Phosphoric acid fuel cells (PAFCs), using natural gas to produce hydrogen as part of integrated system	Efficiency (J/J)	0.36 to 0.38	IPCC (2001)	This is efficiency from natural gas to electricity conversion; Efficiency of 0.50 is possible by 2010; Existing efficiency of 0.80 in cogeneration, expected to rise to 0.90 by 2010
	Fixed cost ($/kW)	1500	IPCC (2001)	This is for integrated system including fuel processor based on natural gas; Fixed cost may to decline to 750 by 2010
Molten carbonate fuel cells (MCFCs), using carbonaceous fuels (e.g. coal, natural gas) and internal reforming to produce hydrogen as part of integral system, and using steam turbine in a bottoming cycle	Efficiency (J/J)	0.65	IPCC (2001)	Not yet introduced commercially due to technical problems like electrode corrosion, sintering of structural fuel cell material, and sensitivity to fuel impurities
Solid oxide fuel cells (SOFCs), with internal reforming (possible due to high temperature) to produce hydrogen, and using by-product heat in a bottoming cycle	Efficiency (J/J)	About 0.65	Estimated based on IPCC (2001)	Not yet introduced commercially due to technical problems related to materials and development of structures
	Fixed cost ($/kW)	1620	IPCC (2001)	Fixed cost expected to decline to 700 in next 2 to3 decades
Hybrid SOFC/CCGT systems, using natural gas to produce hydrogen for fuel cell, coupled with CCGT	Efficiency (J/J)	0.72 to 0.74	IPCC (2001)	Not yet introduced
Existing nuclear power plant	Fixed cost ($/kW)	1700 to 3100	IPCC (2001)	
Pebble bed modular reactor (PBMR -advanced nuclear power plant)	Fixed cost ($/kW)	2090	IPCC (2001)	Still in R&D stage; Uncertainty about cost estimate due to reactor safety and waste disposal
Existing wind power plant	Fixed cost ($/kW)	1000	IPCC (2001)	Fixed cost expected to decline to 635 in next 10-20 years
Solar PV power plant	Fixed cost ($/kW)	4000 to 5000	IPCC (2001)	Fixed cost expected to decline as installed capacity increases
Solar thermal (flat plate solar collectors) power plant	Fixed cost ($/kW)	4000	IPCC (2001)	Fixed cost expected to decline to 2500 by 2030

Note: Also refer to following publications for additional information on efficiencies and costs of new technologies – Blok et al (1995), Brown et al (1998); IEA (1999); IPCC (1996a); Ishitani and Johansson (1996), Penner et al (1999); Perkins (1998); WEC (1995).
* JPY = Japanese Yen in 2002 price (1US$ = 118.05 JPY).

Appendix G: Typical Numbers for Characteristics of Some Removal Processes

The characteristics of removal process and their estimation method are listed in this Appendix. The value provided here can be used as input data of Removal Process table in AIM/Enduse Database System.

G.1 Removal Process for SO$_2$ Emission

G.1.1 Coal washing

Coal washing is the most effective precombustion process to reduce SO$_2$ emission. It involves physical and chemical cleaning of coal. Table G.1 shows the physical coal cleaning's parameters for produced coal in U.S.A. It is estimated with using the value in Kaplan (1994). The assumption in the estimate is that the cite capacity is 500 t/hour and the capacity utilization is 11 hours/day and 365 days/year. The assumption is following Kaplan (1994). Coal washing reduces weight and heat of raw coal. This loss is described as additional energy use to supplement the lack in the model.

Table G.1. Parameters for coal washing

Coal characteristics	Removal rate (%)	Initial cost ($b_p^{0"}$)		Operating cost ($g_p^{0"}$)		Additional energy use (e_p)
		($/GJ)	($/toe)	($/GJ)	($/toe)	
PA Armstrong	41.3%	0.35	14.7	0.10	4.2	5.3%
OH Jefferson	25.4%	0.34	14.1	0.10	4.2	9.9%
WVA Logan	17.4%	0.33	13.6	0.10	4.2	5.3%
IL NO.6	23.2%	0.37	15.4	0.12	5.0	13.6%
MN Rosebud	25.4%	0.52	21.8	0.14	5.7	2.6%
ND Lignite	45.0%	0.62	25.9	0.16	6.7	1.1%

G.1.2 Lime stone injection

SO$_2$ reacts with lime to firm calcium sulfate at high boiler temperatures. With using the principle, sulfur dioxide removal rate of 30-60% can be achieved through the limestone addition. SO$_2$ sorbent such as limestone (CaCO$_3$) or dolomite (CaCO$_3$*MgCO$_3$) is added to the coal pellets fired in stoker boilers or injected into pulverized coal-fired boilers.

Table G.2 shows the limestone injection's parameters estimated with using the value in Cofala and Syri (1998a). It is assumed to estimate fixed cost that the capacity utilization is 5,200 hours per year, boiler size is 10MWth in case below 20MWth, 150MWth in case of 20-300MWth, 500MWth in case over 300MWth, respectively. Operating cost includes the cost of maintenance, administrative overhead, labor cost, sorbent cost and by-product/waste disposal cost, but does not include the cost for increased energy demand. The energy cost is calculated with additional energy use endogenously in the model.

Table G.2. Parameters of limestone injection

Capacity class (MWth)	Fuel for boiler	Removal rate (%)	Initial cost ($b_p^{0"}$)		Operating cost ($g_p^{0"}$)		Additional Energy use (e_p)
			($/GJ)	($/toe)	($/GJ)	($/toe)	
< 20	Brown coal	50%	1.74	72.8	0.15	6.1	0.45%
	Hard coal	50%	1.45	60.7	0.13	5.6	0.45%
	Other solid	50%	1.45	60.7	0.11	4.6	0.45%
	Oil and gas	50%	1.30	54.6	0.13	5.4	0.45%
20-300	Brown coal	50%	2.21	92.7	0.17	6.9	0.45%
	Hard coal	50%	1.85	77.3	0.15	6.3	0.45%
	Other solid	50%	1.85	77.3	0.13	5.3	0.45%
	Oil and gas	50%	1.66	69.5	0.14	6.0	0.45%
> 300	Brown coal	50%	1.81	76.0	0.15	6.3	0.45%
	Hard coal	50%	1.51	63.3	0.14	5.8	0.45%
	Other solid	50%	1.51	63.3	0.11	4.8	0.45%
	Oil and gas	50%	1.36	57.0	0.13	5.5	0.45%

G.1.3 Flue gas desulfurization process

Flue gas desulfurization (FGD) is the chemical process to remove SO_2 from the flue gases. Wet lime or limestone FGD is the representative process of wet FGD. Desulfurization reaction of the process is as follows.

Limestone ; $CaCO_3 + SO_2 \rightarrow CaSO_3 + CaSO_4 + CO_2$
Lime   ; $Ca(OH)_3 + SO_2 \rightarrow CaSO_3 + CaSO_4 + H_2O$

The by-product, gypsum, can be used for a variety of industrial applications. Table G.3 shows wet FGD's parameters estimated with using the value in Cofala and Syri (1998a). The assumption in the estimate and the content of fixed / operating cost are same as in case of lime stone injection.

Wellman-Lord process is the typical economic and technical properties representative for high-efficiency desulfurization techniques. Desulfurization reaction of the process is as follows.

$2NaOH + SO_2 \rightarrow Na_2SO_3 + H_2O$

Table G.4 shows the parameter of Wellman-Lord process as the advanced FGD. This process produces SO2 rich gas instead of gypsum. The gas can be used as raw input in chemical industry.

Table G.3. Parameters of wet flue gas desulfurization process

Capacity class (MWth)	Fuel for boiler	Removal rate (%)	Initial cost ($b_p^{0"}$)		Operating cost ($g_p^{0"}$)		Additional energy use (e_p)
			($/GJ)	($/toe)	($/GJ)	($/toe)	
< 20	Brown coal	95%	6.05	253.2	0.29	12.2	0.90%
	Hard coal	95%	5.04	211.0	0.25	10.6	0.90%
	Other solid	95%	5.04	211.0	0.24	9.9	0.90%
	Oil and gas	95%	4.54	189.9	0.23	9.7	0.90%

Table G.3. Parameters of wet flue gas desulfurization process (continued)

Capacity class (MWth)	Fuel for boiler	Removal rate (%)	Initial cost ($b_p^{0"}$)		Operating cost ($g_p^{0"}$)		Additional energy use (e_p)
			($/GJ)	($/toe)	($/GJ)	($/toe)	
20-300	Brown coal	95%	5.26	220.1	0.26	10.9	0.90%
	Hard coal	95%	4.38	183.4	0.23	9.5	0.90%
	Other solid	95%	4.38	183.4	0.21	8.8	0.90%
	Oil and gas	95%	3.94	165.0	0.21	8.7	0.90%
> 300	Brown coal	95%	4.23	177.3	0.22	9.2	0.90%
	Hard coal	95%	3.53	147.7	0.19	8.0	0.90%
	Other solid	95%	3.53	147.7	0.18	7.4	0.90%
	Oil and gas	95%	3.18	132.9	0.18	7.4	0.90%

Table G.4. Parameters of advanced flue gas desulfurization process

Capacity class (MWth)	Fuel for boiler	Removal rate (%)	Initial cost ($b_p^{0"}$)		Operating cost ($g_p^{0"}$)		Additional energy use (e_p)
			($/GJ)	($/toe)	($/GJ)	($/toe)	
< 20	Brown coal	98%	23.29	974.9	0.94	39.5	1.98%
	Hard coal	98%	19.40	812.4	0.79	33.0	1.98%
	Other solid	98%	19.40	812.4	0.79	33.0	1.98%
	Oil and gas	98%	17.46	731.2	0.71	29.8	1.98%
20-300	Brown coal	98%	12.83	537.3	0.53	22.0	1.98%
	Hard coal	98%	10.69	447.7	0.44	18.4	1.98%
	Other solid	98%	10.69	447.7	0.44	18.4	1.98%
	Oil and gas	98%	9.62	403.0	0.40	16.6	1.98%
> 300	Brown coal	98%	10.12	423.5	0.42	17.5	1.98%
	Hard coal	98%	8.43	352.9	0.35	14.6	1.98%
	Other solid	98%	8.43	352.9	0.35	14.6	1.98%
	Oil and gas	98%	7.59	317.6	0.32	13.2	1.98%

G.2 Removal Process for NO$_x$ Emission

G.2.1 Removal process for stationary sources

There are two types of removal process for NO$_x$ emission from stationary sources. One is combustion modification and the other is flue gas cleaning. The reduction of excess oxygen levels and peak flame temperature are used as the principles of combustion modification. Low-NO$_x$ burner and fluidized bed combustion fall into the category. Selective catalytic reduction (SCR) is widely applied as flue gas cleaning. Denitration reactions of the process are as follows.

$$4NO + 4NH_2 + O_2 \rightarrow 2N_2 + 6H_2O$$
$$6NO_2 + 8NH_3 \rightarrow 7N_2 + 12H_2O$$

Non-selective catalytic reduction (NSCR) needs injection of ammonia or other reducing agents without a catalyst. The reaction is as follows.

$$CO(NH_2)_2 + 2NO + 1/2O_2 \rightarrow 2N_2 + CO_2 + 2H_2O$$

Tables G.5 and G.6 show the parameters of removal process for power plant sector and industrial boilers, respectively. They are estimated by using the values in Cofala and Syri (1998b). It is assumed that capacity utilization is 6,000 hours per year, boiler size is 10MWth in case below 20MWth, 150MWth in case of 20-300MWth, 500MWth in case over 300MWth. Operating cost includes the costs of maintenance, overhead, labor, sorbent, catalyst and by-product/waste disposal, but does not include the cost of increased energy demand. Energy cost is calculated endogenously in the model.

Table G.5. Parameters of removal process for power plant sector

Removal process		Removal rate (%)	Initial cost ($b_p^{0"}$)		Operating cost ($g_p^{0"}$)		additional energy use (e_p)
			($/GJ)	($/toe)	($/GJ)	($/toe)	
Combustion modification (CM)							
< 20	Brown coal	65%	0.81	34.1	0.05	2.0	0.32%
	Hard coal	50%	0.54	22.6	0.03	1.4	0.32%
	Oil and gas	65%	0.29	12.0	0.02	0.7	0.27%
20-300	Brown coal	65%	0.71	29.7	0.04	1.8	0.32%
	Hard coal	50%	0.48	20.0	0.03	1.2	0.32%
	Oil and gas	65%	0.25	10.3	0.01	0.6	0.27%
300 <	Brown coal	65%	0.62	26.1	0.04	1.6	0.32%
	Hard coal	50%	0.40	16.7	0.02	1.0	0.32%
	Oil and gas	65%	0.27	11.4	0.02	0.7	0.27%
Selective catalytic reduction (SCR)							
< 20	Brown coal	80%	1.65	69.3	0.18	7.5	0.32%
	Hard coal	80%	1.38	57.7	0.15	6.4	0.32%
	Oil and gas	80%	0.97	40.5	0.09	3.8	0.27%
20-300	Brown coal	80%	1.32	55.4	0.16	6.7	0.32%
	Hard coal	80%	1.10	46.2	0.14	5.7	0.32%
	Oil and gas	80%	0.78	32.7	0.08	3.3	0.27%
300 <	Brown coal	80%	1.00	42.0	0.14	5.9	0.32%
	Hard coal	80%	0.84	35.0	0.12	5.1	0.32%
	Oil and gas	80%	0.60	24.9	0.07	2.8	0.27%
CM + SCR							
< 20	Brown coal	80%	2.08	87.2	0.19	7.8	0.32%
	Hard coal	80%	1.64	68.8	0.15	6.5	0.32%
	Oil and gas	80%	1.13	47.3	0.09	3.7	0.27%
20-300	Brown coal	80%	1.68	70.4	0.16	6.8	0.32%
	Hard coal	80%	1.33	55.5	0.14	5.7	0.32%
	Oil and gas	80%	0.92	38.4	0.07	3.1	0.27%
300 <	Brown coal	80%	1.30	54.6	0.14	5.9	0.32%
	Hard coal	80%	1.01	42.2	0.12	4.9	0.32%
	Oil and gas	80%	0.75	31.2	0.06	2.7	0.27%

Table G.6. Parameters of removal process for industrial boilers and furnaces

Removal process		Removal rate (%)	Initial cost ($b_p^{0''}$)		Operating cost ($g_p^{0''}$)		Additional energy use (e_p)
			($/GJ)	($/toe)	($/GJ)	($/toe)	
Combustion modification (CM)							
< 20	Solid fuels	50%	0.54	22.6	0.03	1.4	0.27%
	Oil and gas	50%	0.40	16.9	0.02	1.0	0.27%
20-300	Solid fuels	50%	0.48	20.0	0.03	1.2	0.27%
	Oil and gas	50%	0.35	14.6	0.02	0.9	0.27%
300 <	Solid fuels	50%	0.40	16.7	0.02	1.0	0.27%
	Oil and gas	50%	0.28	11.6	0.02	0.7	0.27%
CM + selective catalytic reduction (SCR)							
< 20	Solid fuels	80%	1.64	68.8	0.15	6.2	0.27%
	Oil and gas	80%	1.21	50.5	0.09	4.0	0.27%
20-300	Solid fuels	80%	1.33	55.5	0.13	5.4	0.27%
	Oil and gas	80%	0.98	41.2	0.08	3.4	0.27%
300 <	Solid fuels	80%	1.01	42.2	0.11	4.6	0.27%
	Oil and gas	80%	0.75	31.4	0.07	2.8	0.27%
CM + selective non-catalytic reduction (SNCR)							
< 20	Solid fuels	70%	0.82	34.3	0.10	4.3	0.27%
	Oil and gas	70%	0.59	24.8	0.07	2.7	0.27%
20-300	Solid fuels	70%	0.69	28.7	0.09	3.9	0.27%
	Oil and gas	70%	0.49	20.6	0.06	2.5	0.27%
300 <	Solid fuels	70%	0.55	22.8	0.09	3.6	0.27%
	Oil and gas	70%	0.38	16.1	0.05	2.2	0.27%

G.2.2 Removal process for mobile sources

NO_x emission can be reduced by change in engine design, change in fuel quality after treatment of the exhaust gas by catalytic converters. Table G.7 shows the parameters of removal process for transport sector. They are estimated with using the value in Cofala and Syri (1998b). Operating cost is assumed 11% of fixed cost.

Table G.7. Parameters of removal process for transport sector

Removal process	Removal rate (%)	Initial cost ($b_p^{0''}$)		Operating cost ($g_p^{0''}$)		Additional energy use (e_p)
		($/GJ)	($/toe)	($/GJ)	($/toe)	
Gasoline 4-stroke passenger cars and LDV						
3-way catalytic converter 1992	75%	8.5	358	2.6	107	0
3-way catalytic converter 1996	87%	10.2	429	3.1	129	0
Adv. cnv. with maintenance schemes –EU 2000	93%	24.2	1014	7.3	304	0
Adv. cnv. with maintenance schemes –EU 2005	97%	30.2	1264	9.1	379	0
Diesel passenger cars and LDV						
Combustion modification 1992	31%	5.1	215	1.5	64	0
Combustion modification 1996	50%	9.4	393	2.8	118	0
Adv. CM with maintenance schemes –EU 2000	60%	26.6	1115	8.0	335	0
NO_x converter	80%	35.1	1469	10.5	441	0
Heavy duty vehicles - diesel						
EURO I – 1993	33%	20.5	858	6.1	257	0
EURO II – 1996	43%	61.5	2574	18.4	772	0
EURO III – EU 2000	60%	138.2	5788	41.5	1736	0
EURO IV (NO_x converter)	85%	274.9	11508	82.5	3452	0
Heavy duty vehicle						
Natural gas – catalytic converter	85%	93.9	3933	28.2	1180	0
Gasoline – catalytic converter	85%	93.9	3933	28.2	1180	0
Heavy duty vehicle						
CM + medium vessels	40%	3.9	164	1.2	49	0
CM + large vessels	40%	5.7	237	1.7	71	0
SCR – large vessels	90%	18.0	752	5.4	226	0

Appendix H: Calorific Value, Price, and Emission Coefficients of Energy Kinds

Table H.1. Calorific values and emission coefficients

Energy kind	Calorific value	Emission coefficients	
		CO_2 (t-C/TJ)	SO_2 (kg-SO_2/TJ)
Crude oil	1.0 to 1.023 toe/ton (7.3 to 7.5 barrels/ton)*	IPCC average: 20.0 Japan: 18.66	Japan: 35.1
Natural gas liquids	1.02 toe/ton (1.12 toe/ton for Thailand)	IPCC average: 17.2	
Ethane	1.13 toe/ton	IPCC average: 16.8	
LPG	1.13 toe/ton (1.088 toe/ton for Malaysia) (1.10 toe/ton for China)	IPCC average: 17.2 Japan: 16.32	Japan: 0.8 India: Negligible
Naphtha	1.075 toe/ton (1.054 toe/ton for Malaysia)	IPCC average: 20.0 Japan: 18.16 India: 15.4	Japan: 2.92 India: 111.1
Aviation gasoline	1.07 toe/ton (1.05 toe/ton for Malaysia)	IPCC average: 18.9 Japan: 18.29 India: 18.4	India: 23.2
Motor gasoline	1.07 toe/ton (1.05 toe/ton for Malaysia) (1.03 toe/ton for China)	IPCC average: 18.9 Japan: 18.29 India: 18.6	Japan: 3.8 India: 108.7
Jet gasoline	1. 07 toe/ton	IPCC average: 18.9	
Jet kerosene	1.065 toe/ton (1.032 toe/ton for Malaysia) (1.02 toe/ton for China)	IPCC average: 19.5 Japan: 18.31 India: 18.4	Japan: 52.55
Other kerosene	1.045 toe/ton (1.032 toe/ton for Malaysia) (1.02 toe/ton for China)	IPCC average: 19.6 Japan: 18.51 India: 18.4	Japan: 1.7 India: 116.2
Gas/diesel oil	1.035 toe/ton (1.015 toe/ton for Malaysia) (1.20 toe/ton for China)	IPCC average: 20.2 Japan: 18.72 India: 20.0	Japan: 59.6 India: 465.2
Heavy fuel oil	0.96 toe/ton (0.991 toe/ton for Malaysia) (1.03 toe/ton for China)	IPCC average: 21.1 Japan (type A): 18.90 Japan (type C): 19.54 India: 21.1	Japan (type A): 207.0 Japan (type C): 662.2 India: 1993.6
Lubricants	0.96 toe/ton (1.006 toe/ton for Malaysia)	IPCC average: 20.0	
Bitumen	0.96 toe/ton (0.998 toe/ton for Malaysia)	IPCC average: 22.0	
Paraffin waxes	0.96 toe/ton (1.035 toe/ton for Malaysia)		
Petroleum coke	0.74 toe/ton 0.869 toe/ton for Malaysia)	IPCC average: 27.5	
Other petroleum products	0.96 toe/ton (1.015 toe/ton for Malaysia)	IPCC average: 20.0 Japan: 20.77 India: 18.6	

Table H.1. Calorific values and emission coefficients (contibued)

Fuel	Calorific value	Emission coefficients	
		CO$_2$ (t-C/TJ)	SO$_2$ (kg-SO$_2$/TJ)
Hard coal	0.47 to 0.63 toe/ton[*] (0.50 to 0.63 toe/ton for China)	IPCC average: 25.8 Japan: 24.71 India: 25.6	Japan: 333.0 India: 549.1
Lignite/Brown coal/Sub-bituminous coal	0.20 to 0.42 toe/ton (0.20 to 0.29 toe/ton for most Asian countries)[*]	IPCC average: 26.2 to 27.6 India: 25.6	India: 549.1
Coking coal	0.44 to 0.70 toe/ton[*] (0.55 to 0.85 toe/ton for China)	IPCC average: 25.8 Japan: 23.65 India: 29.4	Japan: 300.6 India: 549.1
Other bitumi-nous coal and anthra-cite	0.44 to 0.70 toe/ton[*]	IPCC average: 25.8 to 26.8 India: 25.6	India: 549.1
Peat	0.20 toe/ton	IPCC average: 28.9	
Coke oven coke and lig-nite coke	0.63 to 0.65 toe/ton[*] (0.99 toe/ton for China)	IPCC average: 29.5 Japan: 29.4 India: 29.4	Japan: 411.7
Peat bri-quettes	0.43 to 0.48 toe/ton[*]	IPCC average: 28.9	
Coke oven gas	0.64 toe/ton	IPCC average: 13.0	Japan: 7.5
Blast furnace gas		IPCC average: 66.0	Japan: 30.5
Charcoal	0.736 toe/ton		
Natural gas	0.901 toe/1000m^3 (0.40 to 0.43 toe/1000m^3 for China) (0.97 toe/1000m^3 for Indone-sia)	IPCC average: 15.3 Japan: 13.47 Japan (town gas): 13.14 India: 14.2	Japan: 1.47 India: Negligible
Solid biomass	Fuelwood: 0.36 toe/ton (0.40 toe/ton for China) Dung: 0.29 toe/ton (0.30 to 0.37 toe/ton for China) Agricultural waste: 0.36 toe/ton (0.30 to 0.37 toe/ton for China) Bagasse: 0.19 toe/ton	IPCC average: 29.9	Japan (fuelwood): 63.2 India (fuelwood): 53.3 India (dung): 42.9 India (agricultural waste): 46.2
Liquid bio-mass	0.65 toe/ton	IPCC average: 20.0	
Gas biomass	Biogas: 0.50 toe/ton for China	IPCC average: 30.6	

Note: Japan's emission factors are based on low heating values; Other numbers are based on high heating values.
[*] These numbers are country-specific. Refer to IEA (2001) for further details.
Sources: IEA (2001); IPCC (2001), IPCC (1996b); Garg and Shukla (2002); JEA (1992, 1998); EDMC (2002)

Table H.2. Crude oil spot prices in 2001 (in US$/barrel)

Source	Price
Brent	24.46
West Texas Intermediate	25.89
West Texas Sour	23.17
Lousiana Light Sweet	25.95
Arab Light	26.75*
Fatah-Dubai	22.81
Iranian Light	22.61
Iranian Heavy	22.08
Urals	23.12
Minas	24.06
Tapis	25.33

* Valid for 2000
Source: IEA (2002)

Table H.3. Oil product spot prices in 2001 (in US$/barrel)

Product	Market		
	USA	Singapore	NW Europe (Rotterdam)
Gasoline	30.96	27.50	28.91
Gas oil	29.78	27.29	29.15
Jet kerosene	31.12	28.31	30.82
Naphtha	-	23.78	23.73
Low sulfur fuel oil	20.68	21.83	19.54
High sulfur fuel oil	17.35	29.37	17.82

Source: IEA (2002)

Table H.4. Prices of enduse energy (in 1000 national currency/toe (NCV))

Country:	Korea	China	India	Indonesia	Japan	Thailand
Exchange rate in 2000 (1US$ =):	'000Won 1.131	'000Yuan 0.00828	'000Rs. 0.0449	'000Rupiah 0.0084	'000Yen 0.1078	'000Bath 0.0401
Exchange rate in 1999 (1US$ =):	-	-	0.0431	0.0079	0.1139	0.0378
Exchange rate in 1998 (1US$ =):	-	-	-	0.0100	0.1309	-
Exchange rate in 1997 (1US$ =):	-	-	-	-	0.1209	-

Energy	Sector	Korea	China	India	Indonesia	Japan	Thailand
Steam coal	Industry	0.094	0.415	$1.21^{**,+}$	$169.1^{\#}$	6.79	1.92^{*}
	Households	-	0.451	-	-	-	-
	Electricity generation	-	-	-	-	$13.33^{\#\#}$	-
Light fuel oil	Industry	0.668	-	12.34^{**}	$305.0^{\#}$	37.12	-
	Households	0.701	-	9.11	338.2^{**}	60.12	18.97
	Electricity generation	-	-	-	-	-	-
High sulfur fuel oil	Industry	0.351	-	8.41^{**}	$362.9^{\#}$	27.49	10.39
	Households	-	-	-	-	-	-
	Electricity generation	0.341	-	8.41^{**}	-	$26.27^{\#\#}$	-
Low sulfur fuel oil	Industry	0.363	-	-	-	35.14	7.08^{**}
	Households	-	-	-	-	-	-
	Electricity generation	-	-	-	-	-	-
Electricity	Industry	0.852	-	41.88^{*}	$2,425.1^{**}$	188.84^{**}	26.51^{*}
	Households	1.068	-	20.26^{*}	$2,253.5^{\#}$	268.72^{*}	27.91^{*}
	Electricity generation	-	-	-	-	-	-
Natural gas	Industry			$1.84^{*,+}$	$883.0^{\#}$	54.22^{*}	4.84^{*}
	Households			-	$111.3^{\#}$	155.00^{*}	-
	Electricity generation			-	-	$28.46^{\#\#}$	-

* Valid for 2000; ** Valid for 1999; $^{\#}$ Valid for 1998; $^{\#\#}$ Valid for 1997
$^{+}$ Price excluding tax (all other prices are inclusive of taxes)
Sources: IEA (2002); IMF (2001)

Appendix I: Units and Conversions

Following tables have been reproduced from IEA (2001).

Table I.1. Conversion factors for energy

To:	TJ	Gcal	Mtoe	MBtu	GWh
From:	multiply by:				
TJ	1	238.8	2.388×10^{-5}	947.8	0.2778
Gcal	4.1868×10^{-3}	1	10^{-7}	3.968	1.163×10^{-3}
Mtoe	4.1868×10^{4}	10^{7}	1	3.968×10^{7}	11630
MBtu	1.0551×10^{-3}	0.252	2.52×10^{-8}	1	2.931×10^{-4}
GWh	3.6	860	8.6×10^{-5}	3412	1

Table I.2. Conversion factors for mass

To:	Kilogram	Ton	Pound
From:	multiply by:		
Kilogram	1	0.001	2.2046
Ton	1000	1	2204.6
Pound	0.454	4.54×10^{-4}	1

Table I.3. Conversion factors for volume

To:	U.S. gallon	U.K. gallon	Barrel	Cubic foot	Litre	Cubic metre
From:	multiply by:					
U.S. gallon	1	0.8327	0.02381	0.1337	3.785	0.0038
U.K. gallon	1.201	1	0.02859	0.1605	4.546	0.0045
Barrel	42.0	34.97	1	5.615	159.0	0.159
Cubic foot	7.48	6.229	0.1781	1	28.3	0.0283
Litre	0.2642	0.220	0.0063	0.0353	1	0.001
Cubic metre	264.2	220.0	6.289	35.3147	1000.0	1

Appendix J: Representation of Technology Systems in AIM-India

Notations

External energy	Device
Internal energy/service	External service

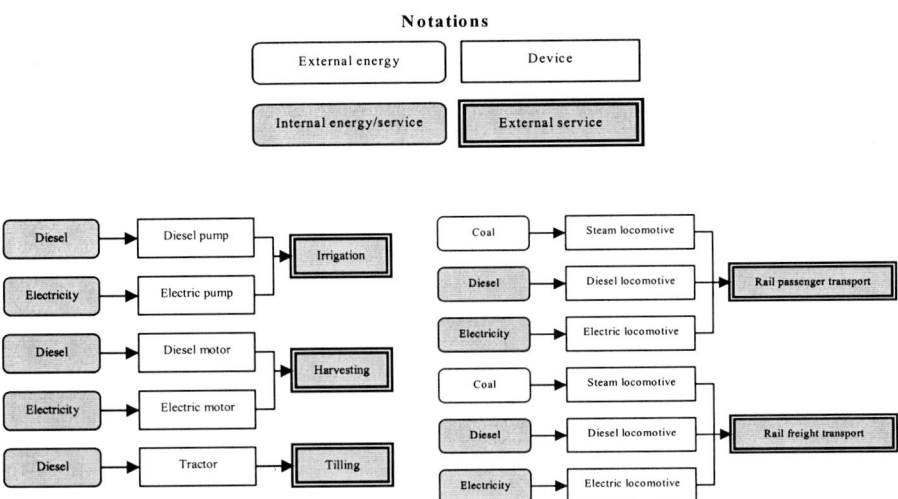

Fig. J.1. Technology systems in agriculture sector

Fig. J.2. Technology systems in rail transport sector

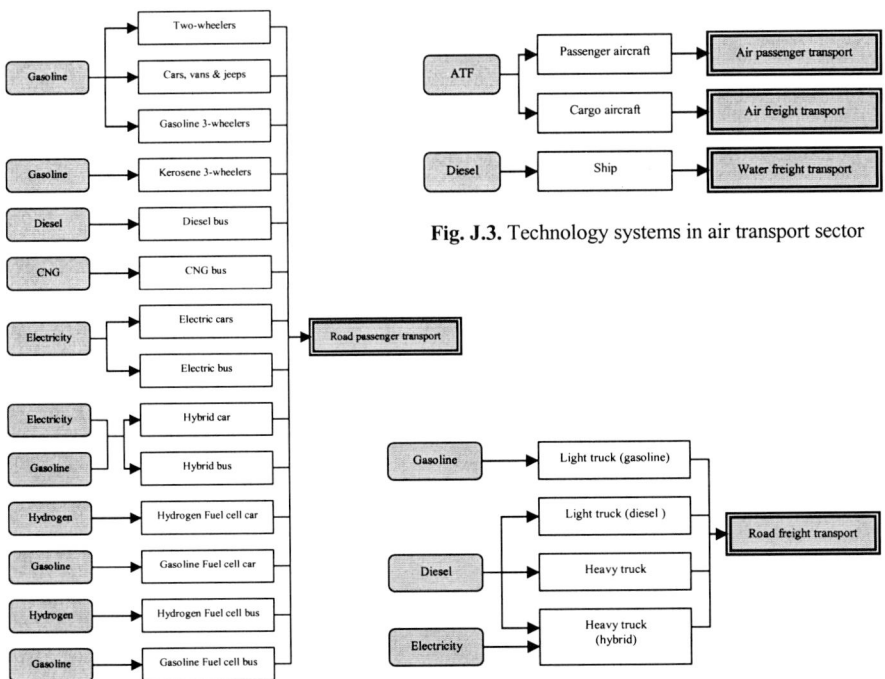

Fig. J.3. Technology systems in air transport sector

Fig. J.4. Technology systems in road transport sector

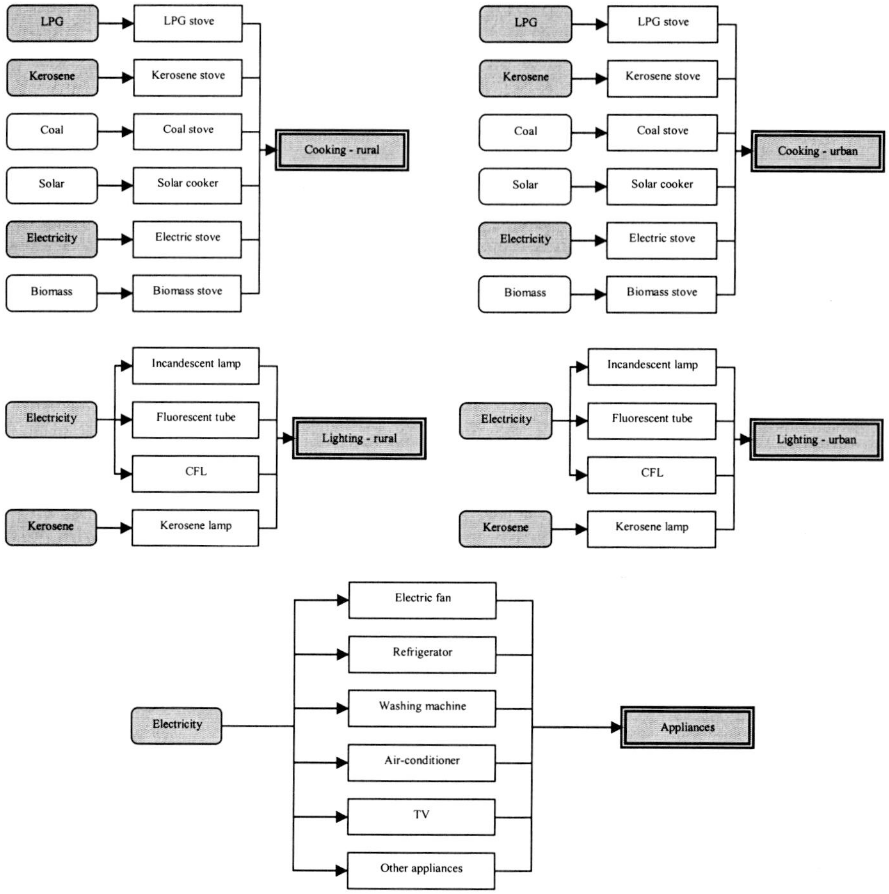

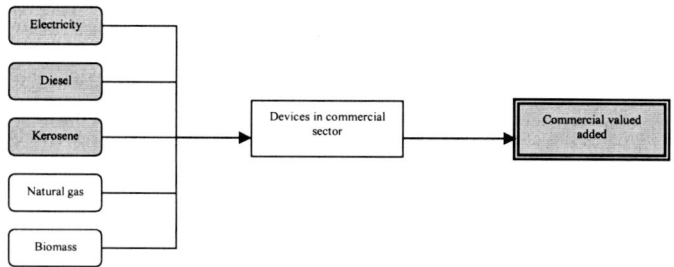

Fig. J.5. Technology systems in residential sector

Fig. J.6. Technology systems in commercial sector

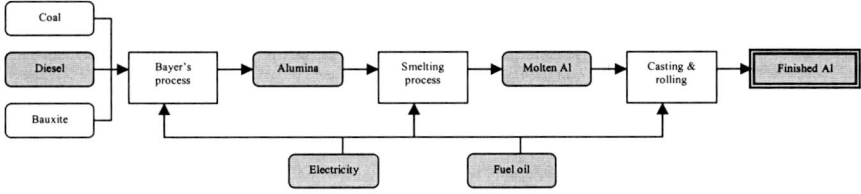

Fig. J.7. Technology systems in aluminum industry

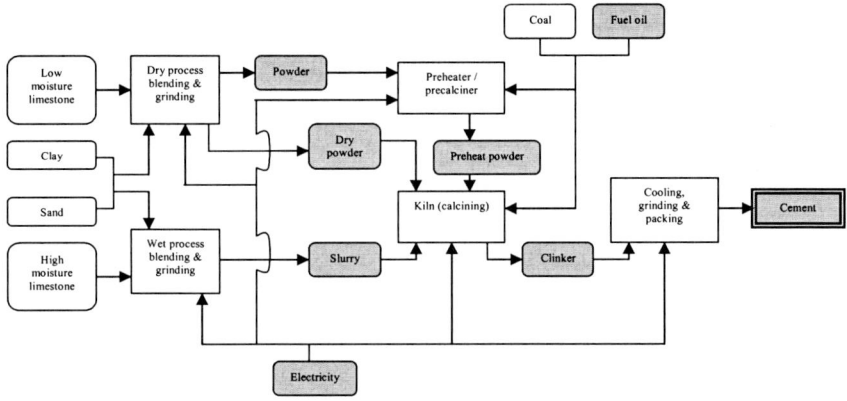

Fig. J.8. Technology systems in cement industry

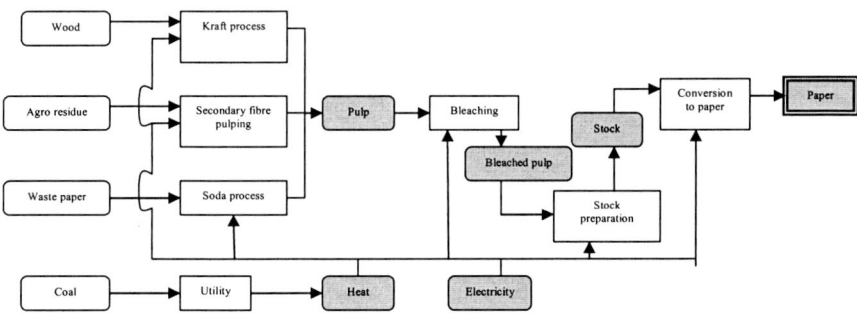

Fig. J.9. Technology systems in pulp & paper industry

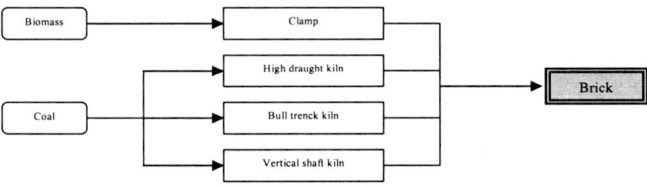

Fig. J.10. Technology systems in brick industry

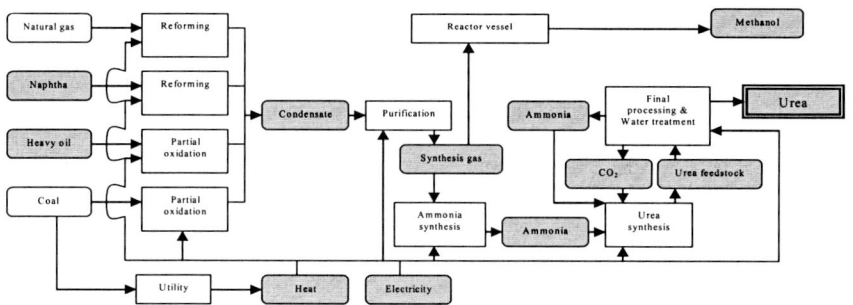

Fig. J.11. Technology systems in nitrogenous fertilizer industry

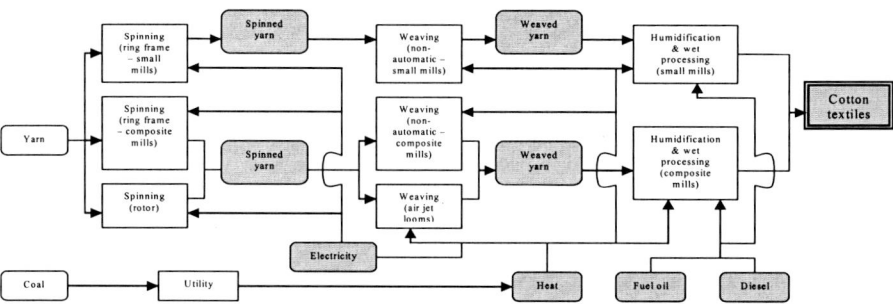

Fig. J.12. Technology systems in cotton textiles industry

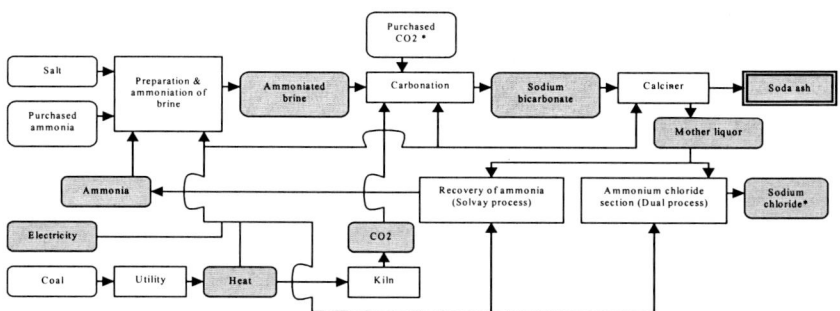

Fig. J.13. Technology systems in soda ash industry

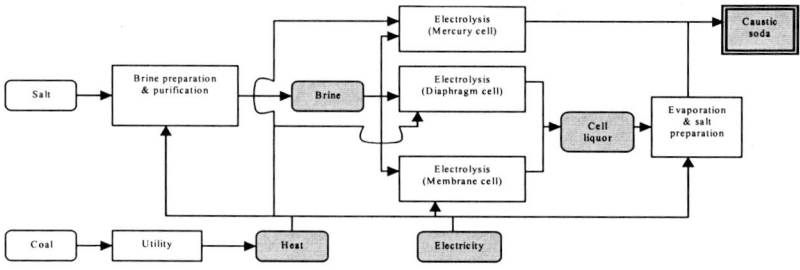

Fig. J.14. Technology systems in caustic soda industry

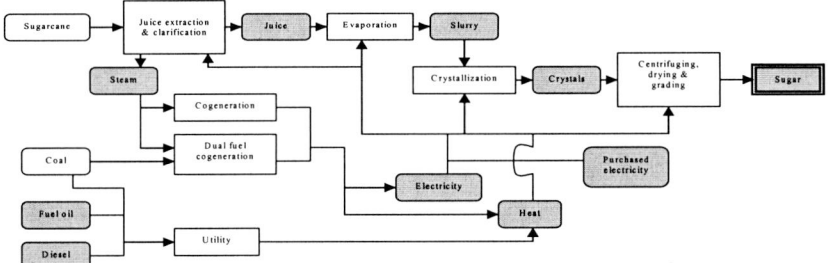

Fig. J.15. Technology systems in sugar industry

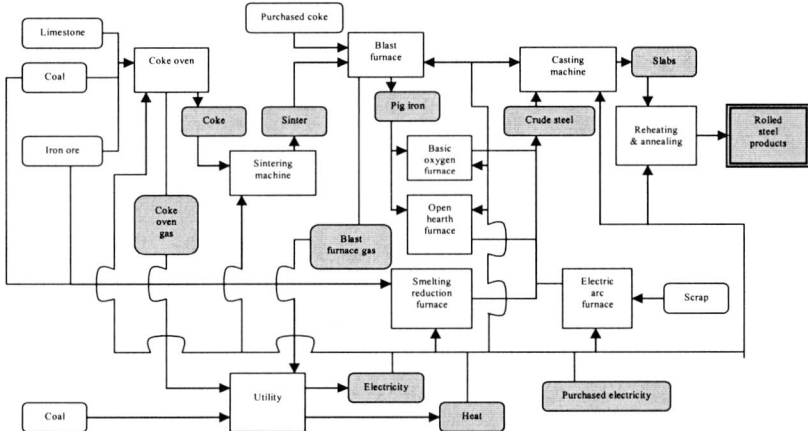

Fig. J.16. Technology systems in iron and steel industry

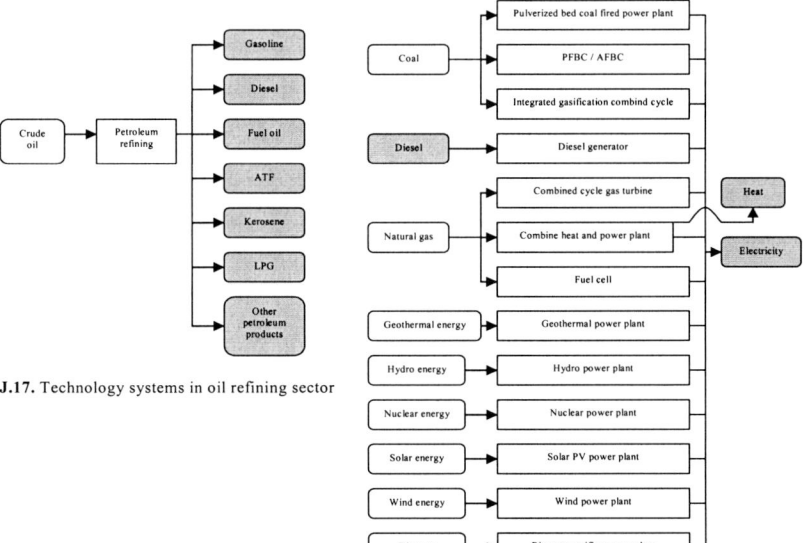

Fig. J.17. Technology systems in oil refining sector

Fig. J.18. Technology systems in electricity sector

Appendix K: Characteristics of Technologies in AIM-India

Device	Device unit	Life (yr.)	Fixed cost (95US$/d.u.)	Service	Service unit	Specific service output (service unit/d.u./yr.)	Energy	Specific energy input (kgoe/d.u./yr.)	NOₓ factor (kg-NOₓ/d.u./yr.)
Agriculture sector									
Tractor	1 tractor	10	5000	Tilling	1000 no	0.001	Diesel	80	0.34
New tractor	1 tractor	10	5000	Tilling	1000 no	0.001	Diesel	60	0.25
Diesel pump for irrigation	1 pump	10	250	Irrigation	toe	0.5	Diesel	1600	6.70
New diesel pump for irrigation	1 pump	10	250	Irrigation	toe	0.5	Diesel	1350	5.65
Electric pump for irrigation	1 pump	10	375	Irrigation	toe	1	Electricity	3000	-
New electric pump (irrigation)	1 pump	10	375	Irrigation	toe	1	Electricity	2500	-
Diesel engine for threshing	1 thresher	10	500	Threshing	toe	0.001	Diesel	3.3	0.01
New diesel engine (threshing)	1 thresher	10	500	Threshing	toe	0.001	Diesel	2.8	0.01
Electric motor for threshing	1 thresher	10	500	Threshing	toe	0.002	Electricity	6.2	-
New electric motor (threshing)	1 thresher	10	500	Threshing	toe	0.002	Electricity	5	-
Aluminium industry									
Bayer's process	1 ton Al	30	1500	Alumina	ton	2.01	Coal	495	6.48
							Electricity	65	
							Fuel oil	157.7	
							Bauxite	5.4	
Improved bayer's process	1 ton Al	30	2000	Alumina	ton	2.01	Coal	360	5.03
							Electricity	55	
							Fuel oil	130	
							Bauxite	5.4	
Energy efficient ALCOA process	1 ton Al	30	2500	Alumina	ton	2.01	Coal	270	4.07
							Electricity	48	
							Fuel oil	120	
							Bauxite	5.4	
Al smelting process (Hall-Heroult)	1 ton Al	20	47000	Molten aluminium	ton	1.02	Alumina	2.01	0.77
							Electricity	1530	
							Fuel oil	107.5	
New Al smelting process (Hall-Heroult with pre-baked anodes)	1 ton Al	20	50000	Molten aluminium	ton	1.02	Alumina	2.01	0.77
							Electricity	1340	
							Fuel oil	80	
New Al smelting process – advanced	1 ton Al	20	51000	Molten aluminium	ton	1.02	Alumina	2.01	0.77
							Electricity	1150	
							Fuel oil	70	
Al casting & rolling (conventional)	1 ton Al	20	1500	Alminium	ton	1	Molten Al	1.02	0.59
							Electricity	125	
							Fuel oil	82	
Al casting & rolling (continuous)	1 ton Al	20	1750	Alminium	ton	1	Molten Al	1.02	0.59
							Electricity	65	
							Fuel oil	82	
New Al casting & rolling (continuous & improved)	1 ton Al	20	2500	Alminium	ton	1	Molten Al	1.02	0.59
							Electricity	50	
							Fuel oil	70	
Brick making industry									
Bull trench kiln – 1	1 million no.	15	4000	Brick	million no	1	Coal	117640	1478
Bull trench kiln – 2	1 million no.	15	3500	Brick	million no	1	Coal	94190	1183
Clamps	1 million no.	25	5000	Brick	million no	1	Biomass	134570	563
High draught kiln	1 million no.	15	6500	Brick	million no	1	Coal	71770	902
vertical shaft brick kiln	1 million no.	15	4000	Brick	million no	1	Coal	57420	721
Caustic soda industry									
Brine preparation & purification	1 ton caustic soda	30	40	Brine	ton	2.1	Salt	3.36	0.42
							Electricity	3.17	
							Heat	12.9	
Improved brine preparation & purification	1 ton caustic soda	30	50	Brine	ton	2.1	Salt	3.36	0.29
							Electricity	2.9	
							Heat	9	
Diaphragm cell electrolysis	1 ton caustic soda	30	100	Cell liquor	ton	1.4	Brine	2.1	-
							Electricity	290	
Mercury cell electrolysis	1 ton caustic soda	30	110	Caustic soda	ton	1	Brine	2.1	-
							Electricity	250	
Membrane cell electrolysis	1 ton caustic soda	30	120	Cell liquor	ton	1.4	Brine	2.1	-
							Electricity	215	
Evaporation & salt separation (diaphragm)	1 ton caustic soda	30	30	Caustic soda	ton	1	Cell liquor	1.4	2.66
							Electricity	76.7	
							Heat	81.7	

d.u. = Device Unit

Device	Device unit	Life (yr.)	Fixed cost (95US$/ d.u.)	Service	Service unit	Specific service output (service unit/d.u./yr.)	Energy	Specific energy input (kgoe/d.u./ yr.)	NOx factor (kg-NOx/ d.u./ yr.)
Evaporation & salt separation (membrane)	1 ton caustic soda	30	30	Caustic soda	ton	1	Cell liquor	1.4	1.79
							Electricity	62	
							Heat	54.6	
Utility in caustic soda plants	1 kgoe heat output	20	0	Heat	kgoe	1	Coal	1.3	-
Cement industry									
Wet process blending & grinding	1 ton clinker	20	1000	Slurry	ton	4	Limestone	1.9	-
							Electricity	2	
Dry process blending & grinding	1 ton clinker	20	1200	Powder	ton	1.5	Limestone	1.49	-
							Electricity	3.5	
Dry process blending & grinding (with pre-calcining)	1 ton clinker	20	1500	Preheated powder	ton	1.3	Limestone	1.49	0.13
							Electricity	4	
							Coal	9.5	
							Fuel oil	1	
Wet process calcining	1 ton clinker	20	4000	Clinker	ton	1	Slurry	4	1.51
							Electricity	4	
							Coal	139	
							Fuel oil	2.3	
Dry process calcining (without pre-calcining)	1 ton clinker	20	4000	Clinker	ton	1	Powder	1.5	0.93
							Electricity	4	
							Coal	85	
							Fuel oil	2.3	
Dry process calcining (after pre-calcining)	1 ton clinker	20	3500	Clinker	ton	1	Preheated powder	1.3	0.82
							Electricity	1.5	
							Coal	65	
							Fuel oil	1.3	
Improved dry process calcining (after pre-calcining)	1 ton clinker	20	3900	Clinker	ton	1	Preheated powder	1.3	0.71
							Electricity	1.5	
							Coal	57	
							Fuel oil	1.3	
Blending, grinding & packing (wet process)	1 ton clinker	20	1000	Cement	ton	1.12	Clinker	1	-
							Electricity	3	
							Fuel oil	2.5	
Blending, grinding & packing (dry process)	1 ton clinker	20	1000	Cement	ton	1.12	Clinker	1	-
							Electricity	4	
							Fuel oil	1.5	
Improved blending, grinding & packing (dry process)	1 ton clinker	20	1200	Cement	ton	1.12	Clinker	1	-
							Electricity	2	
							Fuel oil	1.2	
Nitrogenous fertilizer industry									
Partial oxidation (coal-based)	1 ton urea	30	3000	Synthesis gas	ton	1.5	Coal	3782	40.51
							Electricity	0.238	
							Heat	12	
Partial oxidation (fuel oil-based)	1 ton urea	30	3000	Synthesis gas	ton	1.5	Fuel oil	1780	13.52
							Electricity	5.95	
							Heat	25	
Steam reforming (natural gas-based)	1 ton urea	30	2000	Synthesis gas	ton	1.5	Natural gas	1010	-
							Electricity	4	
Steam reforming (naphtha-based)	1 ton urea	30	2000	Synthesis gas	ton	1.5	Naphtha	1270	-
							Electricity	4.96	
Improved steam reforming (natural gas-based)	1 ton urea	30	2300	Synthesis gas	ton	1.5	Natural gas	900	-
							Electricity	3.5	
Shift conversion & CO2 removal	1 ton urea	30	1000	Purified synthesis gas	ton	1.25	Synthesis gas	1.5	-
							Electricity	66.2	
Improved shift conversion & CO2 removal	1 ton urea	30	1200	Purified synthesis gas	ton	1.25	Synthesis gas	1.5	-
							Electricity	57	
Ammonia synthesis	1 ton urea	30	1500	Ammonia	ton	1.1	Purified synthesis gas	1.25	-
							Electricity	7.14	
Improved ammonia synthesis	1 ton urea	30	1800	Ammonia	ton	1.1	Purified synthesis gas	1.25	-
							Electricity	6.1	
Urea synthesis	1 ton urea	30	1000	Urea	ton	1	Ammonia	1.1	-
							Electricity	97.6	
Improved urea synthesis	1 ton urea	30	1200	Urea	ton	1	Ammonia	1.1	-
							Electricity	75	
Utility in fertilizer plants	1 kgoe heat output	20	0	Heat	kgoe	1	Coal	2.86	-

d.u. = Device Unit

Device	Device unit	Life (yr.)	Fixed cost (95US$/d.u.)	Service	Service unit	Specific service output (service unit/d.u./yr.)	Energy	Specific energy input (kgoe/d.u./yr.)	NO_x factor (kg-NO_x/d.u./yr.)
Soda ash industry									
Preparation, purification & carbonation of Brine (solvay process)	1 ton soda ash	30	100	Sodium bicarbonate	ton	1.4	Salt	1.6	0.05
							Electricity	18.67	
							Heat	3.04	
							Ammonia	0.2	
Preparation, purification & carbonation of Brine (dual process)	1 ton soda ash	30	120	Sodium bicarbonate	ton	1.4	Salt	1.6	0.005
							Electricity	5	
							Heat	0.28	
							Ammonia	0.2	
Preparation, purification & carbonation of Brine (Asahi process)	1 ton soda ash	30	130	Sodium bicarbonate	ton	1.4	Salt	1.6	0.03
							Electricity	22.6	
							Heat	1.74	
							Ammonia	0.2	
Calcination (solvay process)	1 ton soda ash	30	200	Soda ash	ton	1	Sodium bicarbonate	1.4	0.05
				Mother liquor	ton	0.3			
							Electricity	14.5	
							Heat	2.8	
Calcination (dual process)	1 ton soda ash	30	200	Soda ash	ton	1	Sodium bicarbonate	1.4	0.05
				Mother liquor	ton	0.3			
							Electricity	17.4	
							Heat	3	
Calcination (Asahi process)	1 ton soda ash	30	220	Soda ash	ton	1	Sodium bicarbonate	1.4	0.03
				Mother liquor	ton	0.3			
							Electricity	17.6	
							Heat	1.6	
Recovery of ammonia (solvay process)	1 ton soda ash	30	100	Ammonia	ton	0.2	Mother liquor	0.3	0.07
							Electricity	14.5	
							Heat	4.4	
Ammonium chloride section (dual process)	1 ton soda ash	30	120	Ammonia	tn	0.2	Mother liquor	0.3	0.02
							Electricity	99.3	
							Heat	1.2	
Recovery of ammonia (Asahi process)	1 ton soda ash	30	105	Ammonia	tn	0.2	Mother liquor	0.3	0.04
							Electricity	17.6	
							Heat	2.5	
Utility in soda ash plants	1 kgoe heat output	20	0	Heat	kgoe	1	Coal	1.3	-
Pulp and paper industry									
Pulp preparation (kraft process)	1 ton paper	30	150	Pulp	ton	2	Electricity	95	3.45
							Heat	60	
Pulp preparation (soda process)	1 ton paper	30	100	Pulp	ton	2	Electricity	58	3.89
							Heat	120	
Pulp preparation (secondary fiber pulping process)	1 ton paper	30	200	Pulp	ton	2	Electricity	26	1.66
							Heat	45	
Pulp preparation (kraft process & continuous digester)	1 ton paper	30	200	Pulp	ton	2	Electricity	88	3.19
							Heat	52	
Conventional bleaching (large mill)	1 ton paper	30	300	Bleached pulp	ton	1.43	Pulp	2	12.72
							Electricity	73	
							Heat	392	
Conventional bleaching (small mill)	1 ton paper	30	150	Bleached pulp	ton	1.43	Pulp	2	11.73
							Electricity	7	
							Heat	350	
Displacement bleaching	1 ton paper	30	250	Bleached pulp	ton	1.43	Pulp	2	1.62
							Electricity	22	
							Heat	50	
Stock preparation (large mill)	1 ton paper	30	120	Stock	ton	1.1	Bleached pulp	1.43	1.23
							Electricity	106	
							Heat	38	
Stock preparation (small mill)	1 ton paper	30	100	Stock	ton	1.1	Bleached pulp	1.43	1.53
							Electricity	30	
							Heat	47	
Conversion to paper (large mill)	1 ton paper	30	500	Paper	ton	1	Stock	1.1	8.47
							Electricity	134	
							Heat	261	
Conversion to paper (small mill)	1 ton paper	30	300	Paper	ton	1	Stock	1.1	4.87
							Electricity	38	
							Heat	150	
Conversion to paper (large mill with improved evaporator)	1 ton paper	30	550	Paper	ton	1	Stock	1.1	6.17
							Electricity	130	
							Heat	190	
Conversion to paper (small mill with improved evaporator)	1 ton paper	30	350	Paper	ton	1	Stock	1.1	4.38
							Electricity	34	
							Heat	135	
Utility in paper industry	1 kgoe heat output	20	0	Heat	kgoe	1	Coal	1.08	-

d.u. = Device Unit

Device	Device unit	Life (yr.)	Fixed cost (95US$/d.u.)	Service	Service unit	Specific service output (service unit/d.u./yr.)	Energy	Specific energy input (kgoe/d.u./yr.)	NOx factor (kg-NOx/d.u./yr.)
Iron and steel industry									
Small size coke oven	1 ton crude steel	30	6000	Coke Coke gas	toe toe	361.2 12	Coal Electricity	674.9 20.35	9.21
Large size coke oven	1 ton crude steel	30	8000	Coke Coke gas	toe toe	361.2 48	Coal Electricity	600 20.35	8.18
Large size coke oven of Japan type	1 ton crude steel	30	9000	Coke Coke gas	toe toe	361.2 62	Coal Electricity	570 15	7.77
New coking + Coke wetting	1 ton crude steel	30	10000	Coke Coke gas	toe toe	361.2 66	Coal Electricity	550 15	7.50
Large size coke oven + DCQ	1 ton crude steel	30	10500	Coke Coke gas	toe toe	361.2 48	Coal	580	7.91
Large size coke oven of Japan type + DCQ	1 ton crude steel	30	11000	Coke Coke gas	toe toe	361.2 62	Coal	500	6.82
Sintering furnace	1 ton crude steel	20	4000	Sinter	ton	1.15	Coal Oil products Iron Ore	6.21 35 1.03	0.42
Large size sintering furnace	1 ton crude steel	20	4000	Sinter	ton	1.15	Coal Oil products Iron Ore	5.1 35 1.03	0.34
Advanced sintering furnace	1 ton crude steel	20	4000	Sinter	ton	1.15	Coal Oil products Iron Ore	4.6 35 1.03	0.31
Small size blast furnace	1 ton crude steel	20	4000	Pig iron from blast furnace Blast furnace gas	ton toe	1.057 30	Coke Coal Electricity Sinter	358.9 29.54 5.15 1.15	3.89
Large size blast furnace	1 ton crude steel	20	5000	Pig iron from blast furnace Blast furnace gas	tn toe	1.057 72	Coke Coal Electricity Sinter	291.1 55.94 4.7 1.15	3.89
Advanced blast furnace	1 ton crude steel	20	7000	Pig iron from blast furnace Blast furnace gas	tn toe	1.057 102	Coke Coal Electricity Sinter	266.9 54.64 4.3 1.15	2.60
Blast furnace + Wet TRT	1 ton crude steel	20	7500	Pig iron from blast furnace Blast furnace gas	ton toe	1.057 102	Coke Coal Sinter	266.9 51.94 1.15	2.50
Blast furnace + Dry TRT	1 ton crude steel	20	8500	Pig iron from blast furnace Blast furnace gas	ton toe	1.057 102	Coke Coal Sinter	266.9 50.29 1.15	2.40
Blast furnace + Wet TRT + 100kg CPI	1 ton crude steel	20	8700	Pig iron from blast furnace Blast furnace gas	ton toe	1.057 102	Coke Coal Electricity Sinter	176.9 173.94 2.1 1.15	4.60
Blast furnace + Dry TRT + 100kg CPI	1 ton crude steel	20	9000	Pig iron from blast furnace Blast furnace gas	ton toe	1.057 102	Coke Coal Electricity Sinter	176.9 172.29 2.1 1.15	4.50
Blast furnace + Wet TRT + 250kg CPI	1 ton crude steel	20	8900	Pig iron from blast furnace Blast furnace gas	ton toe	1.057 102	Coke Coal Electricity Sinter	107 235.54 2.1 1.15	5.20
Blast furnace + Dry TRT + 250kg CPI	1 ton crude steel	20	9500	Pig iron from blast furnace Blast furnace gas	ton toe	1.057 102	Coke Coal Electricity Sinter	107 233.89 2.1 1.15	5.10
COREX	1 ton crude steel	20	32000	Crude steel	ton	1	Coal Iron Ore	1300 1.39	4.20
DIOS	1 ton crude steel	20	31000	Crude steel	ton	1	Coal Iron Ore	1500 1.39	4.40
Small size ACF	1 ton crude steel	20	4000	Crude steel	ton	1	Electricity Coal Scrap iron	130 95 1	0.34
Large size ACF	1 ton crude steel	20	4500	Crude steel	ton	1	Electricity Coal Scrap iron	110 60 1	0.26

d.u. = Device Unit

Device	Device unit	Life (yr.)	Fixed cost (95US$/d.u.)	Service	Service unit	Specific service output (service unit/d.u./yr.)	Energy	Specific energy input (kgoe/d.u./yr.)	NOₓ factor (kg-NOₓ/d.u./yr.)
DCF	1 ton crude steel	20	5000	Crude steel	ton	1	Electricity / Coal / Scrap iron	100 / 40 / 1	0.18
Advanced ACF	1 ton crude steel	20	6000	Crude steel	ton	1	Electricity / Coal / Scrap iron	100 / 50 / 1	0.19
Advanced DCF	1 ton crude steel	20	7500	Crude steel	ton	1	Electricity / Coal / Scrap iron	90 / 35 / 1	0.16
Small size converter	1 ton crude steel	20	4000	Crude steel	ton	1	Coal / Electricity / Oil products / Pig iron	300 / 40 / 3.5 / 1.057	1.50
Large size converter	1 ton crude steel	20	5000	Crude steel	ton	1	Coal / Electricity / Oil products / Pig iron	250 / 25 / 9 / 1.057	1.10
Converter with ZFG collection	1 ton crude steel	20	6000	Crude steel	ton	1	Coal / Electricity / Oil products / Pig iron	200 / 14.5 / 9 / 1.057	0.90
Open hearth furnace	1 ton crude steel	20	4000	Crude steel	ton	1	Coal / Electricity / Oil products / Pig iron	1300 / 10 / 18 / 1.057	4.20
Flow control 1	1 ton crude steel	1	0.01	Crude steel	ton	1	Crude steel – blast furnace	1	-
Flow control 2	1 ton crude steel	1	0.01	Crude steel	ton	1	Crude steel – electric arc furnace	1	-
Casting machine	1 ton crude steel	20	3000	Slab	ton	1	Coal / Electricity / Crude steel	140 / 0.82 / 1	0.62
Advanced casting	1 ton crude steel	20	3400	Slab	ton	1	Coal / Electricity / Crude steel	70 / 4.22 / 1	0.41
Continuous casting	1 ton crude steel	20	3100	Slab	ton	1	Coal / Electricity / Crude steel	30 / 4.5 / 1	0.20
Advanced continuous casting	1 ton crude steel	20	3700	Slab	ton	1	Coal / Electricity / Crude steel	10 / 2.81 / 1	0.06
Primary rolling machine	1 ton crude steel	20	4000	Finished steel	ton	1	Coal / Electricity / Slab	70 / 13.5 / 1	1.95
Large size primary rolling machine	1 ton crude steel	20	4200	Finished steel	ton	1	Coal / Electricity / Slab	60 / 10 / 1	1.80
Advanced heating furnace	1 ton crude steel	20	4900	Finished steel	ton	1	Coal / Electricity / Slab	50 / 10 / 1	1.60
Direct hot strip mill machine	1 ton crude steel	20	5000	Finished steel	ton	1	Coal / Electricity / Slab	25 / 9 / 1	0.90
Cotton textiles industry									
Spinning (ring frame - small mills)	1000 metres cotton textiles	30	400	Spinned yarn	1000 metres	1.5	Electricity	76	-
Spinning (ring frame - composite mills)	1000 metres cotton textiles	30	400	Spinned yarn	1000 metres	1.5	Electricity	42	-
Spinning (rotor technology)	1000 metres cotton textiles	30	400	Spinned yarn	1000 metres	1.5	Electricity	31	-
Weaving (non-automatic small mills)	1000 metres cotton textiles	30	200	Weaved yarn	1000 metres	1.25	Spinned yarn / Electricity / Heat	1.5 / 4 / 23	1.72
Weaving (improved small mills)	1000 metres cotton textiles	30	250	Weaved yarn	1000 metres	1.25	Spinned yarn / Electricity / Heat	1.5 / 4 / 14.3	1.07

d.u. = Device Unit

Device	Device unit	Life (yr.)	Fixed cost (95US$/d.u.)	Service	Service unit	Specific service output (service unit/d.u./yr.)	Energy	Specific energy input (kgoe/d.u./yr.)	NOₓ factor (kg-NOₓ/d.u./yr.)
Weaving (non-automatic composite mills)	1000 metres cotton textiles	30	200	Weaved yarn	1000 metres	1.25	Spinned yarn	1.5	1.72
							Electricity	18	
							Heat	23	
Weaving (improved composite mills)	1000 metres cotton textiles	30	250	Weaved yarn	1000 metres	1.25	Spinned yarn	1.5	1.07
							Electricity	14	
							Heat	14.3	
Weaving (air jet looms)	1000 metres cotton textiles	30	300	Weaved yarn	1000 metres	1.25	Spinned yarn	1.5	-
							Electricity	21	
Humidification & wet processing (small mills)	1000 metres cotton textiles	30	120	Cotton textiles	1000 metres	1	Weaved yarn	1.25	1.17
							Electricity	15	
							Heat	12	
							Fuel oil	32	
Humidification & wet processing (improved small mills)	1000 metres cotton textiles	30	140	Cotton textiles	1000 metres	1	Weaved yarn	1.25	0.88
							Electricity	13	
							Heat	8.7	
							Fuel oil	25	
Humidification & wet processing (composite mills)	1000 metres cotton textiles	30	120	Cotton textiles	1000 metres	1	Weaved yarn	1.25	1.17
							Electricity	31	
							Heat	12	
							Fuel oil	32	
Humidification & wet processing (improved composite mills)	1000 metres cotton textiles	30	140	Cotton textiles	1000 metres	1	Weaved yarn	1.25	0.88
							Electricity	27	
							Heat	8.7	
							Fuel oil	25	
Utility in textile plants	1 kgoe heat output	20	0	Heat	kgoe	1	Coal	2.3	-
Sugar industry									
Juice extraction & clarification (existing milling & double sulfitation process)	1 ton sugar	30	40	Juice / Steam	ton / kgoe	2 / 200	Sugarcane / Electricity / Heat	3 / 37 / 84	2.73
Juice extraction & clarification (improved sulfitation process)	1 ton sugar	30	45	Juice / Steam	ton / kgoe	2 / 220	Sugarcane / Electricity / Heat	3 / 31 / 80	2.60
Evaporation (existing process)	1 ton sugar	30	40	Sugar slurry	ton	1.5	Juice / Electricity / Heat	2 / 12 / 112	3.70
Evaporation (falling film process)	1 ton sugar	30	45	Sugar slurry	ton	1.5	Juice / Electricity / Heat	2 / 10 / 90	2.92
Crystallizaton (existing process)	1 ton sugar	30	40	Sugar crystals	ton	1.25	Sugar slurry	1.5	4.54
							Electricity	14	
							Heat	140	
Crystallizaton (improved process)	1 ton sugar	30	45	Sugar crystals	ton	1.25	Sugar slurry	1.5	3.98
							Electricity	12	
							Heat	125	
Centrifuging, drying, grading & packing (batch centrifuge)	1 ton sugar	30	20	Sugar	ton	1	Sugar crystals	1.25	1.20
							Electricity	0.9	
							Heat	37	
Centrifuging, drying, grading & packing (continuous centrifuge)	1 ton sugar	30	23	Sugar	ton	1	Sugar crystals	1.25	1.20
							Electricity	0.7	
							Heat	37	
Cogeneration (existing)	1 kgoe steam input	20	0	Heat / Electricity	kgoe / kgoe	1.8 / 0.16	Steam	1	-
Cogeneration (extraction cum back pressure condensing)	1 kgoe steam input	20	0	Heat / Electricity	kgoe / kgoe	1.8 / 0.18	Steam	1	-
Dual fuel cogeneration	1 kgoe steam input	20	0	Heat / Electricity	kgoe / kgoe	1.8 / 0.25	Steam / Coal	1 / 0.2	-
Other industries									
Industry – other – biomass	1	1	20	Ind - others - biomass	toe	1	Bio energy	1000	4.18
Industry - other - commercial fuels	1	1	1000	Industry - others - commercial fuels	index	1	Coal	5532480	98215
							Natural gas	2441680	
							Oil products	1596960	
							Electricity	6211325	

d.u. = Device Unit

Device	Device unit	Life (yr.)	Fixed cost (95US$/ d.u.)	Service	Service unit	Specific service output (service unit/d.u./yr.)	Energy	Specific energy input (kgoe/d.u./ yr.)	NO_x factor (kg-NO_x/ d.u./yr.)
New industry - other - commercial fuels	1	1	700	Industry - others - commercial fuels	index	1	Coal	316240	22553
							Natural gas	2774180	
							Oil products	137872	
							Electricity	9281255	
Commercial sector									
Commercial sector devices - existing mix	1 Million US$95	1	1	Commercial/ service value added	Million US$95	1	Electricity	43600	192.1
							Natural gas	4350	
							Diesel	20540	
							Biomass	23180	
Commercial sector devices - new mix1	1 Million US$95	1	1	Commercial/ service value added	Million US$95	1	Electricity	49390	166.1
							Natural gas	6400	
							Diesel	26060	
							Biomass	10410	
Commercial sector devices - new mix2	1 Million US$95	1	1	Commercial/ service value added	Million US$95	1	Electricity	54680	166.1
							Natural gas	9930	
							Diesel	30240	
							Biomass	4480	
Transport sector									
Cargo aircraft	1 aircraft	15	500000	Air freight	million-km-ton	15	ATF	1335000	16671
New cargo aircraft	1 aircraft	15	500000	Air freight	million-km-ton	15	ATF	1050000	13191
Diesel locomotive	1 train	15	375000	Rail freight	million-km-ton	49	Diesel	200900	10095
Electric locomotive	1 train	15	500000	Rail freight	million-km-ton	50	Electricity	126500	-
New electric locomotive	1 train	15	500000	Rail freight	million-km-ton	50	Electricity	100000	-
Steam locomotive	1 train	15	250000	Rail freight	million-km-ton	50	Coal	348000	4372
Heavy truck	1 truck	10	25000	Road freight	million-km-ton	0.1152	Diesel	4840	162.1
New heavy truck	1 truck	10	25000	Road freight	million-km-ton	0.1152	Diesel	4035	135.2
Light truck	1 truck	10	15000	Road freight	million-km-ton	0.0864	Diesel	4147.2	121.6
New light truck	1 truck	10	15000	Road freight	million-km-ton	0.0864	Diesel	3456	101.3
Ship	1 ship	10	375000	Water freight	million-km-ton	15	Diesel	1200000	75377
New ship	1 ship	10	375000	Water freight	million-km-ton	15	Diesel	900000	56532
Aircraft	1 aircraft	15	500000	Air passenger	million-km-persons	15	ATF	1335000	16671
New aircraft	1 aircraft	15	500000	Air passenger	million-km-persons	15	ATF	1050000	13191
Diesel locomotive	1 train	15	375000	Rail passenger	million-km-persons	49	Diesel	200900	10095
Electric locomotive	1 train	15	500000	Rail passenger	million-km-persons	50	Electricity	126500	-
New electric locomotive	1 train	15	500000	Rail passenger	million-km-persons	50	Electricity	100000	-
Steam locomotive	1 train	15	250000	Rail passenger	million-km-persons	50	Coal	348000	4372
Bus-CNG	1 bus	10	15000	Road passenger	million-km-persons	2.592	Natural gas	10108.8	253.9
New bus-CNG	1 bus	10	45000	Road passenger	million-km-persons	2.592	Natural gas	8553.6	214.9
Bus-diesel	1 bus	10	15000	Road passenger	million-km-persons	2.592	Diesel	12960	325.6
New bus-diesel	1 bus	10	15000	Road passenger	million-km-persons	2.592	Diesel	9849.6	247.5
Electric bus	1 bus	10	45000	Road passenger	million-km-persons	2.592	Electricity	7776	-
Cars, vans & jeeps	1 car	7	6250	Road passenger	million-km-persons	0.027	Gasoline	864	21.7

d.u. = Device Unit

Device	Device unit	Life (yr.)	Fixed cost (95US$/d.u.)	Service	Service unit	Specific service output (service unit/d.u./yr.)	Energy	Specific energy input (kgoe/d.u./yr.)	NO$_x$ factor (kg-NO$_x$/d.u./yr.)
New cars	1 car	7	6250	Road passenger	million-km-persons	0.027	Gasoline	675	16.9
Electric cars	1 car	7	9250	Road passenger	million-km-persons	0.027	Electricity	500	-
Three-wheeler - gasoline	1 vehicle	7	1500	Road passenger	million-km-persons	0.054	Gasoline	1080	27.1
Three-wheeler - kerosene	1 vehicle	7	1500	Road passenger	million-km-persons	0.054	Kerosene	2160	54.3
Two-wheelers	1 scooter	7	500	Road passenger	million-km-persons	0.01296	Gasoline	50.6	1.27
New two-wheelers	1 scooter	7	500	Road passenger	million-km-persons	0.01296	Gasoline	32.4	0.81
Residential sector									
Biomass stove	1 stove	5	2.5	Cooking - rural	toe	0.144	Biomass	1296	5.43
Coal stove	1 stove	5	12.5	Cooking - rural	toe	0.144	Coal	288	1.21
Electric stove	1 stove	5	50	Cooking - rural	toe	0.144	Electricity	158	-
Kerosene stove	1 stove	5	25	Cooking - rural	toe	0.144	Kerosene	216	0.90
LPG stove	1 stove	5	125	Cooking - rural	toe	0.144	LPG	172.8	0.72
New LPG stove	1 stove	5	130	Cooking - rural	toe	0.144	LPG	158	0.66
Solar cooker	1 stove	5	25	Cooking - rural	toe	0.144	Solar	0	-
Incandescent lamp (bulb)	1 bulb	1	0.5	Lighting – rural	billion-lumen-hr	0.001	Electricity	12	-
CFL	1 lamp	1	3.25	Lighting – rural	billion-lumen-hr	0.002	Electricity	5	-
Fl. tube	1 tube	1	3.25	Lighting – rural	billion-lumen-hr	0.0016	Electricity	5.28	-
Kerosene lamp	1 lamp	1	0.75	Lighting – rural	billion-lumen-hr	0.0003	Kerosene	180	0.75
Biomass stove	1 stove	5	2.5	Cooking - urban	toe	0.144	Bio energy	1296	5.43
Coal stove	1 stove	5	12.5	Cooking - urban	toe	0.144	Coal	288	1.21
Electric stove	1 stove	5	50	Cooking - urban	toe	0.144	Electricity	158	-
Kerosene stove	1 stove	5	25	Cooking - urban	toe	0.144	Kerosene	216	0.90
LPG stove	1 stove	5	125	Cooking - urban	toe	0.144	LPG	172.8	0.72
New LPG stove	1 stove	5	125	Cooking - urban	toe	0.144	LPG	158	0.66
Solar cooker	1 stove	5	25	Cooking - urban	toe	0.144	Solar	0	-
Incandescent lamp (bulb)	1 bulb	1	0.5	Lighting – urban	billion-lumen-hr	0.001	Electricity	12	-
CFL	1 lamp	1	3.25	Lighting – urban	billion-lumen-hr	0.002	Electricity	5	-
Fl. tube	1 tube	1	3.25	Lighting – urban	billion-lumen-hr	0.0016	Electricity	5.28	-
Kerosene lamp	1 lamp	1	0.75	Lighting – urban	billion-lumen-hr	0.0003	Kerosene	180	0.75
Fan	1 fan	5	12.5	Fan	1000 no.	0.001	Electricity	15	-
Refrigerator	1 refrigerator	7	300	Refrigerator	1000 no.	0.001	Electricity	80	-
TV	1 tv	7	125	TV	1000 no.	0.001	Electricity	9	-
Washing m/c	1 m/c	7	100	Washing m/c	1000 no.	0.001	Electricity	50	-
Air conditioner	1 a/c	10	1250	Air conditioner	1000 no.	0.001	Electricity	425	-
Other electric appliances	1 appliance	3	25	Other appliances	toe	0.005	Electricity	4	-
Oil refining sector									
Petroleum refinery-1	1 ton crude throughput	30	200	Petroleum products	kgoe	960	Crude Oil	1022	-
Petroleum refinery-2	1 ton crude throughput	30	170	Petroleum products	kgoe	970	Crude Oil	1022	-
Petroleum refinery-3	1 ton crude throughput	30	175	Petroleum products	kgoe	985	Crude Oil	1022	-

d.u. = Device Unit

Device	Device unit	Life (yr.)	Fixed cost (95US$/d.u.)	Service	Service unit	Specific service output (service unit/d.u./yr.)	Energy	Specific energy input (kgoe/d.u./yr.)	NO$_x$ factor (kg-NO$_x$/d.u./yr.)
Electricity generation sector									
Biomass power plant	1 MW	20	950000	Electricity	kgoe	753360	Biomass	2508689	10505
Biomass power plant – advanced	1 MW	20	1050000	Electricity	kgoe	753360	Biomass	1880000	7873
Pulverized bed coal power plant (existing)	1 MW	30	1125000	Electricity	kgoe	753360	Coal	2322000	29171
Pulverized bed coal power plant (improved)	1 MW	30	1740000	Electricity	kgoe	753360	Coal	1680000	21105
PFBC power plant	1 MW	30	1175000	Electricity	kgoe	753360	Coal	2116942	26595
AFBC power plant	1 MW	30	1275000	Electricity	kgoe	753360	Coal	2282681	28677
Diesel generator	1 MW	15	775000	Electricity	kgoe	753360	Diesel	2576491	21578
Gas turbine	1 MW	20	850000	Electricity	kgoe	753360	Natural gas	2184744	13723
CCGT	1 MW	20	600000	Electricity	kgoe	753360	Natural gas	1657392	10411
CCGT – advanced	1 MW	20	650000	Electricity	kgoe	753360	Natural gas	1257392	7898
IGCC power plant (coal)	1 MW	20	2000000	Electricity	kgoe	753360	Coal	1457392	18309
IGCC power plant (biomass)	1 MW	20	2000000	Electricity	kgoe	753360	Biomass	1457392	6103
CHP plant	1 MW	20	2000000	Electricity	kgoe	753360	Natural gas	1600000	10050
				Heat	kgoe	650000			
Proton exchange membrane fuel cell	1 MW	20	1100000	Electricity	kgoe	753360	Natural gas	1650000	10364
Phosphoric acid fuel cell	1 MW	20	1500000	Electricity	kgoe	753360	Natural gas	2000000	12563
Solid oxide fuel cell	1 MW	20	1620000	Electricity	kgoe	753360	Natural gas	1160000	7286
Solid oxide fuel cell coupled with CCGT	1 MW	20	1800000	Electricity	kgoe	753360	Natural gas	1020000	6407
Geothermal power	1 MW	20	1250000	Electricity	kgoe	753360	Geothermal	753360	-
Hydro power plant	1 MW	60	1000000	Electricity	kgoe	753360	Water energy	753360	-
Nuclear power plant	1 MW	40	2200000	Electricity	kgoe	753360	Nuclear	2000000	-
Oil fired power plant	1 MW	30	1250000	Electricity	kgoe	753000	Oil products	2213820	18541
Oil fired power (improved)	1 MW	30	1250000	Electricity	kgoe	753360	Oil products	1695060	14196
Solar PV	1 MW	20	4000000	Electricity	kgoe	753360	Solar	753360	-
Wind power	1 MW	20	1000000	Electricity	kgoe	753360	Wind energy	753360	-

d.u. = Device Unit

Note: Data for existing technologies have been estimated from numerous domestic publications and estimation methodology described in Chap. 4. These are average estimates for India. Data for future technologies are based on international sources. Data for fixed costs of some technologies are not accurate.

Appendix L: Representation of Technology Systems in AIM-Japan

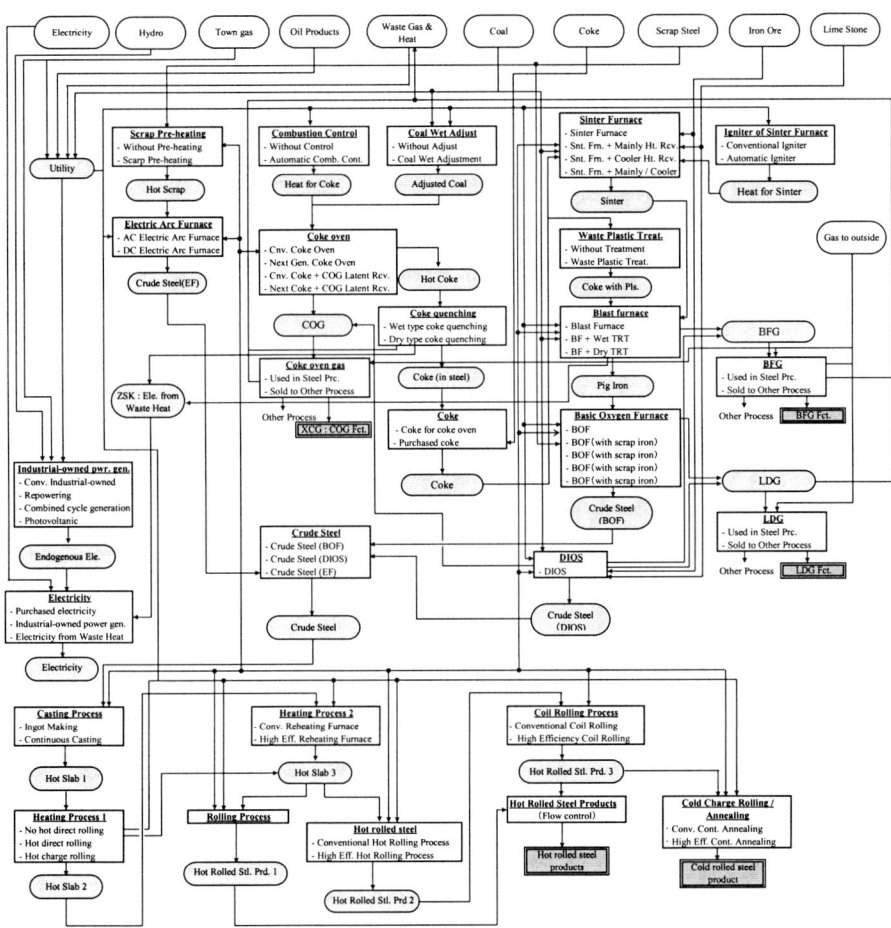

Fig. L.1. Technology system of steel sector in AIM-Japan

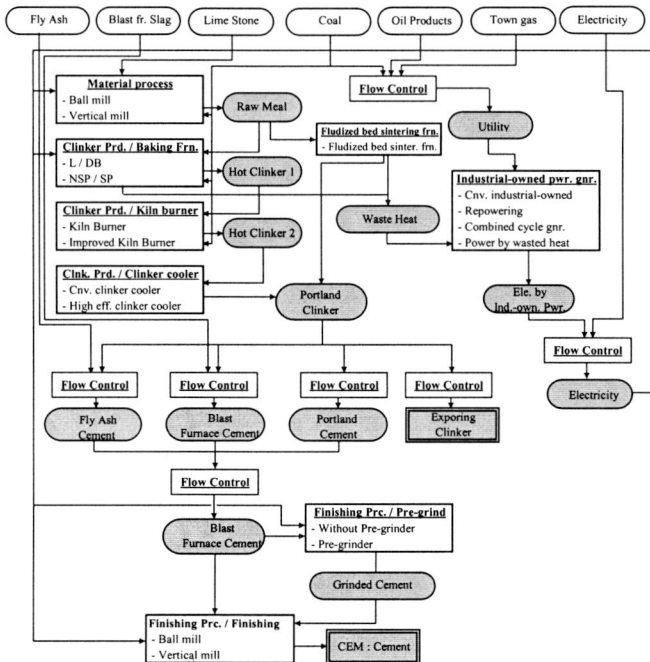

Fig. L.2. Technology system of cement sector in AIM-Japan

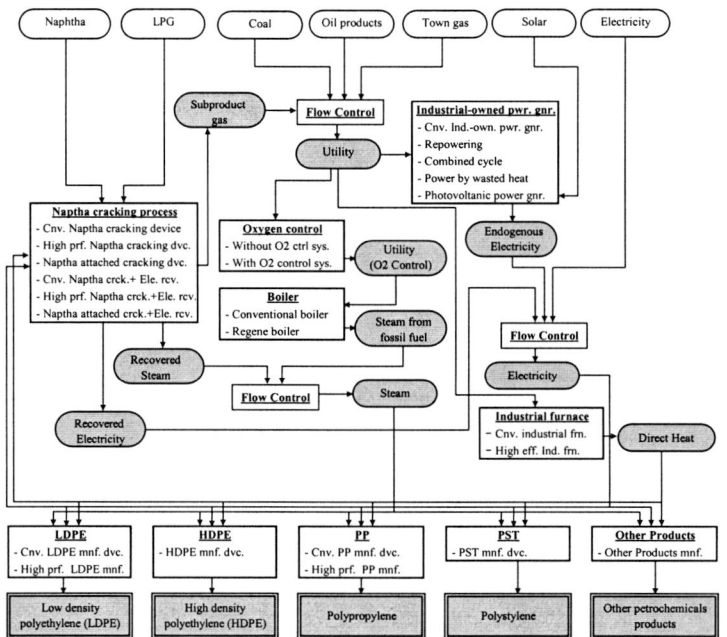

Fig. L.3. Technology system of petrochemicals sector in AIM-Japan

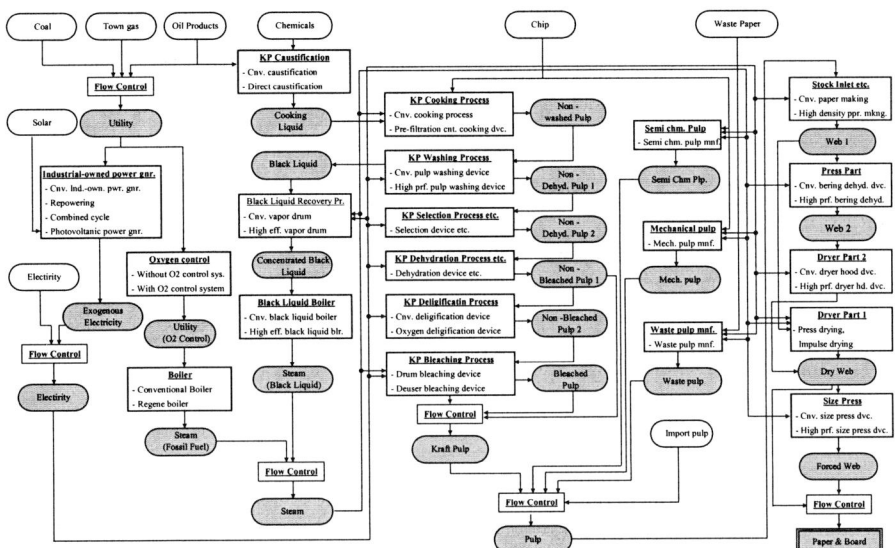

Fig. L.4. Technology system of paper sector in AIM-Japan

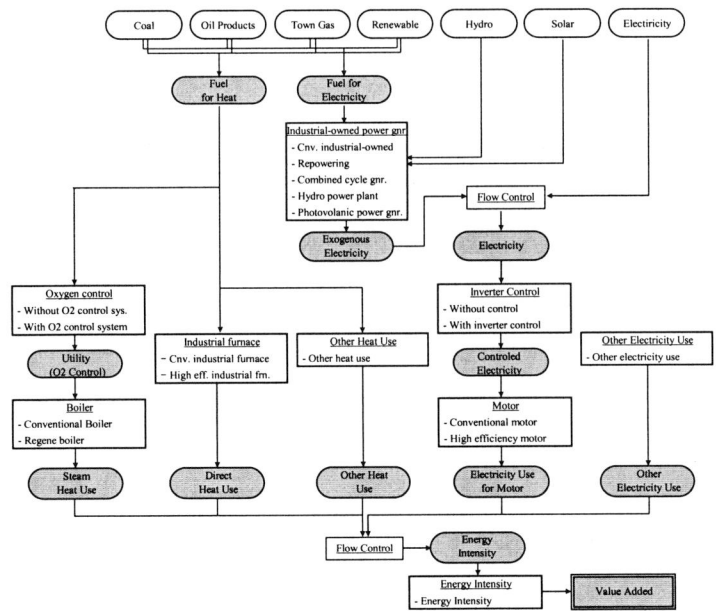

Fig. L.5. Technology system of other industry sector in AIM-Japan

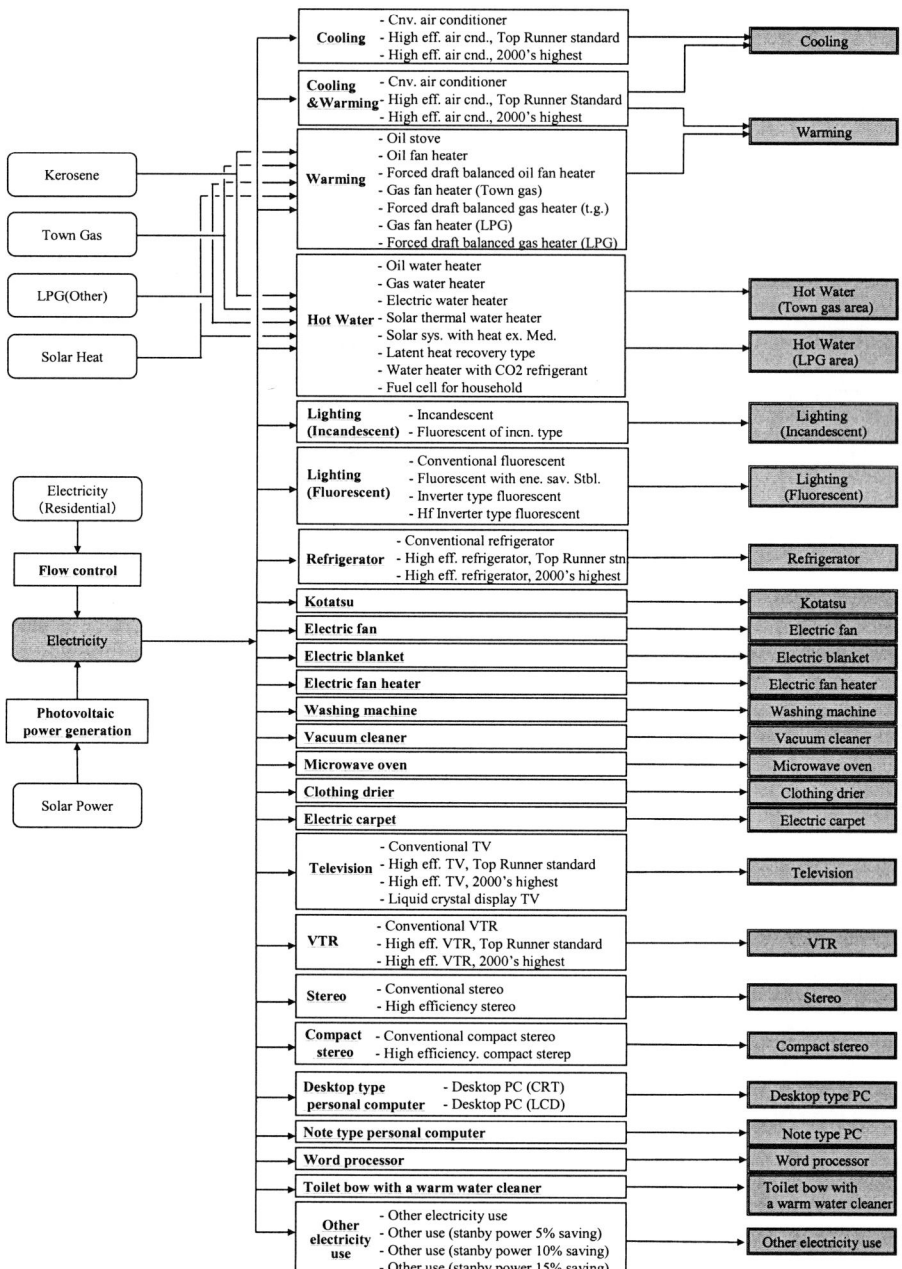

Fig. L.6. Technology system of residential sector in AIM-Japan

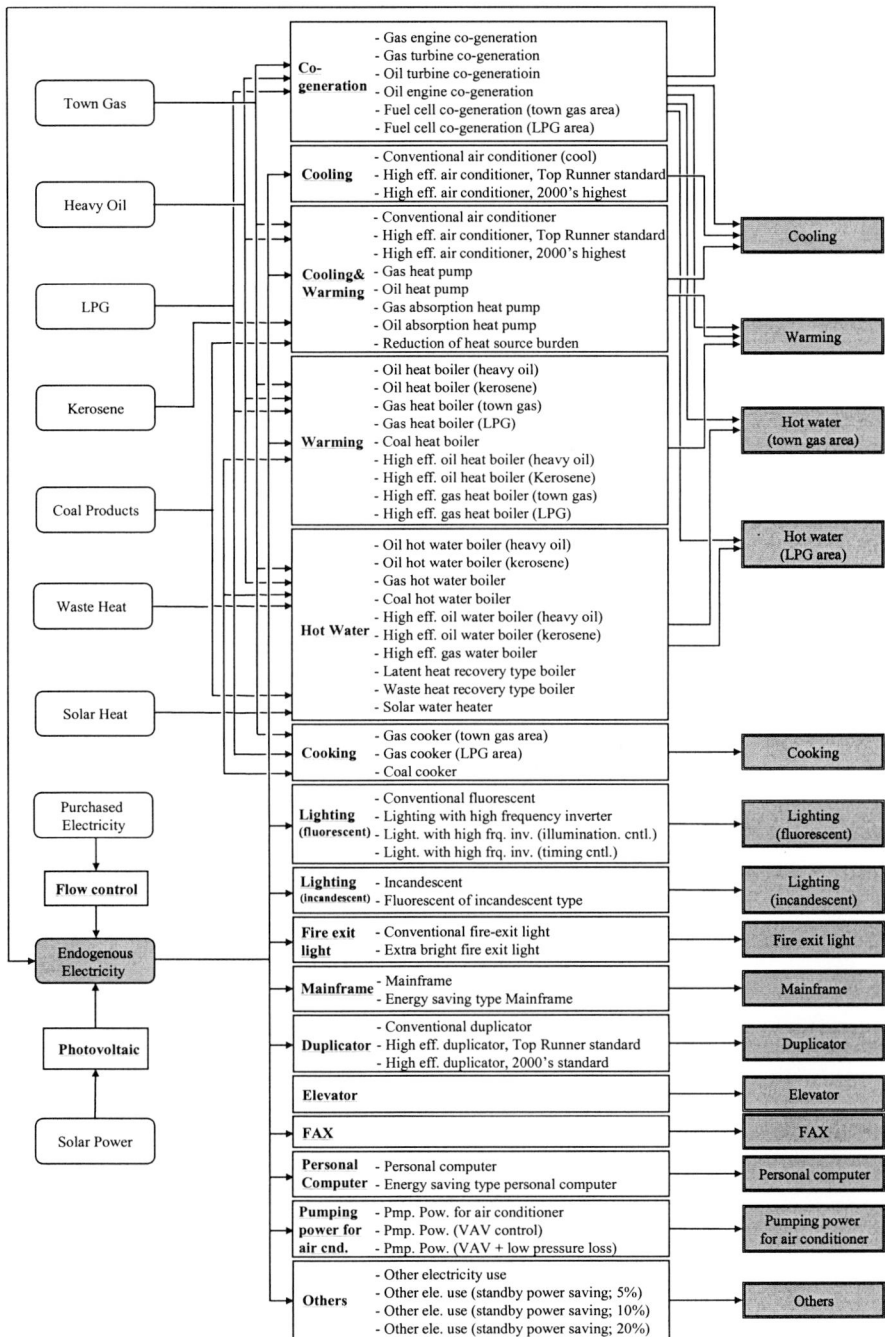

Fig. L.7. Technology system of commercial sector in AIM-Japan

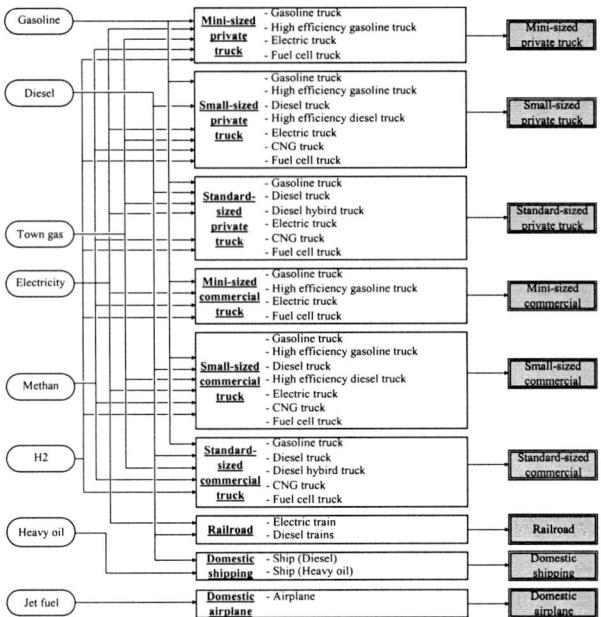

Fig. L.8. Technology system of freight transportation sector in AIM-Japan

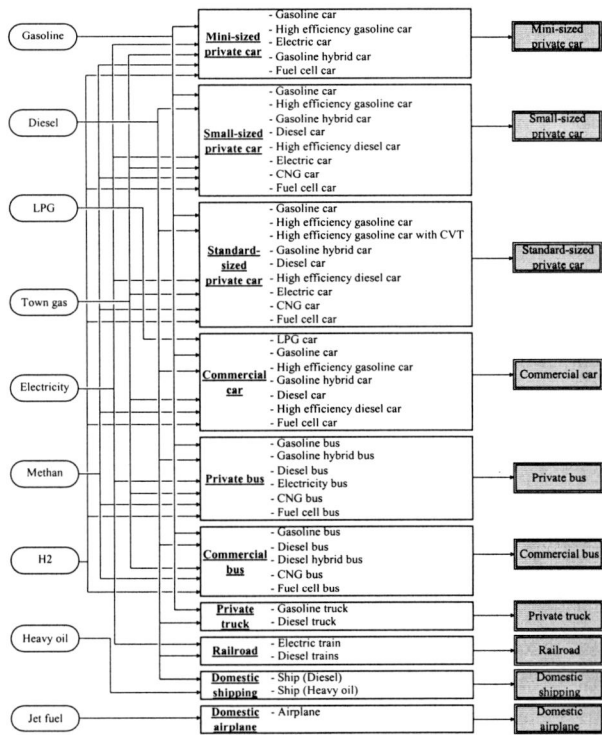

Fig. L.9. Technology system of passenger transportation sector in AIM-Japan

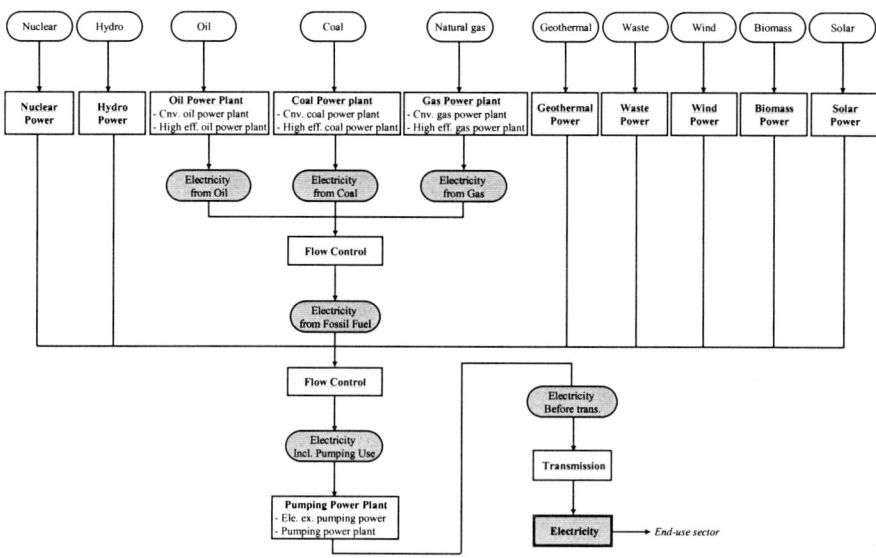

Fig. L.10. Technology system of power generation sector in AIM-Japan

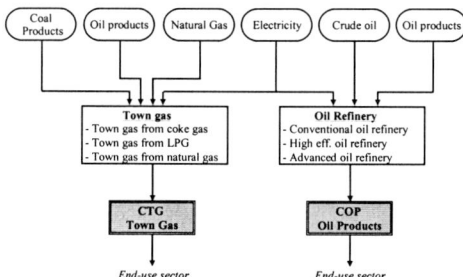

Fig. L.11. Technology system of town gas & oil refinery sector in AIM-Japan

Appendix M Characteristics of Technologies in AIM-Japan

Table M.1. Characteristics of technologies in industrial sector

Energy Device	Device Unit	Fixed Cost (JPY)	Life (Years)	Specific Service		Specific Energy Input		Material Use	NOx (kg/yr. /d.u.)
Industrial Sector - Steel									
Without coal wet adjust eqpt.	1t-c.s.	-	1	Adjusted coke	36.81 ª	Coal	54.41 ª	0.05	0.000
						Utility	0.5 ª	0	
Coke wet adjustment equipment	1t-c.s.	1500	30	Adjusted coke	36.81 ª	Coal	54.41 ª	0.05	0.000
						Utility	0 ª	0	
Without automatic combustion ctrl.	1t-c.s.	-	1	Heat for coke oven	3.74 ª	Utility	3.96 ª	0	0.000
With automatic combustion control	1t-c.s.	100	30	Heat for coke oven	3.74 ª	Utility	3.74 ª	0	0.000
Conventional coke oven	1t-c.s.	5000	30	Hot coke	36.81 ª	Adjusted coke	36.81 ª	0	1.161
				COG	9 ª	Heat for coke oven	3.74 ª	0	
						Electricity	0.23 ª	0	
Next generation coke oven	1t-c.s.	5000	30	Hot coke	36.81 ª	Adjusted coke	36.81 ª	0	0.797
				COG	9 ª	Heat for coke oven	2.96 ª	0	
						Electricity	0.23 ª	0	
Cnv. coke oven + COG latent heat recovery	1t-c.s.	5800	30	Hot coke	36.81 ª	Adjusted coke	36.81 ª	0	0.000
				COG	9 ª	Heat for coke oven	3.74 ª	0	
				Waste gas and heat	0.18 ª	Electricity	0.23 ª	0	
Next gnr. coke oven + COG latent heat recovery	1t-c.s.	5800	30	Hot coke	36.81 ª	Adjusted coke	36.81 ª	0	0.000
				COG	9 ª	Heat for coke oven	2.96 ª	0	
				Waste gas and heat	0.18 ª	Electricity	0.23 ª	0	
Coke wet type quenching	1t-c.s.	5200	30	Coke (in steel)	36.81 ª	Hot coke	36.81 ª	0	0.000
Coke dry type quenching	1t-c.s.	6500	30	Coke (in steel)	36.81 ª	Hot coke	36.81 ª	0	0.000
				Electricity from waste heat	0.29 ª				
				Waste gas and heat	0.77 ª				
COG for steel industry	100Mcal	-	1	COG potential	1 ª	COG	1	0	0.000
				Waste gas and heat	1 ª				
COG for other industries	100Mcal	-	1	COG potential	1 ª	COG	1	0	0.000
				Gas for other process	1 ª	Coal for other process	1		
Coke from coke furnace	100Mcal	-	1	Coke	1 ª	Coke (in steel)	1 ª	0	0.000
Purchased coke	100Mcal	-	1	Coke	1 ª	Coke	1 ª	0	
Conventional igniter of sintering furnace	100Mcal	-	1	Heat for sinter furnace	0.48 ª	Utility	0.88 ª	0	0.000
						Electricity	0.41 ª	0	
Automatic inginter of sintering furnace	100Mcal	100	30	Heat for sinter furnace	0.48 ª	Utility	0.48 ª	0	0.000
						Electricity	0.41 ª	0	
Sintering furnace	1t-c.s.	5000	30	Sinter	1.34 ᵇ	Coal	1.02 ª	0	0.864
						Iron Ore	1.41 ᵇ	1	
						Coke	3.32 ª	0	
						Heat for sinter furnace	0.48 ª	0	
Sintering frn. + Mainly waste heat recovery	1t-c.s.	6500	30	Sinter	1.34 ᵇ	Coal	1.02 ª	0	0.864
				Waste gas and heat	0.27 ª	Iron Ore	1.41 ᵇ	1	
						Coke	3.32 ª	0	
						Heat for sinter furnace	0.48 ª	0	
Sintering frn. + Cooler waste heat recovery	1t-c.s.	8000	30	Sinter	1.34 ᵇ	Coal	1.02 ª	0	0.864
				Waste gas and heat	0.6 ª	Iron Ore	1.41 ᵇ	1	
						Coke	3.32 ª	0	
						Heat for sinter furnace	0.48 ª	0	
Sintering frn. + Mainly/Cooler waste heat recovery	1t-c.s.	9500	30	Sinter	1.34 ᵇ	Coal	1.02 ª	0	0.864
				Waste gas and heat	0.87 ª	Iron Ore	1.41 ᵇ	1	
						Coke	3.32 ª	0	
						Heat for sinter furnace	0.48 ª	0	

Note: Unit with "a"=10⁸cal/year/d.u., Unit with "b"=t/year/d.u., d.u.=device unit, c.s.=crude steel

Table M.1. Characteristics of technologies in industrial sector (continued)

Energy Device	Device Unit	Fixed Cost (JPY)	Life (Years)	Specific Service	Specific Energy Input	Material Use	NOx (kg/yr. /d.u.)
Without waste plastic	1t-c.s.	-	30	Coke with plastic 33.57[a]	Coke 33.57[a]	0	0.000
With waste plastic	1t-c.s.	3000	30	Coke with plastic 33.57[a]	Coke 30.81[a]	0	0.000
Blast furnace	1t-c.s.	5000	30	Pig iron 1.07[b]	Coke with plastic 33.57[a]	0	0.039
				BFG 13.1[a]	Coal 4.71[a]	0	
				Electricity from waste heat 0.001[b]	Sinter 1.34[b]	1	
					Utility 1.81[a]	0	
Blast furnace + Wet top-pressure recovery turbin	1t-c.s.	7100	30	Pig iron 1.07[b]	Coke with plastic 33.57[a]	0	0.000
				BFG 13.1[a]	Coal 4.71[a]	0	
				Electricity from waste heat 0.39[b]	Sinter 1.34[b]	1	
					Utility 1.81[a]	0	
Blast furnace + Dry top-pressure recovery turbin	1t-c.s.	7600	30	Pig iron 1.07[b]	Coke with plastic 33.57[a]	0	0.000
				BFG 13.1[a]	Coal 4.71[a]	0	
				Electricity from waste heat 0.45[a]	Sinter 1.34[b]	1	
					Utility 1.81[a]	0	
BFG for steel industry	100Mcal	-	1	BOG potential 1[a]	BFG 1[a]	0	0.000
				Waste gas and heat 1[a]			
BFG for other industries	100Mcal	-	1	BOG potential 1[a]	BFG 1[a]	0	0.000
				Gas for other process 1[a]	Coal for other process 1		
LDF	1t-c.s.	15000	30	Crude steel (BOF) 1[b]	Pig iron 1.07[b]	1	0.000
				LDG 0.001[a]	Scrap steel 0.07[b]	1	
					Utility 0.18[a]	0	
					Electricity 0.8[a]	0	
LDF + Cnv. LDG recovery	1t-c.s.	16960	30	Crude steel (BOF) 1[b]	Pig iron 1.07[b]	1	0.000
				LDG 2.07[a]	Scrap steel 0.07[b]	1	
					Utility 0.18[a]	0	
					Electricity 0.8[a]	0	
LDF + Closed LDG recovery	1t-c.s.	17100	30	Crude steel (BOF) 1[b]	Pig iron 1.07[b]	1	0.000
				LDG 2.27[a]	Scrap steel 0.07[b]	1	
					Utility 0.18[a]	0	
					Electricity 0.8[a]	0	
LDF + Cnv. LDG recovery + Latent recovery	1t-c.s.	19710	30	Crude steel (BOF) 1[b]	Pig iron 1.07[b]	1	0.000
				LDG 2.07[a]	Scrap steel 0.07[b]	1	
				Waste gas and heat 2.5[a]	Utility 0.18[a]	0	
					Electricity 0.8[a]	0	
LDF + Closed LDG recovery + Latent recovery	1t-c.s.	19850	30	Crude steel (BOF) 1[b]	Pig iron 1.07[b]	1	0.000
				LDG 2.27[a]	Scrap steel 0.07[b]	1	
				Waste gas and heat 2.74[a]	Utility 0.18[a]	0	
					Electricity 0.8[a]	0	
LDG for steel industry	100Mcal	-	1	LDG potential 1[a]	LDG 1[a]	0	0.000
				Waste gas and heat 1[a]			
LDG for other industries	100Mcal	-	1	LDG potential 1[a]	LDG 1[a]	0	0.000
				Gas for other process 1[a]	Coal for other process 1		
Without scrap pre-heat	1t-c.s.	-	1	Hot scrap 1.05[b]	Scrap steel 1.05[b]	1	0.000
					Electricity 0.19[a]	0	
Scrap pre-heat	1t-c.s.	650	30	Hot scrap 1.05[b]	Scrap steel 1.05[b]	1	0.000
					Electricity 0[a]	0	
AC electric furnace	1t-c.s.	4200	30	Crude steel (EF) 1[b]	Hot scrap 1.05[b]	1	0.141
					Electricity 3.73[a]	0	
					Utility 0.6[a]	0	
DC electric furnace	1t-c.s.	5250	30	Crude steel (EF) 1[b]	Hot scrap 1.05[b]	1	0.141
					Electricity 3.41[a]	0	
					Utility 0.6[a]	0	
DIOS	1t-c.s.	22500	30	Crude steel (DIOS) 1[b]	Iron Ore 1.41[b]	1	0.797
				COG 9.02[a]	Coal 60.24[a]	0	
				BFG 13.1[a]	Utility 2.06[a]	0	
				LDG 2.07[a]	Electricity 1.68[a]	0	
Crude steel (LDF)	1t-c.s.	-	1	Crude steel 1[b]	Crude steel (BOF) 1[b]	1	0.000
Crude steel (DIOS)	1t-c.s.	-	1	Crude steel 1[b]	Crude steel (DIOS) 1[b]	1	0.000
Crude steel (Electric furnace)	1t-c.s.	-	1	Crude steel 1[b]	Crude steel (EF) 1[b]	1	0.000
Ingot making	1t-c.s.	5400	30	Hot slab1 0.987[b]	Crude steel 1[b]	1	0.000
					Electricity 0.42[a]	0	
					Utility 1.93[a]	0	
Continuous caster	1t-c.s.	4500	30	Hot slab1 0.987[b]	Crude steel 1[b]	1	0.000
					Electricity 0.14[a]	0	
					Utility 0.13[a]	0	

Note: Unit with "a"=10^8cal/year/d.u., Unit with "b"=t/year/d.u., d.u.=device unit, c.s.=crude steel

Table M.1. Characteristics of technologies in industrial sector (continued)

Energy Device	Device Unit	Fixed Cost (JPY)	Life (Years)	Specific Service		Specific Energy Input		Material Use	NOx (kg/yr. /d.u.)
Without hot charge rolling	1t-c.s.	-	10	Hot slab2	0.987 b	Hot slab1	0.987 b	1	0.000
						Utility	0.9 a	0	
Hot charge rolling	1t-c.s.	1000	30	Hot slab2	0.987 b	Hot slab1	0.987 b	1	0.000
						Utility	0 a	0	
Hot direct rolling	1t-c.s.	2000	30	Hot slab3	0.987 b	Hot slab1	0.987 b	1	0.000
						Utility	0 a	0	
Conventinal hearting furnace	1t-c.s.	2400	30	Hot slab3	0.987 b	Hot slab2	0.987 b	1	0.035
						Utility	2.1 a	0	
High efficiency hearting furnace	1t-c.s.	3000	30	Hot slab3	0.987 b	Hot slab2	0.987 b	1	0.028
						Utility	1.71 a	0	
Rolling process	1t-c.s.	-	1	Hot rolled steel pruducts 1	0.981 b	Hot slab3	0.987 b	1	0.000
						Utility	0.54 a	0	
						Electricity	0.88 a	0	
Conventional hot rolling process	1t-c.s.	5000	30	Hot rolled steel pruducts 2	0.981 b	Hot slab3	0.987 b	1	0.009
						Utility	0.54 a	0	
						Electricity	0.82 a	0	
High efficiency hot rolling process	1t-c.s.	5000	30	Hot rolled steel pruducts 2	0.981 b	Hot slab3	0.987 b	1	0.004
						Utility	0.24 a	0	
						Electricity	0.82 a	0	
Conventional coil rolling	1t-c.s.	5000	30	Hot rolled steel pruducts 3	0.981 b	Hot rolled steel pruducts 2	0.981 b	1	0.000
						Electricity	0.06 a	0	
High efficiency coil rolling	1t-c.s.	5000	30	Hot rolled steel pruducts 3	0.981 b	Hot rolled steel pruducts 2	0.981 b	1	0.000
						Electricity	0 a	0	
Hot rolled sheets	1 t	-	1	Hot rolled steel prod	1 b	Hot rolled steel pruducts 3	1 b	1	0.000
Hot rolled slab	1 t	-	1	Hot rolled steel prod	1 b	Hot rolled steel pruducts 1	1 b	1	0.000
Conventional Continuous Annealing lines	1t-c.s.	8000	30	Cold rolled steel production	1.004 b	Hot rolled steel pruducts 3	0.981 b	1	0.083
						Electricity	1.94 a	0	
						Utility	4.45 a	0	
High Efficiency Continuous Annealing lines	1t-c.s.	10000	30	Cold rolled steel production	1.004 b	Hot rolled steel pruducts 3	0.981 b	1	0.066
						Electricity	1.94 a	0	
						Utility	3.55 a	0	
Purchased electricity	100Mcal	-	1	Electricity	1 a	Electricity	1 a	0	0.000
Industrial-Owned power gnr.	100Mcal	-	1	Electricity	1 a	Endgenous ele.	1 a	0	0.000
Conventional Industrial-owned	100Mcal	1615	1	Endgenous ele.	1 a	Utility	3.23 a	0	0.088
Repowering	100Mcal	1723	1	Endgenous ele.	1 a	Utility	2.56 a	0	0.070
Conbined cycle power plant	100Mcal	2019	1	Endgenous ele.	1 a	Town Gas	2 a	0	0.055
Solar	100Mcal	92180	1	Endgenous ele.	1 a	Solar	2.62 a	0	0.000
Recover	100Mcal	0	1	Endgenous ele.	1 a	Electricity from waste heat	1 a	0	0.000
Coal	100Mcal	-	1	Utility	1 a	Coal	1 a	0	0.000
Oil	100Mcal	-	1	Utility	1 a	Oil Products	1 a	0	0.000
Gas	100Mcal	-	1	Utility	1 a	Town Gas	1 a	0	0.000
Recover	100Mcal	-	1	Utility	1 a	Waste gas and heat	1 a	0	0.000
Industrial Sector - Cement									
Tube mill	1 t	2875	30	Raw meal	1 b	Limestone	1.15 b	1	0.000
						Electricity	0.28 a	0	
Vertical mill	1 t	1917	30	Raw meal	1 b	Limestone	1.15 b	1	0.000
						Electricity	0.22 a	0	
L/DB	1 t	2208	30	Hot clinker 1	1 b	Raw meal	1 b	1	0.952
				Waste heat	0.437 a	Coal	9.72 a	0	
						Electricity	0.257 a	0	
NSP/SP	1 t	2760	30	Hot clinker 1	1 b	Raw meal	1 b	1	0.672
				Waste heat	0.437 a	Coal	6.86 a	0	
						Electricity	0.257 a	0	
Kiln burner	1 t	100	30	Hot clinker 2	1 b	Hot clinker 1	1 b	1	0.000
						Coal	0.02 a	0	
Improved kiln burner	1 t	100	30	Hot clinker 2	1 b	Hot clinker 1	1 b	1	0.000
						Coal	0 a	0	
Fludized bed sintering frn.	1 t	4140	30	Portalnd clinker	1 b	Raw meal	1 b	1	0.570
				Waste heat	0.287 a	Coal	5.82 a	0	
						Electricity	0.257 a	0	

Note: Unit with "a"=10^8cal/year/d.u., Unit with "b"=t/year/d.u., d.u.=device unit, c.s.=crude steel

Table M.1. Characteristics of technologies in industrial sector (continued)

Energy Device	Device Unit	Fixed Cost (JPY)	Life (Years)	Specific Service	Specific Energy Input	Material Use	NOx (kg/yr. /d.u.)
Conventional kiln cooler	1 t	-	30	Portalnd clinker 1 [b]	Hot clinker 2 1 [b]	1	0.000
					Coal 0.35 [a]	0	
High efficiency kiln cooler	1 t	455	30	Portalnd clinker 1 [b]	Hot clinker 2 1 [b]	1	0.000
					Coal 0 [a]	0	
Exported portland clinker	1 t	-	1	Exporting clinker 1 [b]	Portalnd clinker 1 [b]	1	0.000
Mixed portland cement's material	1 t	-	1	Portalnd cement 1 [b]	Portalnd clinker 0.95 [b]	0	0.000
					Blast furnace slag 0.05 [b]	1	
Mixed blast furnace cement's material	1 t	-	1	Blast furnace cement 1 [b]	Portalnd clinker 0.55 [b]	0	0.000
					Blast furnace slag 0.45 [b]	1	
Mixed fly ash cement's material	1 t	-	1	Fly ash cement 1 [b]	Portalnd clinker 0.85 [b]	0	0.000
					Fly ash 0.15 [b]	1	
Flow control (Portland cement)	1 t	-	1	Cement before finish 1 [b]	Portalnd cement 1 [b]	1	0.000
Flow control (Blast furnace cement)	1 t	-	1	Cement before finish 1 [b]	Blast furnace cement 1 [b]	1	0.000
Flow control (Fry ash cement)	1 t	300	1	Cement before finish 1 [b]	Fly ash cement 1 [b]	1	0.000
Without pre-grinder	1 t	-	30	Grinded cement 1 [b]	Cement before finish 1 [b]	1	0.000
					Electricity 0.069 [a]	0	
With pre-grinder	1 t	833	30	Grinded cement 1 [b]	Cement before finish 1 [b]	1	0.000
					Electricity 0 [a]	0	
Ball mill	1 t	2875	30	Cement 1 [b]	Grinded cement 1 [b]	1	0.000
					Electricity 0.354 [a]	0	
Vertical mill	1 t	2917	30	Cement 1 [b]	Cement before finish 1 [b]	1	0.000
					Electricity 0.259 [a]	0	
Other sector	1 t	-	1	Cement others 1 [b]	Electricity 0.063 [a]	1	0.000
Industrial-owned power gnr.	100Mcal	1615	30	Endgenous ele. 1 [a]	Utility 3.23 [a]	0	0.088
Repowering	100Mcal	1723	30	Endgenous ele. 1 [a]	Utility 2.56 [a]	0	0.070
Combind cycle generation	100Mcal	2019	30	Endgenous ele. 1 [a]	Utility 2 [a]	0	0.055
Photovoltanic	100Mcal	92180	30	Endgenous ele. 1 [a]	Solar 2.62 [a]	0	0.000
Heat recovery generation	100Mcal	8640	30	Endgenous ele. 1 [a]	Waste heat 2.62 [a]	0	0.000
Purchased electricity	100Mcal	-	1	Electricity 1 [a]	Electricity 1 [a]	0	0.000
Industrial-owned power gnr.	100Mcal	-	1	Electricity 1 [a]	Endgenous ele. 1 [a]	0	0.000
Coal	100Mcal	-	1	Utility 1 [a]	Coal 1 [a]	0	0.000
Oil	100Mcal	-	1	Utility 1 [a]	Oil Products 1 [a]	0	0.000
Town gas	100Mcal	-	1	Utility 1 [a]	Town Gas 1 [a]	0	0.000
Industrial Sector - Petrochemicals							
Conventional Naptha cracking device	1 t	76800	30	Ethylene 1 [b]	Naptha 439 [a]	0.8	11.897
				Recovered steam 4.66 [a]	LPG 37.5 [a]	0.8	
				Subproduct Gas 95.5 [a]	Direct Heat 0.17 [a]	0	
					Electricity 71.6 [a]	0	
High performance. Naptha cracking device	1 t	96000	30	Ethylene 1 [b]	Naptha 439 [a]	0.8	11.572
				Recovered steam 4.66 [a]	LPG 37.5 [a]	0.8	
				Subproduct Gas 95.5 [a]	Direct Heat 0.17 [a]	0	
					Electricity 56.6 [a]	0	
Naptha attached cracking device	1 t	120000	30	Ethylene 1 [b]	Naptha 439 [a]	0.8	10.430
				Recovered steam 4.66 [a]	LPG 37.5 [a]	0.8	
				Subproduct Gas 95.5 [a]	Direct Heat 0.11 [a]	0	
					Electricity 4.04 [a]	0	
Cnv. Naptha cracking + Electricity recovery	1 t	76800	30	Ethylene 1 [b]	Naptha 439 [a]	0.8	11.897
				Recovered steam 4.66 [a]	LPG 37.5 [a]	0.8	
				Recovered electricity 1.9 [a]	Direct Heat 0.17 [a]	0	
				Subproduct Gas 95.5 [a]	Electricity 71.6 [a]	0	
High prf. Naptha cracking + Electricity recovery	1 t	96000	30	Ethylene 1 [b]	Naptha 439 [a]	0.8	11.572
				Recovered steam 4.66 [a]	LPG 37.5 [a]	0.8	
				Recovered electricity 1.9 [a]	Direct Heat 0.17 [a]	0	
				Subproduct Gas 95.5 [a]	Electricity 56.6 [a]	0	
Naptha attached cracking + Electricity recovery	1 t	120000	30	Ethylene 1 [b]	Naptha 439 [a]	0.8	10.430
				Recovered steam 4.66 [a]	LPG 37.5 [a]	0.8	
				Recovered electricity 1.9 [a]	Direct Heat 0.11 [a]	0	
				Subproduct Gas 95.5 [a]	Electricity 4.04 [a]	0	
Conventional LDPE manufacturing device	1 t	32000	30	LDPE 1 [b]	Direct Heat 10.47 [a]	0	0.271
					Steam 3.38 [a]	0	
					Electricity 0.47 [a]	0	
High performance LDPE manufacturing device	1 t	40000	30	LDPE 1 [b]	Direct Heat 6.6 [a]	0	0.171
					Steam 2.13 [a]	0	
					Electricity 0.3 [a]	0	

Note: Unit with "a"=10^8 cal/year/d.u., Unit with "b"=t/year/d.u., d.u.=device unit

Table M.1. Characteristics of technologies in industrial sector (continued)

Energy Device	Device Unit	Fixed Cost (JPY)	Life (Years)	Specific Service		Specific Energy Input		Material Use	NOx (kg/yr. /d.u.)
Conventional PP manufacturing device	1 t	44000	30	Polypropylene	1 [b]	Direct Heat	5.21 [a]	0	0.132
						Steam	10.79 [a]	0	
						Electricity	0.13 [a]	0	
High performance PP manufacturing device	1 t	55000	30	Polypropylene	1 [b]	Direct Heat	1.72 [a]	0	0.044
						Steam	3.56 [a]	0	
						Electricity	0.04 [a]	0	
HDPE manufacturing device	1 t	-	1	HDPE	1 [b]	Direct Heat	4.77 [a]	0	0.119
						Steam	4.77 [a]	0	
						Electricity	0.04 [a]	0	
Poly stylen manufacturing device	1 t	-	1	Polystylene	1 [b]	Direct Heat	1.9 [a]	0	0.072
						Steam	1.69 [a]	0	
						Electricity	1.01 [a]	0	
Other petro-chemistry products	1 t	-	1	Other petrochemical	1 [b]	Direct Heat	10.47 [a]	0	1.375
						Steam	39.62 [a]	0	
						Electricity	45.07 [a]	0	
Conventional industrial furnace	100Mcal	5000	30	Electricity	1 [a]	Utility	1 [a]	0	0.025
High performance industrial furnace	100Mcal	5000	30	Electricity	1 [a]	Utility	0.7 [a]	0	0.017
Without O2 control system	100Mcal	-	30	Utility (O2 ctrl.)	1.18 [a]	Utility	1.18 [a]	0	0.000
With O2 control system	100Mcal	104	30	Utility (O2 ctrl.)	1.18 [a]	Utility	1.11 [a]	0	0.000
Conventional boiler	100Mcal	1036	30	Steam from fossil	1 [a]	Utility (O2 ctrl.)	1.18 [a]	0	0.025
Regene boiler	100Mcal	1243	30	Steam from fossil	1 [a]	Utility (O2 ctrl.)	1.11 [a]	0	0.023
Steam of fossil fuel	100Mcal	-	1	Steam	1 [a]	Steam from fossil	1 [a]	0	0.000
Recovery steam	100Mcal	-	1	Steam	1 [a]	Recovered steam	1 [a]	0	0.000
Purchased Electricity	100Mcal	-	1	Direct Heat	1 [a]	Electricity	1 [a]	0	0.000
Industrial-Owned power gnr.	100Mcal	-	1	Direct Heat	1 [a]	Endogenous Ele.	1 [a]	0	0.000
Cnv. Industrial-owned power gnr.	100Mcal	1615	30	Endogenous Ele.	1 [a]	Utility	3.23 [a]	0	0.088
Repowering	100Mcal	1723	30	Endogenous Ele.	1 [a]	Utility	2.56 [a]	0	0.070
Conbined cycle generation	100Mcal	2019	30	Endogenous Ele.	1 [a]	Utility	2 [a]	0	0.055
Recovery	100Mcal	-	30	Endogenous Ele.	1 [a]	Recovered electricity	1 [a]	0	0.000
Phitovlotanic	100Mcal	92180	30	Endogenous Ele.	1 [a]	Solar	2.67 [a]	0	0.000
Coal	100Mcal	-	1	Utility	1 [a]	Coal	1 [a]	0	0.000
Oil	100Mcal	-	1	Utility	1 [a]	Oil Products	1 [a]	0	0.000
Gas	100Mcal	-	1	Utility	1 [a]	Town Gas	1 [a]	0	0.000
Subproduct Gas	100Mcal	-	1	Utility	1 [a]	Subproduct Gas	1 [a]	0	0.000
Staticstical Errors	1 t	-	1	Driving force for adjust	1 [b]	Direct Heat	1 [a]	0	0.000
Industrial Sector - Paper									
Mechanical pulp manufacturing device	1 t	-	30	Mechanical pulp	1 [b]	Chips	2.19 [b]	1	0.000
						Steam	0.62 [a]	0	
						Electricity	14.46 [a]	0	
Waste pulp manufacturing device	1 t	-	30	Waste paper pulp	1 [b]	Chips	1.05 [b]	1	0.000
						Steam	0.75 [a]	0	
						Electricity	2.39 [a]	0	
Semi chemical pulp manufacturing device	1 t	-	30	Semi chemical pulp	1 [b]	Chips	2.83 [b]	1	0.000
						Steam	13.35 [a]	0	
						Electricity	6.75 [a]	0	
Conventional caustification	1 t	3000	30	Cooking liquid	1 [b]	Chemicals	1 [b]	1	0.000
						Oil Products	0.49 [a]	0	
Direct caustification	1 t	3000	30	Cooking liquid	1 [b]	Chemicals	1 [b]	1	0.000
						Oil Products	0 [a]	0	
Conventional cooking device	1 t	15238	30	Non-washed pulp	1 [b]	Chips	3.64 [b]	1	0.000
						Cooking liquid	1 [b]	0	
						Steam	8.56 [a]	0	
Pre-filtration continuos cooking device	1 t	18286	30	Non-washed pulp	1 [b]	Chips	3.64 [b]	1	0.000
						Cooking liquid	1 [b]	0	
						Steam	4.54 [a]	0	
Conventional pulp washing device	1 t	2619	30	Non-dehyd. pulp 1	1 [b]	Non-washed pulp	1 [b]	1	0.000
				Black liquid	1 [b]	Electricity	0.31 [a]	0	
						Non-washed pulp	1 [b]	0	
High performance pulp washing device	1 t	3143	30	Non-dehyd. pulp 1	1 [b]	Non-washed pulp	1 [b]	1	0.000
				Black liquid	1 [b]	Electricity	0.04 [a]	0	
						Non-washed pulp	1 [b]	0	
Selection process	1 t	-	30	Non-dehyd. pulp 2	1 [b]	Non-dehyd. pulp 1	1 [b]	1	0.000
						Electricity	1.16 [a]	0	
Conventional vapor drum	1 t	4464	30	Concentrated black liquid	63.9 [a]	Black liquid	1 [b]	1	0.000
						Steam	11.02 [a]	0	
						Electricity	0.74 [a]	0	

Note: Unit with "a"=10^8cal/year/d.u., Unit with "b"=t/year/d.u., d.u.=device unit

Table M.1. Characteristics of technologies in industrial sector (continued)

Energy Device	Device Unit	Fixed Cost (JPY)	Life (Years)	Specific Service	Specific Energy Input	Material Use	NOx (kg/yr. /d.u.)
High performance vapor drum	1 t	5357	30	Concentrated black liquid 63.9[a]	Black liquid 1[b]	1	0.000
					Steam 9.19[a]	0	
					Electricity 0.74[a]	0	
Bleaching process	1 t	-	30	Non-bleached pulp 1 1[b]	Non-dehyd. pulp 2 1[b]	1	0.000
					Electricity 0.52[a]	0	
Conventional delignification device	1 t	3472	30	Non-bleached pulp 2 1[b]	Non-bleached pulp 1 1[b]	1	0.000
					Electricity 0.3[a]	0	
					Oil Products 0.16[a]	0	
Oxygen delignification device	1 t	4167	30	Non-bleached pulp 2 1[b]	Non-bleached pulp 1 1[b]	1	0.000
					Electricity 0.22[a]	0	
					Oil Products 0.13[a]	0	
Drum bleaching device	1 t	17593	30	Bleached pulp 1[b]	Non-bleached pulp 2 1[b]	1	0.000
					Electricity 0.52[a]	0	
					Steam 1.92[a]	0	
Defuser bleaching device	1 t	21111	30	Bleached pulp 1[b]	Non-bleached pulp 2 1[b]	1	0.000
					Electricity 0.34[a]	0	
					Steam 0[a]	0	
Flow Control (Bleaching)	1 t	-	1	Kraft pulp 1[b]	Bleached pulp 1[b]	1	0.000
Flow Control (Non Bleaching)	1 t	-	1	Kraft pulp 1[b]	Non-dehyd. pulp 2 1[b]	1	0.000
Flow Control (Mechanical pulp)	1 t	-	1	Pulp 1[b]	Mechanical pulp 1[b]	1	0.000
Flow Control (Waste)	1 t	-	1	Pulp 1[b]	Waste paper pulp 1[b]	1	0.000
Flow Control (Semi chm. pulp)	1 t	-	1	Pulp 1[b]	Semi chemical pulp 1[b]	1	0.000
Flow Control (Kraft pulp)	1 t	-	1	Pulp 1[b]	Kraft pulp 1[b]	1	0.000
Flow Control (Import pulp)	1 t	-	1	Pulp 1[b]	Import pulp 1[b]	1	0.000
Conventional paper making	1 t	5000	30	Web1 1[b]	Pulp 1[b]	1	0.000
					Electricity 4.18[a]	0	
High density paper making	1 t	5000	30	Web1 1[b]	Pulp 1[b]	1	0.000
					Electricity 2.87[a]	0	
Conventional bearing dehydration device	1 t	2439	30	Web2 1[b]	Web1 1[b]	1	0.000
					Steam 14.79[a]	0	
High performance bearing dehydration device	1 t	2927	30	Web2 1[b]	Web1 1[b]	1	0.000
					Steam 9.68[a]	0	
Conventional dryer hood device	1 t	1543	30	Dry Web 1[b]	Web2 1[b]	1	0.000
					Electricity 0.78[a]	0	
High performance dryer hood device	1 t	1852	30	Dry Web 1[b]	Web2 1[b]	1	0.000
					Electricity 0.46[a]	0	
Press drying, Impulse drying	1 t	4779	30	Dry Web 1[b]	Web1 1[b]	1	0.000
					Electricity 0.46[a]	0	
					Steam 5.68[a]	0	
Conventional size press device	1 t	4630	30	Forced Web 1[b]	Dry Web 1[b]	1	0.000
					Steam 6.24[a]	0	
High performance size press device	1 t	5556	30	Forced Web 1[b]	Dry Web 1[b]	1	0.000
					Steam 0.91[a]	0	
Flow Control	1 t	-	1	Paper & Board 1[b]	Forced Web 0.53[b]	1	0.000
					Dry Web 0.47[b]	0	
Purchased Electricity	100Mcal	-	1	Electricity 1[a]	Electricity 1[a]	0	0.000
Industrial-owned power gnr.	100Mcal	-	1	Electricity 1[a]	Exogeneous Elec. 1[a]	0	0.000
Cnv. Industrial-owned power gnr.	100Mcal	1615	30	Exogeneous Elec. 1[a]	Utility 3.23[a]	0	0.088
Cnv. Ind.-owned + Repowering	100Mcal	1723	30	Exogeneous Elec. 1[a]	Utility 2.56[a]	0	0.070
Conbined cycle generation	100Mcal	2019	30	Exogeneous Elec. 1[a]	Utility 2[a]	0	0.055
Phitovlotanic	100Mcal	92180	30	Exogeneous Elec. 1[a]	Solar 2.62[a]	0	0.000
Without O2 control system	100Mcal	-	30	Utility (O2 ctrl.) 1.18[a]	Utility 1.18[a]	0	0.000
With O2 control system	100Mcal	104	30	Utility (O2 ctrl.) 1.18[a]	Utility 1.11[a]	0	0.000
Conventional boiler	100Mcal	1036	30	Steam (Fossil fuel) 1[a]	Utility (O2 ctrl.) 1.18[a]	0	0.025
Regene boiler	100Mcal	1243	30	Steam (Fossil fuel) 1[a]	Utility (O2 ctrl.) 1.11[a]	0	0.023
Conventional black liquid boiler	100Mcal	500	30	Steam (Black liquid) 0.5[a]	Concentrated black liquid 1[b]	0	0.000
High eff. black liquid boiler	100Mcal	500	30	Steam (Black liquid) 0.7[a]	Concentrated black liquid 1[b]	0	0.000
Coal	100Mcal	-	30	Utility 1[a]	Coal 1[a]	0	0.000
OPR	100Mcal	-	30	Utility 1[a]	Oil Products 1[a]	0	0.000
Gas	100Mcal	-	30	Utility 1[a]	Town Gas 1[a]	0	0.000
Steam (Fossil fuel)	100Mcal	-	1	Steam 1[a]	Steam (Fossil fuel) 1[a]	0	0.000
Steam (Black liquid)	100Mcal	-	1	Steam 1[a]	Steam (Black liquid) 1[a]	0	0.000
Other	100Mcal	-	1	Paper & Board 1[b]	Electricity 1[a]	0	0.000

Note: Unit with "a"=10^8 cal/year/d.u., Unit with "b"=t/year/d.u., d.u.=device unit

Table M.1. Characteristics of technologies in industrial sector (continued)

Energy Device	Device Unit	Fixed Cost (JPY)	Life (Years)	Specific Service		Specific Energy Input		Material Use	NOx (kg/yr. /d.u.)
Industrial Sector - Others									
Energy / Add value	1Myen	-	1	Value added	1 c	Energy use	137.2 a	1	0.000
Vapor	100Mcal	-	1	Energy use	1 a	Steam heat use	1 a	0	0.000
Direct Heat	100Mcal	-	1	Energy use	1 a	Direct heat use	1 a	0	0.000
Other heat	100Mcal	-	1	Energy use	1 a	Other heat use	1 a	0	0.000
Electricity for power	100Mcal	-	1	Energy use	1 a	Ele. use for motor	1 a	0	0.000
Other electricity	100Mcal	-	1	Energy use	1 a	Other electricity use	1 a	0	0.000
Without O2 control system	100Mcal	-	1	Utility (O2 Ctrl.)	1.18 a	Fuel for heat	1.18 a	0	0.000
With O2 control system	100Mcal	104	30	Utility (O2 Ctrl.)	1.18 a	Fuel for heat	1.11 a	0	0.000
Conventional boiler	100Mcal	1036	30	Steam heat use	1 a	Utility (O2 ctrl.)	1.18 a	0	0.025
Regene boiler	100Mcal	1243	30	Steam heat use	1 a	Utility (O2 ctrl.)	1.11 a	0	0.023
Conventional industrial furnace	100Mcal	5000	30	Direct heat use	1 a	Fuel for heat	1 a	0	0.028
High performance industrial furnace	100Mcal	5500	30	Direct heat use	1 a	Fuel for heat	0.7 a	0	0.020
Other Heat	100Mcal	-	1	Other heat use	1 a	Fuel for heat	1 a	0	0.000
Without inverter	100Mcal	-	1	Controled electricity	1 a	Value added	1 a	0	0.000
Inverter	100Mcal	1646	30	Controled electricity	1 a	Value added	0.65 a	0	0.000
Conventinal motor	100Mcal	1651	30	Ele. use for motor	1 a	Ctrl. electricity	1 a	0	0.000
High performance motor	100Mcal	2229	30	Ele. use for motor	1 a	Ctrl. electricity	0.97 a	0	0.000
Other electricity	100Mcal	-	1	Other electricity use	1 a	Value added	1 a	0	0.000
Industrial-owned power geration	100Mcal	-	1	Value added	1 a	Endogenous ele.	1 a	0	0.000
Purchased Electricity	100Mcal	-	1	Value added	1 a	Electricity	1 a	0	0.000
Cnv. Industrial-owned power gnr.	100Mcal	1615	30	Endogenous ele.	1 a	Fuel for Electricity	3.23 a	0	0.088
Cnv. Ind.-owned + Repowering	100Mcal	1723	30	Endogenous ele.	1 a	Fuel for Electricity	2.56 a	0	0.070
Conbined cycle generation	100Mcal	2019	30	Endogenous ele.	1 a	Fuel for Electricity	2 a	0	0.055
Water power	100Mcal	-	30	Endogenous ele.	1 a	Hydro power	2.616 a	0	0.000
Photovlotanic	100Mcal	92180	30	Endogenous ele.	1 a	Solar	2.616 a	0	0.000
Coal fuel for heat	100Mcal	-	1	Fuel for heat	1 a	Coal	1 a	0	0.000
Oil fuel for heat	100Mcal	-	1	Fuel for heat	1 a	Oil Products	1 a	0	0.000
Gas fuel for heat	100Mcal	-	1	Fuel for heat	1 a	Town Gas	1 a	0	0.000
New energy for heat	100Mcal	-	1	Fuel for heat	1 a	New Energy	1 a	0	0.000
Coal fuel for electricity	100Mcal	-	1	Fuel for electricity	1 a	Coal	1 a	0	0.000
Oil fuel for electricity	100Mcal	-	1	Fuel for electricity	1 a	Oil Products	1 a	0	0.000
Gas fuel for electricity	100Mcal	-	1	Fuel for electricity	1 a	Town Gas	1 a	0	0.000
New energy for electricity	100Mcal	-	1	Fuel for electricity	1 a	New Eenergy	1 a	0	0.000

Note: Unit with "a"=10^8 cal/year/d.u., Unit with "b"=t/year/d.u., d.u.=device unit

Table M.2. Characteristics of technologies in residential and commercial sectors

Energy Device	Device Unit	Fixed Cost (10³JPY)	Life (Years)	Specific Service		Specific Energy Input (10⁸cal/yr./d.u.)		Material Use	NOx (kg/yr./d.u.)
Residential Sector									
Conventional air conditioner (cool)	1 unit	171	10	Cooling	13.26 [a]	Electricity	4.74	0	0.00
High efficiency air conditioner, Top Runner Standard (cool)	1 unit	180	10	Cooling	13.26 [a]	Electricity	4.74	0	0.00
High efficiency air conditioner, 2000's highest (cool)	1 unit	216	10	Cooling	13.26 [a]	Electricity	2.41	0	0.00
Conventional air conditioner (cool&warm)	1 unit	190	10	Warming Cooling	38.75 [a] 13.26 [a]	Electricity	17.18	0	0.00
High efficiency air conditioner, Top Runner Standard	1 unit	200	10	Warming Cooling	38.75 [a] 13.26 [a]	Electricity	17.18	0	0.00
High efficiency air conditioner, 2000's highest (cool&warm)	1 unit	240	10	Warming Cooling	38.75 [a] 13.26 [a]	Electricity	8.92	0	0.00
Oil stove	1 unit	3	8	Warming	23.60 [a]	Kerosene	25.49	0	0.07
Oil fan heater	1 unit	20	8	Warming	34.33 [a]	Kerosene Electricity	37.05 1.58	0 0	0.63
Forced draft balanced oil fan heater	1 unit	118	8	Warming	34.97 [a]	Kerosene	40.58	0	0.62
Gas fan heater	1 unit	50	8	Warming	21.46 [a]	Town Gas	21.46	0	0.15
Forced draft balanced gas heater	1 unit	133	8	Warming	42.91 [a]	Town Gas	51.49	0	0.75
Gas fan heater	1 unit	50	8	Warming	21.46 [a]	LPG	21.46	0	0.18
Forced draft balanced gas heater	1 unit	133	8	Warming	42.91 [a]	LPG	51.49	0	0.85
Oil water heater	1 unit	244	15	Hot water (t.g. area)	25.28 [a]	Kerosene	38.90	0	0.97
Gas water heater	1 unit	156	15	Hot water (t.g. area)	25.28 [a]	Town Gas	33.71	0	0.63
Electric water heater	1 unit	240	15	Hot water (t.g. area)	25.28 [a]	Electricity	28.09	0	0.00
Solar thermal water heater	1 unit	190	10	Hot water (t.g. area)	12.50 [a]	Solar Heat	12.50	0	0.00
Solar system with heat exchange media	1 unit	550	33	Hot water (t.g. area)	19.50 [a]	Solar Heat	19.50	0	0.00
Latent heat recovery type	1 unit	190	15	Hot water (t.g. area)	25.28 [a]	Town Gas	26.61	0	0.50
Water heater with CO2 refrigerant	1 unit	400	15	Hot water (t.g. area)	25.28 [a]	Electricity	8.43	0	0.16
Fuel cell for household	1 unit	500	20	Hot water (t.g. area) Electricity	26.73 [a] 23.39 [a]	Town Gas	66.83	0	0.00
Oil water heater	1 unit	244	15	Hot water (LPG area)	25.28 [a]	Kerosene	38.90	0	0.97
Gas water heater	1 unit	156	15	Hot water (LPG area)	25.28 [a]	LPG	33.71	0	0.72
Electric water heater	1 unit	240	15	Hot water (LPG area)	25.28 [a]	Electricity	28.09	0	0.00
Solar thermal water heater	1 unit	190	10	Hot water (LPG area)	12.50 [a]	Solar Heat	12.50	0	0.00
Solar system with heat exchange media	1 unit	550	33	Hot water (LPG area)	19.50 [a]	Solar Heat	19.50	0	0.00
Latent heat recovery type	1 unit	190	15	Hot water (LPG area)	25.28 [a]	LPG	26.61	0	0.57
Water heater with CO2 refrigerant	1 unit	400	15	Hot water (LPG area)	25.28 [a]	Electricity	8.43	0	0.18
Fuel cell for household	1 unit	500	20	Hot water (LPG area) Electricity	26.73 [a] 23.39 [a]	LPG	66.83	0	0.00
Incandescent	1 unit	0.2	1	Light (Incandescent)	0.50 [a]	Electricity	0.50	0	0.00
Fluorescent of incandescent type	1 unit	2.1	6	Light (Incandescent)	0.50 [a]	Electricity	0.17	0	0.00
Conventional fluorescent	1 unit	14	7	Light (Fluorescent)	0.19 [a]	Electricity	0.19	0	0.00
Fluorescent with enery saving stabilizer	1 unit	18	7	Light (Fluorescent)	0.19 [a]	Electricity	0.17	0	0.00
Inverter type fluorescent	1 unit	18	7	Light (Fluorescent)	0.19 [a]	Electricity	0.16	0	0.00
Hf Inverter type fluorescent	1 unit	23	7	Light (Fluorescent)	0.19 [a]	Electricity	0.14	0	0.00
Conventional refrigerator	1 unit	170	6	Refrigerator	1.00 [b]	Electricity	7.24	0	0.00
High efficiency refrigerator, Top Runner Standard	1 unit	180	6	Refrigerator	1.00 [b]	Electricity	7.24	0	0.00
High efficiency refrigerator, 2000's highest	1 unit	183	6	Refrigerator	1.00 [b]	Electricity	3.51	0	0.00
Kotatsu	1 unit	10	6	Kotatsu	1.00 [b]	Electricity	1.63	0	0.00
Fan	1 unit	10	6	Fan	1.00 [b]	Electricity	0.15	0	0.00
Electric blanket	1 unit	10	6	Electric blanket	1.00 [b]	Electricity	0.48	0	0.00
Electric fan heater	1 unit	10	6	Electric fan heater	1.00 [b]	Electricity	0.83	0	0.00
Washing machine	1 unit	10	6	Washing machine	1.00 [b]	Electricity	0.47	0	0.00
Vacuum cleaner	1 unit	10	6	Vacuum cleaner	1.00 [b]	Electricity	1.14	0	0.00
Microwave oven	1 unit	10	6	Microwave oven	1.00 [b]	Electricity	1.03	0	0.00
Clothing drier	1 unit	10	6	Clothing drier	1.00 [b]	Electricity	4.44	0	0.00
Electric carpet	1 unit	10	6	Electric carpet	1.00 [b]	Electricity	2.63	0	0.00

Note: Unit with "a"=10⁸cal/year/d.u., Unit with "b"=number of device unit/year/d.u., d.u.=device unit

Table M.2. Characteristics of technologies in residential and commercial sectors (continued)

Energy Device	Device Unit	Fixed Cost (10³JPY)	Life (Years)	Specific Service		Specific Energy Input (10⁸cal/yr./d.u.)		Material Use	NOx (kg/yr. /d.u.)
Conventional TV	1 unit	50	6	Television	1.00 [b]	Electricity	2.20	0	0.00
High efficiency TV, Top Runner Standard	1 unit	51	6	Television	1.00 [b]	Electricity	2.20	0	0.00
High efficiency TV, 2000's highest	1 unit	52	6	Television	1.00 [b]	Electricity	1.44	0	0.00
Liquid crystal display TV	1 unit	100	6	Television	1.00 [b]	Electricity	0.92	0	0.00
Conventional VTR	1 unit	20	6	VTR	1.00 [b]	Electricity	0.60	0	0.00
High efficiency VTR, Top Runner Standard	1 unit	21	6	VTR	1.00 [b]	Electricity	0.60	0	0.00
2000's highest efficiency VTR	1 unit	21	6	VTR	1.00 [b]	Electricity	0.09	0	0.00
Conventional stereo	1 unit	50	6	Stereo	1.00 [b]	Electricity	0.85	0	0.00
High efficiency stereo	1 unit	52	6	Stereo	1.00 [b]	Electricity	0.25	0	0.00
Conventional combination tape recorder and radio	1 unit	20	6	Compact Stereo	1.00 [b]	Electricity	0.30	0	0.00
High efficiency combination tape recorder and radio	1 unit	21	6	Compact Stereo	1.00 [b]	Electricity	0.17	0	0.00
Conventional combination tape recorder and radio	1 unit	20	6	Compact Stereo	1.00 [b]	Electricity	0.30	0	0.00
High efficiency combination tape recorder and radio	1 unit	21	6	Compact Stereo	1.00 [b]	Electricity	0.17	0	0.00
Desktop personal computer (CRT)	1 unit	200	5	Desktop type PC	1.00 [b]	Electricity	0.75	0	0.00
Desktop personal computer (LCD)	1 unit	250	5	Desktop type PC	1.00 [b]	Electricity	0.40	0	0.00
Note type personal computer	1 unit	10	5	Note type PC	1.00 [b]	Electricity	0.08	0	0.00
Word processor	1 unit	10	5	Word processor	1.00 [b]	Electricity	0.04	0	0.00
Toilet bow with a warm water cleaner	1 unit	10	6	Toilet bow with a warm water cleaner	1.00 [b]	Electricity	0.21	0	0.00
Other electricity use	1 unit	50	6	Other electricity use	1.00 [b]	Electricity	4.36	0	0.00
Other ele. use (standby power saving; 5%)	1 unit	51	6	Other electricity use	1.00 [b]	Electricity	4.25	0	0.00
Other ele. use (standby power saving; 10%)	1 unit	52	6	Other electricity use	1.00 [b]	Electricity	4.13	0	0.00
Other ele. use (standby power saving; 20%)	1 unit	53	6	Other electricity use	1.00 [b]	Electricity	3.90	0	0.00
Photovoltaic power generation	1 unit	3,000	20	Electricity	31.54 [a]	Solar Power	70.96	0	0.00
Flow control	1 unit	0	1	Electricity	1.00	Electricity	1.00	0	0.00
Commercial Sector									
Gas engine co-generation	1kW	200	30	Warming	8.45 [a]	Town Gas	58	0	177.4
				Cooling	8.45 [a]				
				Hot water (t.g. area)	11.27 [a]				
				Electricity	14.53 [a]				
Gas turbine co-generation	1kW	220	30	Warming	14.84 [a]	Town Gas	104	0	17.7
				Cooling	14.84 [a]				
				Hot water (t.g. area)	19.79 [a]				
				Electricity	22.16 [a]				
Oil turbine co-generatioin	1kW	220	30	Warming	14.84 [a]	Heavy Oil	104	0	29.9
				Cooling	14.84 [a]				
				Hot water (LPG area)	19.79 [a]				
				Electricity	22.16 [a]				
Oil engine co-generation	1kW	200	30	Warming	7.47 [a]	Heavy Oil	78	0	302.1
				Cooling	7.47 [a]				
				Hot water (LPG area)	9.97 [a]				
				Electricity	27.10 [a]				
Fuel cell co-generation (t.g. area)	1kW	450	30	Warming	14.91 [a]	Town Gas	124	0	0.0
				Cooling	14.91 [a]				
				Hot water (t.g. area)	19.88 [a]				
				Electricity	49.71 [a]				
Fuel cell co-generation (LPG area)	1kW	450	30	Warming	14.91 [a]	LPG	124	0	0.0
				Cooling	14.91 [a]				
				Hot water (LPG area)	19.88 [a]				
				Electricity	49.71 [a]				
Conventional air conditioner (cool)	1 unit	1,800	20	Cooling	173 [a]	Electricity	69	0	0.00
High efficiency air conditioner, Top Runner Standard (cool)	1 unit	1,800	20	Cooling	173 [a]	Electricity	69	0	0.00
High efficiency air conditioner, 2000's highest (cool)	1 unit	1,900	20	Cooling	173 [a]	Electricity	43	0	0.00

Note: Unit with "a"=10⁸cal/year/d.u., Unit with "b"=number of device unit/year/d.u., d.u.=device unit

Table M.2. Characteristics of technologies in residential and commercial sectors (continued)

Energy Device	Device Unit	Fixed Cost (10³JPY)	Life (Years)	Specific Service		Specific Energy Input (10⁸cal/yr./d.u.)		Material Use	NOx (kg/yr. /d.u.)
Conventional air conditioner	1 unit	1,850	20	Warming	311 [a]	Electricity	180	0	0.00
(cool&warm)				Cooling	173 [a]				
High efficiency air conditioner,	1 unit	1,850	20	Warming	311 [a]	Electricity	180	0	0.00
Top Runner Standard				Cooling	173 [a]				
High efficiency air conditioner,	1 unit	2,280	20	Warming	311 [a]	Electricity	112	0	0.00
2000's highest (cool&warm)				Cooling	173 [a]				
Gas heat pump	1 unit	1,900	20	Warming	311 [a]	Town Gas	336	0	86.8
				Cooling	173 [a]	Electricity	10	0	
Oil heat pump	1 unit	2,200	20	Warming	311 [a]	Kerosene	336	0	90.0
				Cooling	173 [a]	Electricity	10	0	
Gas absorption heat pump	1 unit	58,100	30	Warming	12100 [a]	Town Gas	29,740	0	14.7
				Cooling	16300 [a]	Electricity	135	0	
Oil absorption heat pump	1 unit	60,000	30	Warming	12100 [a]	Heavy Oil	29,740	0	21.9
			30	Cooling	16300 [a]	Electricity	232	0	
Reduction of heat source burden	1000m²	3,100	20	Warming	240.2 [a]	New Energy	0	0	0.0
			20	Cooling	31.2 [a]				
Oil heat boiler (heavy oil)	1 unit	1,186	20	Warming	240 [a]	Heavy Oil	304	0	24.0
Oil heat boiler (kerosene)	1 unit	1,186	20	Warming	240 [a]	Kerosene	304	0	20.3
Gas heat boiler (town gas)	1 unit	1,186	20	Warming	240 [a]	Town Gas	278	0	18.3
Gas heat boiler (LPG)	1 unit	1,186	20	Warming	240 [a]	LPG	278	0	24.3
Coal heat boiler	1 unit	1,186	20	Warming	240 [a]	Coke	304	0	70.5
Oil hot water boiler (heavy oil)	1 unit	1,186	20	Hot water (t.g. area)	240 [a]	Heavy Oil	304	0	21.4
Oil hot water boiler (kerosene)	1 unit	1,186	20	Hot water (t.g. area)	240 [a]	Kerosene	304	0	18.4
Gas hot water boiler (town gas)	1 unit	1,186	20	Hot water (t.g. area)	240 [a]	Town Gas	278	0	15.2
Coal hot water boiler	1 unit	1,186	20	Hot water (t.g. area)	240 [a]	Coke	304	0	27.9
Latent heat recovery type boiler	1 unit	1,245	20	Hot water (t.g. area)	240 [a]	Town Gas	253	0	15.2
Solar water heater	1 unit	1,010	20	Hot water (t.g. area)	30.6 [a]	Solar Heat	31	0	0.0
Waste heat recovery type boiler	1 unit	1,186	20	Hot water (t.g. area)	240 [a]	New Energy	304	0	0.0
Oil hot water boiler (heavy oil)	1 unit	1,186	20	Hot water (LPG area)	240 [a]	Heavy Oil	304	0	21.4
Oil hot water boiler (kerosene)	1 unit	1,186	20	Hot water (LPG area)	240 [a]	Kerosene	304	0	18.4
Gas hot water boiler (LPG)	1 unit	1,186	20	Hot water (LPG area)	240 [a]	LPG	278	0	22.8
Coal hot water boiler	1 unit	1,186	20	Hot water (LPG area)	240 [a]	Coke	304	0	27.9
Latent heat recovery type boiler	1 unit	1,305	20	Hot water (LPG area)	240 [a]	LPG	253	0	22.8
Waste heat recovery type boiler	1 unit	1,010	20	Hot water (LPG area)	30.6 [a]	Solar Heat	31	0	0
Gas cooker	100m²	100	6	Cooking	1 [a]	Town Gas	17	0	28.78
Coal cooker	100m²	100	6	Cooking	1 [a]	Coke	17	0	52.67
Gas cooker	100m²	100	6	Cooking	1 [a]	LPG	17	0	29.93
Conventional fluorescent	100m²	484	11	Light (fluorescent)	1 [a]	Electricity	51	0	0
Lighting with high frequency inverter	100m²	484	11	Light (fluorescent)	1 [a]	Electricity	41	0	0
Light. with high frq. Inv. (ill. cntl.)	100m²	553	11	Light (fluorescent)	1 [a]	Electricity	26	0	0
Light. with high frq. Inv. (tim. cntl.)	100m²	671	11	Light (fluorescent)	1 [a]	Electricity	22	0	0
Incandescent	100m²	40	1	Light (incandescent)	1 [a]	Electricity	51	0	0
Fluorescent of incandescent type	100m²	250	7	Light (incandescent)	1 [a]	Electricity	17	0	0
Conventional fire-exit light	100m²	39	10	Fire exit light	1 [a]	Electricity	2	0	0
Extra bright fire exit light	100m²	41	10	Fire exit light	1 [a]	Electricity	1	0	0
Mainframe	100m²	610	6	Mainframe	1 [a]	Electricity	2	0	0
Energy saving type Mainframe	100m²	611	6	Mainframe	1 [a]	Electricity	2	0	0
Conventional duplicator	100m²	30	6	Duplicator	1 [a]	Electricity	2	0	0
High eff. dupl., Top Runner standard	100m²	30	6	Duplicator	1 [a]	Electricity	2	0	0
2000's highest efficiency duplicator	100m²	33	6	Duplicator	1 [a]	Electricity	2	0	0
Elevator	100m²	690	17	Elevator	1 [a]	Electricity	2	0	0
FAX	100m²	10	6	FAX	1 [a]	Electricity	0	0	0
Personal computer	100m²	20	6	Personal Computer	1 [a]	Electricity	1	0	0
Energy saving type PC	100m²	22	6	Personal Computer	1 [a]	Electricity	0	0	0
Pumping power for air conditioner	100m²	100	6	Pumping power for air conditioner	1 [a]	Electricity	11	0	0
Pumping power (VAV control)	100m²	119	6	Pumping power for air conditioner	1 [a]	Electricity	8	0	0
Pumping power (VAV + low pressure loss)	100m²	135	6	Pumping power for air conditioner	1 [a]	Electricity	6	0	0

Note: Unit with "a"=10⁸cal/year/d.u., Unit with "b"=number of device unit/year/d.u., d.u.=device unit

Table M.2. Characteristics of technologies in residential and commercial sectors (continued)

Energy Device	Device Unit	Fixed Cost (10^3 JPY)	Life (Years)	Specific Service		Specific Energy Input (10^8 cal/yr./d.u.)		Material Use	NOx (kg/yr. /d.u.)
Other electricity use	100m^2	100		Others	1 ^{a}	Electricity	21	0	
Other ele. use (standby power saving; 5%)	100m^2	106		Others	1 ^{a}	Electricity	20	0	0
Other ele. use (standby power saving; 10%)	100m^2	115	6	Others	1 ^{a}	Electricity	19	0	0
Other ele. use (standby power saving; 20%)	100m^2	123	6	Others	1 ^{a}	Electricity	17	0	0
Photovoltaic power generation	1kW	1,700	6	Electricity	10.512 ^{a}	Solar Power	24	0	0
Flow control	1	-	6	Electricity	1 ^{a}	Electricity	1	0	0

Note: Unit with "a"=10^8cal/year/d.u., Unit with "b"=number of device unit/year/d.u., d.u.=device unit

Table M.3. Characteristics of technologies in transportation sector

Energy Device	Device Unit	Fixed Cost (10³JPY)	Life (Years)	Specific Service (100 person-km/yr/d.u.)		Specific Energy Input (10⁸cal/yr./d.u.)		Material Use	NOx (kg/yr. /d.u.)
Passenger Transporation									
Mini car (gasoline)	1 unit	748	10	Mini private car	98	Gasoline	51.3	0	1.01
Mini car (gasoline, high eff.)	1 unit	785	10 0	Mini private car	98	Gasoline	43.0	0	1.01
Mini car (electricity)	1 unit	4,000	10	Mini private car	98	Electricity	11.7	0	
Mini car (gasoline hybrid)	1 unit	5,000	10 0	Mini private car	98	Gasoline	23.6	0	
Mini car (fuel cell, gasoline)	1 unit	5,000	10	Mini private car	98	Gasoline	25.2	0	
Mini car (fuel cell, town gas)	1 unit	5,000	10	Mini private car	98	Town Gas	25.7	0	
Mini car (fuel cell, methane)	1 unit	5,000	10	Mini private car	98	Methanol	24.1	0	
Mini car (fuel cell, hydrogen)	1 unit	5,000	10	Mini private car	98	Hydrogen	21.6	0	
Small car (gasoline)	1 unit	1,527	10	Small private car	152	Gasoline	84.5	0	1.68
Small car (diesel)	1 unit	1,464	10	Small private car	152	Lighting	100.7	0	6.14
Small car (diesel, high eff.)	1 unit	1,537	10	Small private car	152	Lighting	84.1	0	6.14
Small car (gasoline, high eff.)	1 unit	1,603	10	Small private car	152	Gasoline	62.6	0	1.68
Small car (electricity)	1 unit	4,950	10	Small private car	152	Electricity	14.6	0	
Small car (compressed town gas)	1 unit	2,000	10	Small private car	152	Town Gas	83.3	0	
Small car (gasoline hybrid)	1 unit	2,150	10	Small private car	152	Gasoline	42.3	0	
Small car (fuel cell, Hydrogen)	1 unit	5,000	10	Small private car	152	Hydrogen	35.5	0	
Small car (fuel cell, gasoline)	1 unit	5,000	10	Small private car	152	Gasoline	41.4	0	
Small car (fuel cell, town gas)	1 unit	5,000	10	Small private car	152	Town Gas	42.3	0	
Small car (fuel cell, methane)	1 unit	5,000	10	Small private car	152	Methanol	39.7	0	
Standard car (gasoline)	1 unit	3,020	8	Regular private car	152	Gasoline	123.5	0	1.68
Standard car (diesel)	1 unit	2,990	8	Regular private car	152	Lighting	147.1	0	6.14
Standard car (diesel, high eff.)	1 unit	3,140	8	Regular private car	152	Lighting	122.8	0	6.14
Standard vhcl. (gsln. drct. inj.)	1 unit	3,096	8	Regular private car	152	Gasoline	95.0	0	1.68
Standard vhcl. (electricity)	1 unit	9,540	8	Regular private car	152	Electricity	16.7	0	
Standard vhcl. (natural gas)	1 unit	6,415	8	Regular private car	152	Town Gas	128.1	0	
Standard vhcl. (CVT)	1 unit	3,035	8	Regular private car	152	Gasoline	112.2	0	
Standard vhcl. (gasoline hybrid)	1 unit	5,000	8	Regular private car	152	Gasoline	56.8	0	
Standard vhcl. (fuel cell, gsl.)	1 unit	5,000	8	Regular private car	152	Gasoline	60.5	0	
Standard vhcl. (fuel cell, gas)	1 unit	5,000	8	Regular private car	152	Town Gas	61.7	0	
Standard vhcl. (fuel cell, mthn.)	1 unit	5,000	8	Regular private car	152	Methanol	58.0	0	
Standard vhcl. (fuel cell, H2)	1 unit	5,000	8	Regular private car	152	Hydrogen	51.9	0	
Commercial car (LPG)	1 unit	1,780	10	Commercial car	479	LPG	905.4	0	11.06
Commercial car (gasoline)	1 unit	3,020	10	Commercial car	479	Gasoline	603.2	0	11.06
Commercial car (diesel)	1 unit	1,812	10	Commercial car	479	Lighting	448.7	0	40.35
Commercial car (diesel, high eff.)	1 unit	1,903	10	Commercial car	479	Lighting	374.6	0	40.35
Commercial car (gasoline drct. in.)	1 unit	3,096	10	Commercial car	479	Gasoline	464.0	0	11.06
Commercial car (gasoline hybrid)	1 unit	4,252	10	Commercial car	479	Gasoline	277.5	0	
Commercial car (fuel cell, gsln.)	1 unit	5,000	10	Commercial car	479	Gasoline	295.6	0	
Commercial car (fuel cell, gas)	1 unit	5,000	10	Commercial car	479	Town Gas	301.6	0	
Commercial car (fuel cell, mthn.)	1 unit	5,000	10	Commercial car	479	Methanol	283.5	0	
Commercial car (fuel cell, H2)	1 unit	5,000	10	Commercial car	479	Hydrogen	253.4	0	
Private bus (gasoline)	1 unit	2,052	12	Private bus	1403	Gasoline	147.7	0	19.38
Private bus (diesel)	1 unit	4,430	12	Private bus	1403	Lighting	198.5	0	54.11
Private bus (electricity)	1 unit	25,000	12	Private bus	1403	Electricity	63.4	0	
Private bus (CNG)	1 unit	16,660	12	Private bus	1403	Town Gas	538.4	0	
Private bus (gasoline hybrid)	1 unit	14,050	12	Private bus	1403	Gasoline	165.5	0	
Private bus (fuel cell, gasoline)	1 unit		12	Private bus	1403	Gasoline	97.3	0	
Private bus (fuel cell, methane)	1 unit		12	Private bus	1403	Methanol	93.3	0	
Private bus (fuel cell, hydrogen)	1 unit		12	Private bus	1403	Hydrogen	83.4	0	
Commercial bus (diesel)	1 unit	14,100	14	Commercial bus	7381	Lighting	1,266.0	0	177.06
Commercial bus (CNG)	1 unit	22,815	14	Commercial bus	7381	Town Gas	4,742.8	0	
Commercial bus (diesel hybrid)	1 unit	20,000	14	Commercial bus	7381	Lighting	1,055.0	0	
Commercial bus (fuel cell, gsln.)	1 unit		14	Commercial bus	7381	Gasoline	620.3	0	
Commercial bus (fuel cell, mthn.)	1 unit		14	Commercial bus	7381	Methanol	595.0	0	
Commercial bus (fuel cell, H2)	1 unit		14	Commercial bus	7381	Hydrogen	531.7	0	
Private truck (gasoline)			1	Private truck	1	Gasoline	0.4	0	
Private truck (diesel)			1	Private truck	1	Lighting	0.4	0	
Railroad (electricity)			1	Railroad	1	Electricity	0.0	0	
Railroad (diesel)			1	Railroad	1	Lighting	0.0	0	0.015
Ship (diesel)			1	Domestic shipping	1	Heavy Oil	0.4	0	0.286
Ship (hot valve)			1	Domestic shipping	1	Lighting	0.4	0	0.286
Airplane			1	Domestic airline	1	Jet fuel	0.5	0	0.056

d.u.=device unit

Table M.4. Characteristics of technologies in transportation sector (continued)

Energy Device	Device Unit	Fixed Cost (10^3JPY)	Life (Years)	Specific Service (100 ton-km/yr./d.u.)		Specific Energy Input (10^8cal/yr./d.u.)		Material Use	NOx (kg/yr. /d.u.)
Freight Transportaion									
Mini prv. truck (gasoline)	1 unit	953	10	Mini prv. truck	1.5	Gasoline	13.9	0	4.65
Mini prv. truck (gasoline, high eff.)	1 unit	1,001	10	Mini prv. truck	1.5	Gasoline	12.5	0	
Mini prv. truck (electricity)	1 unit	3,100	10	Mini prv. truck	1.5	Electricity	4.9	0	
Mini prv. truck (CNG)	1 unit	1,950	10	Mini prv. truck	1.5	Town Gas	32.4	0	
Mini prv. truck (fuel cell, gasoline)	1 unit	5,000	10	Mini prv. truck	1.5	Gasoline	6.8	0	
Mini prv. truck (fuel cell, methane)	1 unit	5,000	10	Mini prv. truck	1.5	Methanol	6.5	0	
Small prv. truck (gasoline)	1 unit	1,279	10	Small prv. trick	15	Gasoline	103.0	0	7.70
Small prv. truck (gasoline,high eff.)	1 unit	1,343	10	Small prv. trick	15	Gasoline	92.5	0	7.70
Small prv. truck (diesel)	1 unit	1,264	10	Small prv. trick	15	Lighting	131.6	0	12.55
Small prv. truck (diesel,high eff.)	1 unit	1,327	10	Small prv. trick	15	Lighting	121.3	0	12.55
Small prv. truck (electricity)	1 unit	11,230	10	Small prv. trick	15	Electricity	95.2	0	
Small prv. truck (CNG)	1 unit	2,130	10	Small prv. trick	15	Town Gas	316.5	0	
Small prv. truck (fuel cell, gasoline)	1 unit	5,000	10	Small prv. trick	15	Gasoline	64.5	0	
Small prv. truck (fuel cell, methane)	1 unit	5,000	10	Small prv. trick	15	Methanol	61.9	0	
Standard prv. truck (gasoline)	1 unit	1,619	12	Regular prv. truck	277	Gasoline	119.1	0	18.88
Standard prv. truck (diesel)	1 unit	2,924	12	Regular prv. truck	277	Lighting	335.6	0	60.85
Standard prv. truck (electricity)	1 unit	20,000	12	Regular prv. truck	277	Electricity	124.5	0	
Standard prv. truck (CNG)	1 unit	4,755	12	Regular prv. truck	277	Town Gas	1,129.5	0	
Standard prv. truck (diesel hybrid)	1 unit	12,000	12	Regular prv. truck	277	Lighting	335.6	0	
Standard prv. truck (fuel cell, gasoline)	1 unit	99,999	12	Regular prv. truck	277	Gasoline	164.4	0	
Standard prv. truck (fuel cell, methane)	1 unit	99,999	12	Regular prv. truck	277	Methanol	157.7	0	
Mini cmm. truck (gasoline)	1 unit	953	10	Mini cmm. truck	31	Gasoline	211.1	0	16.66
Mini cmm. truck (gasoline, high eff.)	1 unit	1,001	10	Mini cmm. truck	31	Gasoline	189.6	0	16.66
Mini cmm. truck (electricity)	1 unit	3,100	10	Mini cmm. truck	31	Electricity	75.0	0	
Mini cmm. truck (CNG)	1 unit	1,950	10	Mini cmm. truck	31	Town Gas	493.7	0	
Mini cmm. truck (fuel cell, gasoline)	1 unit	5,000	10	Mini cmm. truck	31	Gasoline	103.4	0	
Mini cmm. truck (fuel cell, methane)	1 unit	5,000	10	Mini cmm. truck	31	Methanol	99.2	0	
Small cmm. truck (gasoline)	1 unit	1,533	10	Small cmm. trick	117	Gasoline	229.8	0	14.97
Small cmm. truck (gasoline, high eff.)	1 unit	1,610	10	Small cmm. trick	117	Gasoline	206.4	0	14.97
Small cmm. truck (diesel)	1 unit	1,715	10	Small cmm. trick	117	Lighting	338.7	0	24.40
Small cmm. truck (diesel, high eff.)	1 unit	1,801	10	Small cmm. trick	117	Lighting	312.1	0	24.40
Small cmm. truck (electricity)	1 unit	11,230	10	Small cmm. trick	117	Electricity	244.9	0	
Small cmm. truck (CNG)	1 unit	2,130	10	Small cmm. trick	117	Town Gas	814.3	0	
Small cmm. truck (fuel cell, gasoline)	1 unit	5,000	10	Small cmm. trick	117	Gasoline	166.0	0	
Small cmm. truck (fuel cell, methane)	1 unit	5,000	10	Small cmm. trick	117	Methanol	159.2	0	
Standard cmm. truck (diesel)	1 unit	4,600	12	Standard cmm. truck	2276	Lighting	1,425.5	0	203.80
Standard cmm. truck (CNG)	1 unit	9,333	12	Standard cmm. truck	2276	Town Gas	3,809.8	0	
Standard cmm. truck (diesel hybrid)	1 unit	12,000	12	Standard cmm. truck	2276	Lighting	1,425.5	0	
Standard cmm. truck (fuel cell, gasln.)	1 unit	99,999	12	Standard cmm. truck	2276	Gasoline	698.5	0	
Standard cmm. truck (fuel cell, mthn.)	1 unit	99,999	12	Standard cmm. truck	2276	Methanol	670.0	0	
Railroad (electricity)		9,999	1	Railroad	1	Electricity	0.1	0	
Railroad (diesel)		9,999	1	Railroad	1	Lighting	0.1	0	0.026
Ship (diesel)		9,999	1	Domestic shipping	1	Heavy Oil	0.2	0	0.173
Air plane		9,999	1	Domestic airline	1	Jet fuel	5.5	0	0.663

d.u.=device unit, prv.=private, cmm.=commercial
* Mini vechicle: Overall length 3.4m or less, Overall width 1.48m or less, Overall height 2.0m or less, Displacement 660cc
* Small vechicle: Overall length 4.7m or less, Overall width 1.7m or less, Overall height 2.0m or less, Displacement 2000cc
* Standard vechicle: Larger than small vechicle

Table M.5. Characteristics of technologies in energy conversion sector

Energy Device	Device Unit	Fixed Cost (10³JPY)	Life (Years)	Specific Service (10⁸cal/yr./d.u)		Specific Energy Input (10⁸cal/yr./d.u.)		Material Use	NOx (kg/yr./d.u.)
Electricity Generation									
Transmission loss	1		1	Electricity sales	1	Ele. Before trasmission	1	0	0.00
Ele. Ex. Pumping-up	1		1 0	Ele. Before trasmissi	1	Ele. Incl. Pumping	1	0	0.00
Pumping-up power plant	1 kW	75.34	40	Ele. Before trasmissi	75.34	Ele. Incl. Pumping	166	0	0.00
Nuclear power plant	1kW	430	40 0	Ele. Incl. Pumping	72.03	Nuclear Power	197	0	0.00
hydro power plant	1kW	882	40	Ele. Incl. Pumping	74.96	Hydro Power	197	0	0.00
Flow control	1		1	Ele. Incl. Pumping	1	Ele. Gerated from fossil fuel	1	0	0.00
Geothermal power plant	1kW		40	Ele. Incl. Pumping	69.31	Geothermal Power	197.1	0	0.00
Waste power plant	1kW		40	Ele. Incl. Pumping	67.81	Waste Power	197.1	0	0.00
Wind power plant	1kW		40	Ele. Incl. Pumping	67.81	Wind Power	197.1	0	0.00
Biomass power plant	1kW		40	Ele. Incl. Pumping	67.81	Biomass Power	197.1	0	0.00
Solar power plant	1kW		40	Ele. Incl. Pumping	75.34	Solar Heat	197.1	0	0.00
Flow control (oil)	1		1	Ele. Gerated from fossil fuel	1	Ele. By oil plant	1	0	0.00
Flow control (coal)	1		1	Ele. Gerated from fossil fuel	1	Ele. By coal plant	1	0	0.00
Flow control (gas)	1		1	Ele. Gerated from fossil fuel	1	Ele. By gas plant	1	0	0.00
Conventional oil plant	1kW	190	40	Ele. By oil plant	71.57	Oil	190.3	0	5.94
Hi performance oil plant	1kW	190	40	Ele. By oil plant	71.57	Oil	178.9	0	5.58
Conventional coal plant	1kW	289	40	Ele. By coal plant	71.57	Coal	180.4	0	12.81
Hi performance coal plant	1kW	289	40	Ele. By coal plant	71.57	Coal	178.9	0	12.70
Conventional gas plant	1kW	208	40	Ele. By gas plant	73.08	Natural Gas	179.4	0	2.69
Hi performance NG plant	1kW	208	40	Ele. By gas plant	73.08	Natural Gas	149.1	0	2.24
Town Gas									
Town gas from coke	1		1	Town Gas Sales	1	Coal Products	1.26	0.797	0.00
						Electricity	0.01	0	
Town gas from LPG	1		1	Town Gas Sales	1	LPG	1.26	0.797	0.00
						Electricity	0.01	0	
Town gas from natural gas	1		1	Town Gas Sales	1	Natural Gas	1.00	0.996	0.00
						Electricity	0.01	0	
Oil Refinery									
Oil refining (conventional)	1		30	Oil Products Sales	1	Oil	1.00	1	0.001
						Electricity	0.00	0	
						Oil Products	0.05	0	
Oil refining (new)	1		30	Oil Products Sales	1	Oil	1.00	1	0.001
						Electricity	0.00	0	
						Oil Products	0.05	0	
Oil refining (advanced)	1		30	Oil Products Sales	1	Oil	1.00	1	0.001
						Electricity	0.00	0	
						Oil Products	0.04	0	

d.u.=device unit

Index